INNOVATION ON EDUCATION AND SOCIAL SCIENCES

PROCEEDINGS OF THE INTERNATIONAL JOINT CONFERENCE ON ARTS AND HUMANITIES (IJCAH 2021), SURABAYA, INDONESIA, 2 OCTOBER 2021

Innovation on Education and Social Sciences

Edited by

Irena Maureen, Elly Imah, Siti Savira, Syafi'ul Anam, Masilva Mael & Lina Hartanti
Universitas Negeri Surabaya, Indonesia

Routledge
Taylor & Francis Group

Routledge is an imprint of the Taylor & Francis Group, an informa business

Typeset in Times New Roman by MPS Limited, Chennai, India

Library of Congress Cataloging-in-Publication Data
A catalog record has been requested for this book

First published 2022
by Routledge
2 Park Square, Milton Park, Abingdon, Oxon OX14 4RN
and by Routledge
605 Third Avenue, New York, NY 10158
e-mail: enquiries@taylorandfrancis.com
www.routledge.com – www.taylorandfrancis.com

ISBN: 978-1-032-20752-0 (Hbk)
ISBN: 978-1-032-20753-7 (Pbk)
ISBN: 978-1-003-26506-1 (eBook)
DOI: 10.1201/9781003265061

Innovation on Education and Social Sciences – Maureen et al (Eds)

Table of contents

Innovation on Education and Social Sciences – Maureen et al (Eds)

Foreword

The COVID-19 pandemic is still taking a growing toll on many countries, including Indonesia. We have had 4,2 million cases until the day of the conference — and many of our closest relatives, teachers, students, and friends have perished. In terms of education, the pandemic has inflicted the most havoc on educational systems in modern years. From preschool to university, everyone was impacted.

The International Joint Conference of Art and Humanities 2021 was organized by Universitas Negeri Surabaya (UNESA) – Indonesia as an effort to share stories and experiences between educators and those who care about education, especially in this time of crisis. This prestigious event was coordinated and assisted by AIPI – the Artificial Intelligence and Scientific Publication Unit of UNESA and the Scientific Committee of the conference.

We would like to express our gratitude to all keynote speakers and authors who generously shared their knowledge and experience. We also want to thank all of the conference participants who are actively participated in the discussions — we had attendees from Iraq, China, India, Bangladesh, the Philippines, and Thailand, as well as fellow educators from Indonesia.

This book is an attempt to elicit thoughts and actions so that we might support each other in facing this pandemic and then grow and thrive together once it has passed.

The Editors

Innovation on Education and Social Sciences – Maureen et al (Eds)
© 2022 copyright the Author(s), ISBN 978-1-032-20752-0
Open Access: www.taylorfrancis.com, CC BY-NC-ND 4.0 license

Organization

Steering Committee

Prof. Dr. Nurhasan, M.Kes.
Universitas Negeri Surabaya, Indonesia

Prof. Dr. Bambang Yulianto, M.Pd.
Universitas Negeri Surabaya, Indonesia

Andrzej Cirocki, Ph.D.
York University, UK

Jonnie Hill, Ph.D.
USA

Dr. Timothy Laurie
University of Technology Sydney, Australia

Prof. Chia Hua Kuo, Ph.D., FACSM, CTSSN.
University of Taipei, Taipei

Prof. Yongming Li, Ph.D.
Shanghai University of Sport, China

Prof. Robert Newton, Ph.D., AEP, CSCS*D, FESSA, FNSCA.
Edith Cowan University, Australia

Dr. Cem Tinaz
Istanbul Bilqi University, Turkey

Pratiwi Retnaningdyah, Ph.D.
Universitas Negeri Surabaya, Indonesia

Syafi'ul Anam, Ph.D.
Universitas Negeri Surabaya, Indonesia

Prof. Nining Widyah Kusnanik, M.Appl.Sc.
Universitas Negeri Surabaya, Indonesia

Organizing Committee

Dr. Irena Y. Maureen, S.Pd. M.Sc.
Siti Ina Savira, S.Psi. M.Ed.Cp.
Aditya Chandra Setiawan, S.Pd., M.Pd.
Ainur Rifqi, S.Pd., M.Pd.
Damajanti Kusuma Dewi, S.Psi., M. Si.
Lina Purwaning Hartanti, M.EIL.
Yunan Fathur Rahman, M.A.
Tsuroyya, M.A.
Laily Maulida Septiana Harti, S.S., M.AppLing.
Vivi Ervina Dewi, S.Pd., M.Pd.
Dr. Nurhayati, ST.,MT.
Bayu Agung Pramono, S.Pd., M.Kes.
Hijrin Fitroni, S.Or., M.Pd
Yetty Septiani Mustar, S.KM., M.P.H.
Muhammad Dzul Fikri, S.Or., M.Pd.
Nilam Ade Pangestu

Innovation on Education and Social Sciences – Maureen et al (Eds)

Acknowledgement

Organizing a conference is no simple task. It takes many hours, many months, of effort by many people working quietly behind the scenes. We want to thank:

- Universitas Negeri Surabaya for making this conference happen
- The UNESA Unit of Artificial Intelligence and Scientific Publication
- The keynote speakers, the authors, and the participants of the conference
- The steering committee and the national committee from different faculties of UNESA
- and also the CRC-Press/Balkema Publisher

Curriculum, instruction, and pedagogy

Innovation on Education and Social Sciences – Maureen et al (Eds)
© 2022 copyright the Author(s), ISBN 978-1-032-20752-0
Open Access: www.taylorfrancis.com, CC BY-NC-ND 4.0 license

Learning concrete structures using IT-based interactive media

Suprapto, Theodorus Wiyanto Wibowo & Luthfiyah Nurlaela
Universitas Negeri Surabaya, Surabaya, Indonesia

ABSTRACT: Concrete structure is a compulsory course in Civil Engineering. The problem that occurs is that students do not understand the basics of structural analysis. A media that connects fundamental structural analysis with concrete structures have not yet been developed. The purpose of this research is to develop interactive IT-based learning media in this subject. This research is development research. The data collection was conducted with testing and questionnaires that were then analyzed by using quantitative descriptive techniques. The results of this study are as follows: (1) the results of student responses get an average score of 45.1 and are in the very good category; (2) the student learning outcomes test has an average value of 74.84; (3) the concrete structure of learning media that needs to be improved is the appearance of the color composition to make it more attractive.

1 INTRODUCTION

One of the compulsory courses in civil engineering is the concrete structure course. The concrete structure course is a course that provides understanding and skills in the calculation of reinforced concrete structures in buildings such as plates, beams, and columns. Measures in concrete structures start from loading, reaction, dimensions of the structure, calculation of reinforcement, to detailed drawings of reinforcement sketches.

The learning of concrete structures uses a learning model adapted to the subject or material to be delivered. The beginning of learning at the first meeting until the third meeting usually uses a direct learning model or introduces new material to students. Learning on advanced material uses a problem-based learning model; the lecturer gives an analysis case for planning a concrete structure, either slab, beam, or column, then students are asked to solve some specific cases. The learning method used also adjusts the subject matter or material. The material was delivered using lecture and demonstration methods to provide students with the basic concepts of concrete structures. Discussion and question-and-answer methods are used when solving problems in the analysis and planning of concrete structures.

The assessment arrangements in this course consist of participation score (20%), assignments score (30%), mid-term test (20%), and final test (30%). The participation score is made up from attendance plus an activity score during classes (active in discussions, questions and answers, presentations, and expressing opinions according to the material). Assignments are given at the end of each subject to provide enrichment or practice in solving a concrete structure case. The topics of the assignments are the subject of one-way and two-way slabs, beams with flexural reinforcement and shear reinforcement, short columns, and slender columns. The mid-term written test was conducted in the eighth week, while the final test was conducted at the end of the semester.

During the pandemic, the teaching-learning process was carried out through distance learning to tackle the spread of COVID-19. The process received several complaints, especially in topics that required demonstrations, like "concrete structures". The basis for calculating reinforcement in concrete structures is fundamental structure analysis (shear, axial, and moment forces). The Mechanical Engineering course is the basis before taking the Concrete Structure course.

The problem is that the basics of the structural analysis course are separated by one to two semesters from the Concrete Structure course. Students in the basics of structural analysis course do not understand the implementation of their calculations on concrete structures, so in the Concrete Structures course, the lecturer must relate again to the material that has been obtained in the Engineering Mechanics course. Students in the Concrete Structure course do not understand how to determine reinforcement, which must be based on the value of a moment, shear, and axial (D, M, and N planes), all three of which are obtained from engineering mechanics calculations.

The learning of the Concrete Structure course always repeats the material for the Mechanical Engineering course. Students who cannot determine moments, shears, and axial (D, M, and N planes) will not be able to continue to calculate plans in concrete structures, so instructional media are needed to facilitate the understanding of concrete structures associated with engineering mechanics, especially on the subject of reinforced concrete columns.

So far, media that can increase understanding to implement engineering mechanics with concrete structures have not been developed. Based on these problems, one solution is to develop interactive learning media that can connect Engineering Mechanics with Concrete Structures, for example, making programs with programming languages, i.e., visual basic, Autoware, or other suitable programming languages.

Instructional media are assistive tools in learning both inside and outside the classroom. Learning media consist of physical and non-physical media or software. Learning media are used in the context of communication and interaction between educators and students as an intermediary from the source of information to the recipient. Therefore, learning media are any tools, environments, and all forms of activities that are conditioned to increase knowledge, change attitudes, and instill skills for students who use these media (Barr & Linebarger, 2010; Fenstermacher et al., 2010).

Learning media are all that can be used to send messages from teachers to students so that they can increase students' interest in learning. Learning media can help the teaching and learning process. Media are commonly used in learning. The media used range from conventional to modern. Conventional media include books, film projectors, radio, television, and cassettes. Modern media are computers, DVDs, the Internet, and smartphones (Linebarger, 2015).

Learning media are used to facilitate the interaction of educators and students in the teaching and learning process. Learning media are something that are used to assist in the learning process. Learning media must be able to attract the attention of students. Good learning media consist of audio and visual sources. Good media must be able to make learning more fun and be able to reach all students (Hidayati & Wuryandari, 2013b).

Learning needs media support, such as audio, video, animation, and so on (Hidayati & Wuryandari, 2013b). Learning media that can increase student interest in subjects need to be developed. Attractive learning media can help students learn, so they can improve their learning outcomes (Linebarger, 2015). The development of learning media can be done by utilizing the latest information technology. Information technology can be used to create interesting learning media (A. El Saddik, 2001).

Learning media must be prepared and planned properly. The learning media used must follow the subject matter to be delivered. Media selection is very important because it will affect success in learning. Learning objectives will be achieved with the help of effective media (Michael & Jones, 2014). Learning media can be the solution in delivering material related to basic and pedagogical theories. The success of the media depends on the characteristics of students and subjects (Abdulrahaman et al., 2020).

Students can learn actively and independently with learning media, but there is still control from the educator. Students must be controlled and reminded of the task at hand. Educators must be ready if there are questions and discussions from students. Learning media must follow technological developments, so they must be developed continuously and periodically (Sangsawang, 2015).

Information and Communication Technology (ICT) is an important aspect in developing digital learning media consisting of text, audio, video, and so on. ICT can combine several technologies that provide information in the best format, package, and size (Guan et al., 2018). Appropriate use of ICT in teaching changes the learning environment from teacher-centered to learner-centered. The shift from teaching to learning creates student-centered learning where the teacher is a facilitator in learning (Coleman et al., 2016).

Related research on the development of learning media resulted in discovering that the low achievement of students in the concrete structure course was due to the low understanding of students. To overcome this problem, interactive learning media were made to improve student understanding (Widodo & Raharjo, 2008). The use of interactive multimedia in the learning process has an impact; those students who are more active in asking questions, are more interested in the media used, and have a better interest in learning (Crengu, 2014). This is supported by the increase in the average learning outcomes (pre-test and post-test). Good media are seen from the quality of the material, appearance, and other media equipment. Virtual-based learning media can be used well because it is easy and interesting to use in learning. Good learning media can improve the quality of learning (Gde et al., 2020).

Based on the description above, it can be concluded that learning media is a tool in the communication or delivery of material from educators to students. Learning media can be hardware or software. Good learning media can provide convenience in delivering material, making learning interesting and fun. Media must also be accessible by students easily anytime and anywhere. Good learning media can facilitate learning and be able to improve student learning outcomes.

Based on the description above, this research takes the title "Learning Concrete Structures Using IT-Based Interactive Media". The aims of this study are as follows: (1) to determine the feasibility of the media that can connect engineering mechanics with concrete structures; (2) to determine student responses to learning with interactive media in the Mechanical Engineering with Concrete Structures course; and (3) to find out the understanding of the concept of mechanical engineering and concrete structures by students by applying learning using interactive media.

2 METHOD

This research is development research. The development carried out is the development of the concrete structure learning media. The 4D model used in the development of this instructional media has the stages of define, design, development, and dissemination. This research is only up to development with limited trials. The media modeling can be seen in Figure 1. This media can be accessed at http://pembelajaran-strukturbeton.unesa.ac.id.

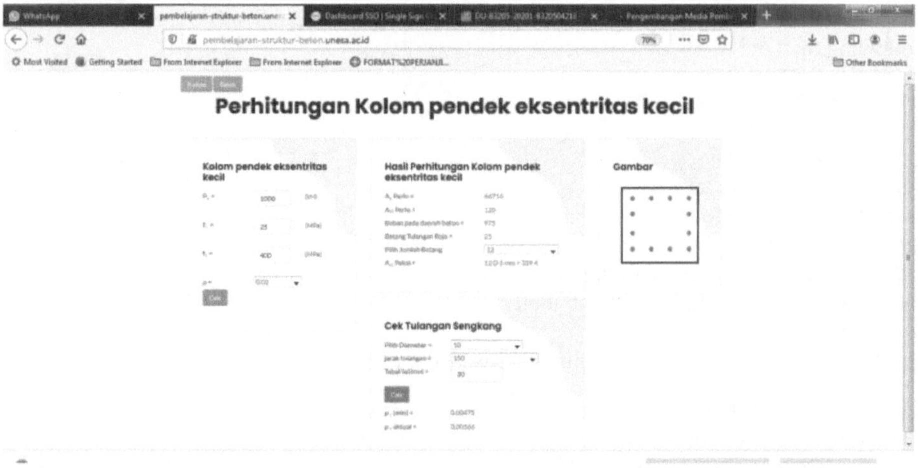

Figure 1. Column concrete structure media.

The data collection used a questionnaire response to the use of learning media and written tests. The data analysis technique uses the quantitative descriptive analysis technique. The data that has been obtained are then analyzed and described in the form of tables and graphs, to allow arrival at the conclusions.

The assessment of the feasibility of learning media is carried out by comparing the score of the assessment results with the criterion score on the variables based on the normal curve. The analysis steps are as follows:

- Calculate the total score of each sample on each variable
- Calculate the maximum score for each variable, with the formula for maximum score = Number of instrument items × highest score
- Calculate the minimum score for each variable, that is, the minimum score formula = Number of instrument items × lowest score
- Calculate the ideal Mean (Mi) with the formula = (maximum score + minimum score)
- Calculate the ideal standard deviation (SDi) with the formula = 1/6 (maximum score – minimum score)

Determine the assessment criteria for each variable based on the range of scores on the normal curve as shown in Table 1.

Table 1. Criteria assessment of each variable.

Score Range	Category
$X > Mi + 1,5\ SDi$	Very good
$Mi + 0,5\ SDi < X \leq Mi + 1,5\ SDi$	Good
$Mi - 0,5\ SDi < X \leq Mi + 0,5\ SDi$	Quite good
$Mi - 1,5\ SDi < X \leq Mi - 0,5\ SDi$	Not good
$X \leq Mi - 1,5\ SDi$	Very Poor

3 RESULTS AND DISCUSSION

3.1 *Student Response*

Student response analysis was obtained by filling out a questionnaire using respondents from the class under study. The questionnaire that was distributed to respondents contained statements of several aspects while using learning media. The questionnaire consisted of 10 statements concerning the following matters:

- Media display layout
- Media display quality
- Clarity of text in media
- Color composition in media
- Ease of media operation
- Ease of accessing media
- Ease of understanding the material on the media
- The suitability of the media with the material
- Use of media in learning
- Clarity of the media in providing illustrations of the material

The assessment of student responses to the concrete structure learning media was carried out by comparing the average scores with the assessment criteria. The assessment of student responses on the questionnaire sheet is contained in 10 statement items. The analysis is done by first determining the highest ideal score, lowest ideal score, ideal mean, and ideal standard deviation. The highest ideal score and the lowest ideal score are found as follows.

$$\begin{aligned}\text{Highest ideal score} &= \text{Number of instrument items x highest score}\\ &= 10 \times 5 = 50\\ \text{Lowest ideal score} &= \text{Number of instrument items x lowest score}\\ &= 10 \times 1 = 10\end{aligned}$$

After knowing the highest ideal score and the lowest ideal score, the ideal mean (Mi) and ideal standard deviation (SDi) can be determined.

$$\begin{aligned}\text{Mi} &= \tfrac{1}{2}\,(\text{Highest ideal score} + \text{Lowest ideal score})\\ &= \tfrac{1}{2}\,(50 + 10)\\ &= 30\\ \text{SDi} &= 1/6\,(\text{Highest ideal score} + \text{Lowest ideal score})\\ &= 1/6\,(50 - 10)\\ &= 6,67\end{aligned}$$

Based on the ideal mean value and the ideal standard deviation, it can be determined by the categorization of the assessment of the concrete structure learning media. The categorization can be seen in Table 2 below.

Table 2. Student response assessment criteria.

Score Range	Category
$X > 40$	Very good
$33{,}3 < X \leq 40$	Good
$26{,}7 < X \leq 33{,}3$	Quite good
$20 < X \leq 26{,}7$	Not good
$X \leq 20$	Very Poor

The results of the responses to 33 students got a total score of 1489, with an average number of 45.1. The summary of student response data can be seen in Table 3 and a comparison of the number of student response scores can be seen in Figure 2.

Table 3. Data description of student response to concrete structure learning media.

No	Min	Max	Total Value	Average
1	3	5	148	4,48
2	3	5	146	4,42
3	3	5	154	4,67
4	2	5	136	4,12
5	3	5	151	4,58
6	1	5	143	4,33
7	3	5	150	4,55
8	3	5	156	4,73
9	3	5	158	4,79
10	3	5	147	4,45
Total	27	50	1489	45,12

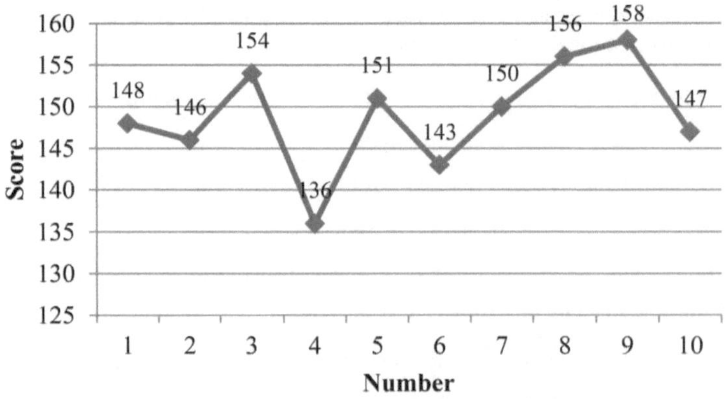

Figure 2. Student response to concrete structure learning media.

Based on Table 3. the average score of student responses is 45.1, so that according to Table 2 student responses to the application of concrete structure learning media are included in the very good category. The highest score in statement number 9 is that this media is useful in learning, then statement number 8 is that the media is by the material. The lowest value in statement number 4 is the color composition in the media.

Based on the results of these responses, it was found that according to students the concrete structure media followed the material and was useful in understanding the material of concrete structures. The concrete structure media is still less attractive, especially in the color composition. These results are from research that explains that good media can be seen from the quality of the material, appearances, and other media equipment. Learning media must be easy and interesting to use in learning (Gde et al., 2020).

3.2 *Student test results*

Student cognitive test results are quiz test results indicated by the scores obtained by students after taking the test through the assessment sheet. The results of students' cognitive tests can be seen in Table 4.

Table 4. Student cognitive learning test results.

Number of Respondents	32
Total score	2395
Average	74,84375
Minimum Value	60
Maximum Value	90

Based on the data in Table 4, it is found that the average student score is 74.84, which is above the graduation standard limit value of 70, but there is still one student with a score of 60. The impacts include students being more active in asking questions, being more interested in the media used, and having a better interest in learning, which increase the average acquisition of learning outcomes. This is in line with the findings of Hidayati and Wuryandari (2013a) that learning media can improve students' understanding with the help of media with actual construction forms. The concrete structure media that is made easily accessible and by the material makes it easier for students to understand the material. However, of appearance and color composition are considered lacking by students, and are parts that must be improved to get better results.

4 CONCLUSIONS

4.1 *Conclusions*

Based on the results of research and discussion, it can be concluded as follows.

- The results of student responses to the concrete structure learning media got a total score of 1489, with an average of 45.1, which falls in the very good category.
- Student learning outcomes test after the application of concrete structure learning media had an average score of 74.84, a minimum score of 60, and a maximum value of 90.
- In terms of feasibility, the concrete structure learning media is feasible to use, and what needs to be improved is the appearance of the color composition to make it more attractive.

4.2 *Suggestions*

Based on the conclusions above, several things can be suggested, as follows:

- Students like learning with interesting media, so they feel happy in learning, for that each subject needs to be have interesting media and make it easier for students to learn.
- The institution should provide training in making media, to improve the quality of learning media for each lecturer.
- Students should make good use of the advances in technology and media that have been provided with positive things, such as accessing learning materials on the web, e-learning, journals, and e-books, and so on.
- Future research should be conducted with more in-depth research on concrete structure learning media in other ranges of material and with a more attractive appearance.

REFERENCES

A. El Saddik. (2001). *Interactive multimedia learning*. Springer.

Abdulrahaman, M. D., Faruk, N., Oloyede, A. A., Surajudeen-bakinde, N. T., & Olawoyin, L. A. (2020). Multimedia tools in the teaching and learning processes?: A systematic review. *Heliyon*, *6*(11), e05312. https://doi.org/10.1016/j.heliyon.2020.e05312

Barr, R., & Linebarger, D. L. (2010). Introduction to the special issue: content and context in infant-directed media. *Infant and Child Development*, *19*(6), 553–556.

Coleman, L. O., Gibson, P., Cotten, S. R., Howell-Moroney, M., & Stringer, K. (2016). Integrating computing across the curriculum: the impact of internal barriers and training intensity on computer integration in the elementary school classroom. *J. Educ. Comput. Res*, *54*(2), 275–294.

Fenstermacher, S. K., Barr, R., Garcia, A., Salerno, K., Shwery, C., Pempek, T. A., Calvert, S., & Linebarger, D. L. (2010). Infant-directed media: an analysis of product information and claims. *Infant and Child Development*, *19*(6), 557–576.

Gde, A., Mulyono, W., & Karim, N. (2020). *Virtual Learning-Based Media to Improve Learning Quality*.

Guan, N., Song, J., & Li, D. (2018). On the advantages of computer multimedia-aided English teaching. *Procedia Comput. Sci*, *131*, 727–732.

Hidayati, N., & Wuryandari, A. I. (2013a). Media Design for Learning Indonesian in Junior High School Level. *Procedia – Social and Behavioral Sciences*, *67*(November 2011), 490–499. https://doi.org/10.1016/j.sbspro.2012.11.354

Hidayati, & Wuryandari. (2013b). Media design for learning Indonesian in junior high school level. *Procedia – Soc. Behav. Sci*, *6*, 490–499.

Linebarger. (2015). Educational media: potentials for learning, Int. Encycl. *Soc. Behav. Sci. Second Ed*, *6*, 246–254.

Michael, & Jones. (2014). *Educational Media and Technology Yearbook*. Springer.

Rideout, V., Foehr, U. G., & Roberts, D. F. (2010). *Generation M2: Media in the Lives of 8–18-Year-Olds*. Kaiser Family Foundation, Menlo Park, CA.

Rideout, V., & Robb, M. B. (2020). *The Common Sense census: Media use by kids age zero to eight.* Common Sense Media.
Sangsawang, T. (2015). Instructional Design Framework for Educational Media. *Procedia – Social and Behavioral Sciences*, *176*, 65–80. https://doi.org/10.1016/j.sbspro.2015.01.445
Widodo, & Raharjo. (2008). Internet-Based Interactive Learning to Improve Student Competency Achievement in the Concrete Structure Course I. *Jurnal Pendidikan Teknik Sipil Dan Perencanaan*.

Innovation on Education and Social Sciences – Maureen et al (Eds)

Development of hybrid project-based learning in State University of Surabaya

Mustaji, Siti Masitoh & Hirnanda Dimas Pradana
Educational Technology Department, Universitas Negeri Surabaya, Surabaya, Indonesia

ABSTRACT: All aspects of life are connected to the internet in the industrial revolution 4.0 era. All information can be accessed quickly and easily without regard for space, distance, or time. Students can learn with anything, anywhere, and at any time. Lecturers are not the only source of information. Today's online learning environment makes it challenging to incorporate project-based learning in order to produce usable products. Project-based learning is a type of learning that applies to both online and offline learning environments. This research aims to create a viable and effective hybrid project-based learning environment that will improve learning outcomes. The ADDIE model is used to develop products. An expert validated the feasibility model, and the effectiveness model was validated using independent t-test samples and the SPSS program. Based on the results of expert assessment, the following feasibility results are known: (1) expert assessment related to learning model of 96.87%; (2) expert assessment related to learning materials of 95.34%; and (3) expert assessment related to learning design of 97.39%. Experts gave an overall rating of 95.78%. As a result, a value of 96.53% was obtained from all expert trials. The findings indicate that hybrid project-based learning is both feasible and effective in terms of enhancing learning outcomes. Hybrid Project-Based Learning (HPjBL) provides numerous benefits to students and educators. Academic research is increasingly supporting the use of HPjBL to engage students, reduce absenteeism, enhance collaborative learning skills, and boost academic performance.

1 INTRODUCTION

Information technology and communication are advancing at a breakneck pace at the moment. Today's education requires the inclusion of internet services as a resource. Internet service is a global network comprising numerous computer networks that enable any computer to be connected to the internet by anyone, at any time, and from anywhere to perform data transmission and information retrieval (Warsita, 2008). Students can use the internet to access and analyze available information. The ability to access, organize, integrate, evaluate, and create information through communication technology is becoming increasingly important in the world of education—the industrial revolution era 4.0 "forces" the educational system to establish technology-enhanced learning processes (Hamimah et al., 2020).

Project-based learning is a type of learning that is now being researched intensively in relation to 21st century skills. Project-based learning is a cutting-edge learning methodology that uses digital technologies integrated into existing educational institutions. The core principle behind the project-based learning model is "learning by doing," which employs a project-based approach to make activities more meaningful and suitable for real-world application (Barron et al., 1998). Project-based learning is a type of education that involves students solving project-based problems by creating artifacts (Koh, Herring, & Hew, 2010). Project assignments are another term for project-based learning. It is defined by the use of questions or issues to entice students to participate in design, problem-solving, decision-making, and investigative activities that lead to a product or presentation (Tan & Chapman, 2016). Students engage in problem-solving activities such as posing and clarifying problem questions, debating ideas, making predictions, designing plans

DOI 10.1201/9781003265061-2

and experiments, collecting and analyzing data, drawing conclusions, communicating their ideas and findings to others, posing new questions, and creating artifacts in project-based learning (Huang, Spector, Yang, & Garnar, 2019). Project assignments are a defining feature of this learning model. They motivate students to engage in activities by developing their knowledge through problem-solving searches that result in production.

The goal of this research is to establish project-based learning via the use of online information and communication technologies. The potential for online education to make the learning process more student-centered, inventive, and adaptable is enormous. Online learning is described as a learning experience that takes place in a synchronous or asynchronous environment and involves the use of various devices (such as mobile phones, laptops, and other computers) that have internet connection (Dhawan, 2020). When online learning is combined with project-based learning models, the combined approach is called Hybrid Project-Based Learning (HPjBL). HPjBL consists of a single class group and enables communication with a broad audience via information search or even across disciplines (Krauss & Boss, 2013). It will then enable students to search for additional information via the internet, thereby expanding their knowledge through problem-solving and project development using internet technology.

When students are learning online, they find it difficult to construct work in lectures. The lack of a learning paradigm that promotes project-based learning in online education is the source of this problem. As a result, we need online learning innovation that allows students to generate goods as a result of their education. Students are productive persons, thus a model that accommodates their demands for generating or manufacturing usable products is essential. Furthermore, online education provides more freedom and encourages students to seek self-directed learning (Sadikin & Hamidah, 2020).

The use of HPjBL is expected to improve learning outcomes. Learning outcomes are an essential part of learning HPjBL and are expected to improve student outcomes. Learning outcomes are critical components of learning (Sudjana & Ibrahim, 2009). The term "learning outcomes" refers to behavioral changes in the cognitive, affective, and psychomotor domains (Dimyati & Mudjiono, 2006). Additionally, it was mentioned that the learning outcomes are the result of learning interaction. Students' ability level has a significant impact on the learning outcomes they achieve. According to Clark, 70% of student learning outcomes in school are determined by students' abilities and 30% by the environment (Sudjana & Rivai, 2001).

Along with the two factors mentioned previously, additional factors contribute to students' learning outcomes, most notably the approach to learning factor. It relates to students' learning efforts, which include strategies and methods for learning. This research will assess students' success in learning planning courses by examining various explanations for learning outcomes, particularly in creating learning implementation plans and teaching simulation videos. With all of the constraints and complications, lecturers are required to fully implement online learning despite all of the complications that accompany it (Wahyuni, Aziz, Wargadinata, & Efiyanti, 2021).

HPjBL implementation is expected to accommodate space and time constraints consistent with the role of digital technology in supporting learning and pedagogy. Students can use more ideas, propose more actions, and contextualize them using the HPjBL model. As a result, project-based learning in a virtual environment has the potential to foster more meaningful learning. Additionally, it will make it easier for teachers to keep track of all student activities within the online environment provided. Lecturers should be more inventive in their learning resources to enhance the distance learning process, which can be accomplished by developing learning resources. The previous statement is reinforced by research findings that the learning process must be conducted effectively via online education (Sintema, 2020). Each project concludes with a reflection stage that provides students with an opportunity to reflect.

2 METHOD

This study applied the research and development method. The procedure for development follows the ADDIE model (analyze, design, develop, implement, and evaluate) (Branch, 2009).

The analysis stage is where the requirements for designing learning are analyzed. The design stage is used to create any necessary models, materials, or hybrid project-based learning devices. At the development stage, a hybrid project-based learning model was developed. The implementation stage involves testing model products in both synchronous, and asynchronous online modes with research subjects enrolled in programming courses related to the design of the learning system, specifically graduate students at Surabaya State University. Trials of implementation had a pre-experiment and a single group pretest–posttest design. Implementation of knowledge was based on the stages described by Han and Bhattcharya (Dewi, 2020; Warsono & Hariyanto, 2012).

Specifically, there are five stages: (1) planning, (2) design and development, (3) monitoring, (4) peer review, and (5) revision. The evaluation stage gathers data through knowledge tests and product assessment sheets from learning system design courses. In order to determine the model's feasibility, an expert test learning model, expert learning materials, and expert test learning design are conducted using the structured interview method. In order to determine the efficacy of hybrid project-based online learning, independent t-test samples were used. In control classes conventional models are used, and in experimental classes hybrid project-based learning models are used. Independent test statistics for t-test samples are calculated using the SPSS program.

a. Analysis. The analysis section conducts a needs analysis, an audience analysis, a topic analysis, and a task analysis (Moradmand, Datta, & Oakley, 2014).
b. Design. This design stage is the planning stage to implement online learning by creating a conducive learning environment. This design stage is where the developer will plan the project he is developing (development of a product). Unfortunately, numerous products fail or do not work because the design is not mature or ready but is still compelled to continue (Pradana & Kuswandi, 2018).
c. Develop. The development stage involves the process of integrating technology, management, and educational processes. Additionally, learning designers have begun developing and producing high-quality educational materials, beginning with design, instructional materials, and learning (Aldoobie, 2015).
d. Implementation. The implementation stage is a critical step in putting the developed learning products into action (Budiarta, 2014).
e. Evaluation. The final stage is to conduct an evaluation, the results of which will indicate whether the media is valid or invalid (Gusmida & Islami, 2017). This stage occurs following the distribution of online learning to students. The lecturer administers pretest and posttest questions to students to ascertain their ability levels.

3 RESULTS

3.1 *Hybrid project-based learning feasibility*

Measuring the feasibility is conducted through an expert review of the learning model, an expert review of the learning materials, and an expert review of the learning design using the structured interview method. Table 1 contains data on the model's feasibility.

Table 1. Hybrid project-based learning feasibility.

No	Feasibility	Percentage
1	Learning Model	96.87
2	Learning Materials	95.34
3	Learning Tools	97.39
	Mean	96.53

The following feasibility results were found as a result of the expert assessment: (1) expert assessment related to the learning model was 96.87%, (2) expert assessment related to the learning materials was 95.34%, and (3) expert assessment related to the learning design was 97.39%. Experts gave an overall rating of 95.78%. As a result, a percentage value of 96.53% was obtained from all expert trials. If the calculation's results are converted to the score criteria specified in Arikunto (2013), the percentage is in the category of very good and feasible. HPjBL is an innovation of the learning model in the 21st century in response to challenges in education to produce a relevant learning model for students today that has a different way of learning compared with generations in the previous era. The novelty of this model is that in the model syntax, there is a peer review stage between students that is done as the basis for improving student work. In general, HPjBL stages use reviews and evaluations from lecturers as the basis for improving student work. This peer review stage is also one of the efforts to improve learning outcomes.

3.2 *Effectiveness of hybrid project-based learning*

Once deemed worthy by experts, the model's effectiveness in improving learning outcomes is determined in the subsequent implementation stage. HPjBL implementation consists of six stages, which are detailed below. These stages must be carried out coherently to develop the student's capacity to comprehend the learning material. Additionally, students will understand what it means to be a student capable of developing marketable products.

The first phase is the Plan, in which a lecturer establishes a discussion forum to discuss lecture contracts, raises fundamental questions about the importance of designing the learning system to be developed, determines the theme of the learning materials to be designed and developed, and agrees on schedules and deadlines for working on the learning system development project. At this stage, a preliminary understanding of the learning system's design is developed that is easily comprehendible by students—the student comprehension of the learning system's design.

The second phase is development and design. At this stage, lecturers assist students in creating products, developing learning systems, and designing them. While students design and develop until the learning system is complete, they also create products that adhere to the theory of learning and self-learning. Students have already developed a design for the product they will develop at this point. The students' designed learning system adheres to established theories of learning.

Monitoring is stage three. At this stage, lecturers monitor the upload process for the learning system's design of students' work due to their learning. At the same time, students submit their assigned work by the agreed-upon deadline. Students have created products that have been validated by the lecturer and are ready to be uploaded to the designated location.

Peer review is stage four. Lecturers observe the peer review process and student activity as part of the learning process at this stage. As part of the learning process, students provide feedback and evaluations on one another's work during peer review activities. In addition, students provide feedback and suggestions on their classmates' work at this stage. From now on, this work will serve as a guide for students as they revise their products in order to create high-quality, publishable works.

Revisions occur in stage six. At this point, lecturers assist students in revising their work in response to peer reviewer feedback and suggestions. When peer reviewers make suggestions, students revise their work/products. Students revise their products during this stage in order to make them better. The outcomes of student-led product revisions will result in high-quality and publishable products.

Publish at stage six. At this point, lecturers monitor students' progress toward publication and assign grades in addition to providing feedback on students' work. Students revise their works in light of peer review recommendations. At this point, it is feasible to disseminate student-created products.

3.3 *Calculation results of pretest control class and experimental class*

At the start of the experiment on 10 March 2021, students were divided into control and experimental classes, with each group consisting of 24 students. The purpose of the two pretest average similarity tests in the control class and this experiment is to ascertain whether or not students in both classes have some initial abilities in common. The two-test average pretest similarity test is based on an independent t-test sample. The following conclusions were drawn from the pretest results for the control and experimental groups as in Tables 2–4.

Table 2. Pretest average (mean) score test results control and experiment classes.

	Group	N	Mean	Std. Deviation	Std. Error Mean
Value	Experiment	41	73,5122	3,95678	,61795
	Control	41	73,4146	4,51096	,70449

Table 3. Control and experiment class pretest results.

		Levene's Test for Equality of Variances		t-test for Equality of Means						
									95% Confidence Interval of the Difference	
		F	Sig.	T	df	Sig. (2-tailed)	Mean Difference	Std. Error Difference	Lower	Upper
Value	Equal variances assumed	,815	,369	,104	80	,917	,09756	,93711	–1,76734	1,96246
	Equal variances not assumed			,104	78,664	,917	,09756	,93711	–1,76783	1,96295

According to the calculations, the abovementioned average score (mean) for the pretest control class (73.41) and the experimental class (73.51) is greater than the known average score (mean). Then, using the asymp column, the calculation result with t-test yielded a significance value. Sig (2-tailed) of 0.917 or a significance of >0.05 (0.917 > 0.05), the pretest value in this case represented by the value of the message design course indicates that the initial ability in the experimental and control classes is identical. Thus, the initial ability pretest value is known, and the initial capabilities of the control and experiment classes are identical. After establishing that both classes have the same initial ability, the next step is to provide learning in both classes with the same materials, teachers, facilities, and infrastructure; the only difference in learning between the two groups is that control classes are treated according to conventional models, whereas experimental classes are treated according to experimental models.

3.4 *Results of calculations posttest control class and experimental class*

After learning with HPjBL in experimental and conventional classes or control classes, both classes were be given a posttest. Posttest data on learning outcomes derived from student-created product assessments via HPjBL. Each student produces work and is evaluated on the learning system's understanding and product design projects. After collecting posttest data from both classes, two posttest averages were calculated in the control and experiment classes to ascertain the extent of the HPjBL's influence. This two-test average posttest similarity employs an independent t-test sample.

The following results were obtained from the posttest results provided for the experiment class and the control class:

Table 4. Average score test result (mean) posttest control and experiment classes.

	Group	N	Mean	Std. Deviation	Std. Error Mean
Value	Experiment	41	83,9512	4,19494	,65514
	Control	41	75,4146	2,63605	,41168

Table 5. Test result t posttest control and experiment classes.

		Levene's Test for Equality of Variances		t-test for Equality of Means						
									95% Confidence Interval of the Difference	
		F	Sig.	T	Df	Sig. (2-tailed)	Mean Difference	Std. Error Difference	Lower	Upper
Value	Equal variances assumed	17,357	,000	11,033	80	,000	8,53659	,77375	6,99677	10,07640
	Equal variances not assumed			11,033	67,329	,000	8,53659	,77375	6,99231	10,08086

According to the mean calculation in Table 5, the average score (mean) for the control class posttest is 75.41, while the experiment class has an average score of 83.95. Then, using the asymp column, the calculation result from the t-test yielded a significance value of Sig 0.00 (2-tailed) or significance 0.05 (0.00 0.05). Because the only differentiator in the learning process is the HPjBL treatment, it is the treatment that accounts for the difference in posttest values between the control and experiment groups. Posttest results in experimental classes using HPjBL and control classes using conventional learning models or traditional learning, in general, indicate that HPjBL effectively improves learning outcomes.

HPjBL provides numerous benefits to students and educators. Academic research is increasingly supporting the use of HPjBL to engage students, reduce absenteeism, enhance collaborative learning skills, and boost academic performance (Dewi, 2020). Additionally, HPjBL benefits students in the following ways: (1) increased attendance, increased self-reliance, and a more positive attitude toward learning (Thomas, 2000); (2) academic benefits are comparable to or superior to those generated by alternative models, with students involved in the project taking greater ownership of their learning than they do during more traditional classroom activities (Boaler, 1997); (3) opportunities for the development of sophisticated skills such as critical thinking, problem-solving, collaboration, and communication; (4) increased access to a broader range of educational opportunities in the classroom, as well as strategies for engaging culturally diverse students (Railsback, 2002). HPjBL positively affected student learning and grades because it enhanced students' critical thinking and problem-solving abilities, increased peer-to-peer discussion opportunities during group HPjBL activities, and increased students' interest and motivation to learn through practical engineering applications (Magnus et al., 2020). Based on the findings of research and literature from various sources, the astounding conclusion that HPjBL can improve learning outcomes can be drawn.

4 CONCLUSION

HPjBL development products are deserving of being used as a model for enhancing learning outcomes. It has been demonstrated by the expert and user evaluations of each component of the development product, specifically the learning models, learning materials, and learning tools, that the value of all model components is sufficient to justify their use in learning.

HPjBL can aid in the improvement of educational outcomes. It was accomplished by comparing the posttest scores of the control and experimental groups, which revealed that learning with HPjBL produced significantly better results than conventional learning. HPjBL is a type of blended learning that can be used with other courses that share similar characteristics. When distributing this model more widely, it is necessary to analyze learner characteristics and learning objectives to develop more effective learning models. Additional variables such as learning motivation, learning independence, and a variety of other pertinent variables may be assessed by other researchers. In this model, interactions are suggested to be combined with synchronous interactions to create real-time interactions, a novel concept in online learning models.

REFERENCES

Aldoobie, N. (2015). ADDIE Model. *American International Journal of Contemporary Research*, 5(6).

Arikunto, S. (2010). *Prosedur Penelitian Suatu Pendekatan Praktik*. Rineka Cipta.

Barron, B. J. S. S., Schwartz, D. L., Vye, N. J., Moore, A., Petrosino, A., Zech, L., & Bransford, J. D. (1998). Doing with understanding: Lessons from research on problem-and project-based learning. *Journal of the Learning Sciences*, 7(3–4), 271–311. https://doi.org/10.1080/10508406.1998.9672056

Boaler, J. (1997). *Experiencing School Mathematics: Teaching Styles, Sex, and Setting*. Open University Press.

Branch, R. M. (2009). Instructional design: The ADDIE approach. In *Encyclopedia of Evolutionary Psychological Science* (Vol. 722). Springer Science & Business Media. https://doi.org/10.1007/978-3-319-19650-3_2438

Budiarta, I. wayan. (2014). Pengembangan Multimedia Interaktif Model Addie Untuk Meningkatkan Motivasi Belajar Sejarah Siswa Kelas X-1 Semester Genap Di Sman 1 Sukasada, Buleleng, Bali. *Jurnal Pendidikan Sejarah*, 2(1), 1–12. https://ejournal.undiksha.ac.id/index.php/JJPS/article/view/3620

Dewi, U. (2020). *Model Project Based Online Learning* (1st ed.). CB Beta Aksara.

Dhawan, S. (2020). Online learning: A panacea in the time of COVID-19 crisis. *Journal of Educational Technology Systems*, 49(1), 5–22. https://doi.org/10.1177/0047239520934018

Dimyati, & Mudjiono. (2006). *Belajar dan Pembelajaran*. Rineka Cipta.

Gusmida, R., & Islami, N. (2017). The development of learning media for the kinetic theory of gases using the ADDIE model with augmented reality. *Journal of Educational Sciences*, 1(1), 1. https://doi.org/10.31258/jes.1.1.p.1-10

Hamimah, H., Arlis, S., Arwin, A., Chandra, C., Anita, Y., Kenedi, A., & Kharisma, A. (2020). Thinking the most convenient analysis of alpha generation by using social science story digital books. *Ýlköðretim Online*, 19, 78–86. https://doi.org/10.17051/ilkonline.2020.654895

Huang, R., Spector, J. M., Yang, J., & Garnar, A. W. (2019). Educational technology: a primer for the 21st century. In *Springer*. Springer. https://doi.org/10.4324/9780203735657-8

Koh, J. H. L., Herring, S. C., & Hew, K. F. (2010). Project-based learning and student knowledge construction during asynchronous online discussion. *The Internet and Higher Education*, 13(4), 284–291. https://doi.org/10.1016/j.iheduc.2010.09.003

Krauss, Jj., & Boss, S. (2013). Thinking Through Project-Based Learning: Guiding Deeper Inquiry. In *Library Media Connection* (Vol. 32, Issue 4). SAGE Publications.

Magnus, D. D. M., Carbonera, L. F. B., Pfitscher, L. L., Farret, F. A., Bernardon, D. P., & Tavares, A. A. (2020). An Educational Laboratory Approach for Hybrid Project-Based Learning of Synchronous

Machine Stability and Control: A Case Study. *IEEE Transactions on Education*, 63(1), 48–55. https://doi.org/10.1109/TE.2019.2956909
Moradmand, N., Datta, A., & Oakley, G. (2014). The Design and Implementation of an Educational Multimedia Mathematics Software?: Using ADDIE to Guide Instructional System Design. *The Journal of Applied Instructional Design*, 4(1), 37–49. http://www.jaidpub.org/wp-content/uploads/2014/10/Moramand-Datta-Oakley.pdf
Pradana, H. D., & Kuswandi, D. (2018). *Augmented Reality Learning Materials for Motion Picture Making Subject*, 6(3), 108–114.
Railsback, J. (2002). *Project-Based Instruction: Creating Excitement for Learning.*
Sadikin, A., & Hamidah, A. (2020). Pembelajaran Daring di Tengah Wabah Covid-19. *Biodik*, 6(2), 109–119. https://doi.org/10.22437/bio.v6i2.9759
Sintema, E. J. (2020). Effect of COVID-19 on the performance of grade 12 students: Implications for STEM education. *Eurasia Journal of Mathematics, Science and Technology Education*, 16(7), 1–6. https://doi.org/10.29333/EJMSTE/7893
Sudjana, N., & Ibrahim. (2009). *Penelitian dan penilaian pendidikan*. Sinar Baru Algensindo.
Sudjana, N., & Rivai, A. (2001). Penggunaan media pengajaran dalam PBM. *Bandung: Sinar Baru Algensindo*.
Tan, J. C. L., & Chapman, A. (2016). *Project-Based Learning for Academically-Able Students: Hwa Chong Institution in Singapore*. SensePublishers.
Thomas, J. W. (2000). *A review of research on project-based learning*.
Wahyuni, E. N., Aziz, R., Wargadinata, W., & Efiyanti, A. Y. (2021). Investigation of Primary School Teacher Readiness in Online Learning during the Covid-19 Pandemic. *Madrasah: Jurnal Pendidikan Dan Pembelajaran Dasar*, 13(2), 97–113. https://doi.org/10.18860/mad.v13i2.11343
Warsita, B. (2008). *Teknologi pembelajaran landasan dan aplikasinya*. Rineka Cipta.
Warsono, H., & Hariyanto, M. S. (2012). Pembelajaran aktif teori dan asesmen. *Bandung: PT Remaja Rosdakarya*.

Innovation on Education and Social Sciences – Maureen et al (Eds)

Profile of changes in adaptive physical education learning during the Covid-19 pandemic

Erick Burhaein
Doctoral Program of Sports Education, School Postgraduates Studies, Universitas Pendidikan Indonesia, Indonesia

Beltasar Tarigan, Dian Budiana & Yudy Hendrayana
Faculty of Sports and Health Education, Universitas Pendidikan Indonesia, Indonesia

Diajeng Tyas Pinru Phytanza
Faculty of Education, Department of Special Education, Universitas Negeri Yogyakarta, Indonesia

ABSTRACT: The COVID-19 pandemic has an uneven and measurable impact on adaptive physical education (APE) learning given the different geographical conditions and capacities of people in Indonesia. The purpose of this study is to describe the dynamics of learning implementation in pandemic times. The methodology used in this research is descriptive with qualitative approaches and literature studies. Participants in this study were 2 APE teachers from West Java and Bali provinces in Indonesia. The interview instrument has been validated for each item through expert judgment by nine experts. The reliability of the instrument was tested with an Alpha Cronbach value of 0.913. Data analysis is done using interpretive descriptive methods, through an analysis cycle in the form of interactive cycles. Based on the results of the study, it can be concluded that learning has changed from before the pandemic, i.e., offline to online is considered the best solution today. But various problems and obstacles of learning APE online and distance learning (DL) appear in its implementation. Therefore, it takes knowledge about the implementation and appropriate strategies to achieve APE learning success. Therefore, this study has implications for teachers' knowledge of implementation and strategies in achieving APE success on the DL during pandemic times. The authors suggest further researchers review the implementation and learning strategies of APE during the COVID-19 pandemic within the region and with a wider number of respondents.

1 INTRODUCTION

Today the global world is shocked by a new type of virus that has never been encountered before and is widespread globally. The Covid-19 virus, or what we usually call the coronavirus, was designated as a global pandemic by WHO in September 2021 (Wu & Zha, 2020; Purwanto, Lumintuarso, & Burhaein, 2021; Putra, Purwanto, & Burhaein, 2021). The Covid-19 outbreak is a type of epidemic that spreads very high and fast. This epidemic attacks the human immune (Li et al., 2020; Simpson et al., 2020; Woods et al., 2020) and respiratory systems (Castañeda-Babarro et al., 2020; Chen et al., 2020; Wei et al., 2020). Learning has undergone massive changes since the pandemic. Changes have occurred in some of the main aspects of learning by students. Some of the main aspects of learning are affective (Bland et al., 2021; Ko et al., 2020; UNESCO, 2020), psychomotor (Alrashed et al., 2021; Karina, Maryanti, & Wulandary, 2021), and cognitive (Brom et al., 2020; Fiorenzato et al., 2021; Miners, Kehoe, & Love, 2020).

DOI 10.1201/9781003265061-3

During the pandemic, the implementation of education in Indonesia was carried out by online distance learning (DL), this includes APE learning that is changing to online. Due to the pandemic, the implementation of education in Indonesia is carried out online, especially adapted physical education (APE) (Burhaein, 2020; Nurulfa et al., 2021; Phytanza, Burhaein, & Pavlovic, 2021; Yunus & Rezki, 2020). This causes APE learning that cannot be carried out evenly and measurably, given the different geographical conditions of Indonesia and the different abilities of the people, so online learning is not ready (Burhaein, Tarigan, & Phytanza, 2020; Direktorat Pembinaan PK-LK Republik Indonesia, 2013). It is undeniable that school-age children in remote villages experience difficulties because of the minimal information infrastructure and technology (Mahmood, 2021). The learning process carried out in schools is a public policy. The existence of online learning instructions or DL of APE is undoubtedly a big problem for the community. To overcome this, the world of education certainly requires recovery and appropriate resources to overcome these problems. DL is intended to avoid social contact to prevent the transmission of Covid-19 (Burhaein, 2021; Yunitasari & Hanifah, 2020).

The results of the study of the effectiveness of online learning during a pandemic show that the use of e-learning systems was no longer something novel. It is just that not all schools have implemented this system. Moreover, in remote areas or villages, not all families/parents can fulfill these facilities and infrastructure, considering the uneven economic status (Adedoyin & Soykan, 2020). So the learning process based on e-learning, commonly called online, is not delivered perfectly and optimally.

Furthermore, there is a scarcity of APE learning studies during pandemics in Indonesia. This scarcity is characterized by the study of physical education learning in non-disabled children. Previous research on DL on physical education for elementary school (SD) students (Putra, 2020), junior high school (SMP) (Simbolon et al., 2021), and high school (SMA) (Sulaksono & Wibowo, 2021) has showed several things, including students feel encouraged to study remotely without adequate facilities and infrastructure at home; and students also do not have a culture of DL because so far, the learning system has been done face-to-face/offline. Students are used to being in the school environment to interact with their friends, play and have fun and face-to-face with teachers, in addition, the school is closed for too long so the children are bored, a factor that causes students to lose their social soul. At school, they can play and interact with their friends, but at home they cannot and only study alone with their parents.

Based on the problem gap above, a research question was formulated: How did the APE learning change during the pandemic? Therefore, this study aims to describe the dynamics of change in the implementation of adaptive physical education learning during the pandemic.

2 METHODS

This study uses a descriptive method with a qualitative approach and literature study. The qualitative research method is a naturalistic research method because the research is carried out in natural conditions (natural settings), and the data collected is analyzed qualitatively (Burhaein, Phytanza, & Demirci, 2020). Qualitative research is considered appropriate to describe the current situation, namely online learning during a pandemic.

The type of data collected is secondary data in the form of the results of previous experiences from various sources, namely articles, library sources, and documents, as well as elaborating sentences which are answers from research subjects conducted by interviews. The participants in this study were 2 APE teachers, each in the Provinces of West Java and Bali. The research instrument is an interview guide whose data collection was carried out via the mobile WhatsApp application. This interview instrument has previously been validated for each item through expert judgment by nine experts (Ayre & Scally, 2014; Lawshe, 1975, 1985). Then reliability was tested and gave an Alpha Cronbach value of 0.913. The guidelines for interviews conducted with teachers can be seen in Table 1.

Table 1. Interview guidelines.

No.	Question
1.	How is the implementation of learning during the pandemic?
2.	What are applications used in online learning?
3.	How is online learning using the application?
4.	How are the learning outcomes of students while using the online learning media?
5.	In your opinion, is online media used effectively?
6.	What are the obstacles faced during online learning?

Data analysis was conducted to interpret the research data been collected through interviews. Data analysis is a crucial step in interpretation and making a research conclusion. Data analysis was carried out using interpretive descriptive methods, through an analysis cycle in an interactive cycle (Miles, Huberman, & Saldaña, 1994, 2014). The interactive cycle consists of data collection, data display, data reduction, and conclusion visualization. A flowchart of the qualitative data analysis is shown in Figure 1.

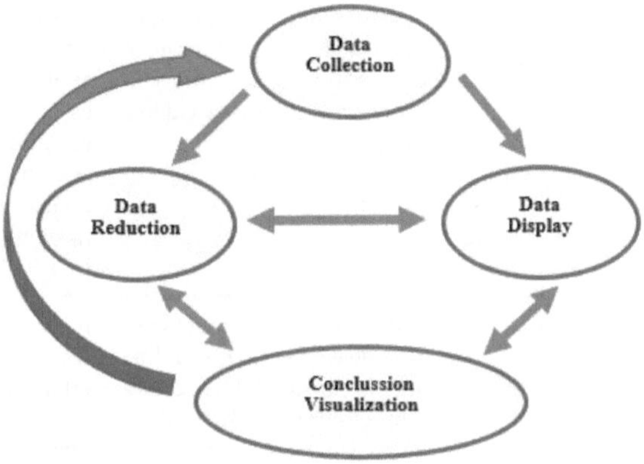

Figure 1. Flowchart qualitative data analysis.

3 RESULTS

Adaptive physical education (APE) learning during the COVID-19 pandemic is currently being carried out online. Through online learning, it is hoped that students will continue to gain knowledge as usual so that they do not miss out on the material obtained. However, online learning is not entirely welcomed by students (Table 2).

Table 2. Interview result.

No.	Questions	APE Teacher (1)	APE Teacher (2)
1.	How is the implementation of learning during the pandemic?	Before the pandemic, APE learning was carried out offline or face-to-face. The current condition is that all learning including APE learning is "forced" to be done online. [A1, T1]	The situation when the COVID-19 pandemic occurred caused school learning to be carried out online from March 2020 until now. So that APE learning is also carried out online or distance learning (DL) [A1, T2]

(continued)

Table 2. Continued.

No.	Questions	APE Teacher (1)	APE Teacher (2)
2.	What are applications used in online learning?	The platforms used during distance learning are Zoom meeting, Google Classroom, and WhatsApp (WA) [A2, T1]	The platforms used during distance learning are WhatsApp (WA) & Telephone (only occasionally). [A2, T2]
3.	How is online learning using the application?	APE learning during a pandemic requires us to keep our distance. So DL/online is the best way at this time that is adapted to various existing conditions. [A3, T1]	Online learning cannot be carried out and face-to-face learning cannot be carried out. They are learning through assignments delivered via Telephone to parents. The role of parents is very high in success, but the fact is that parents do not always play an active role in fostering their children's learning. [A3, T2]
4.	How are the learning outcomes of students while using the online learning media?	Learning outcomes are occasionally seen with live performance through motion observations in the zoom or sent motion videos. [A4, T1]	Learning outcomes are not effective on an online platform/application. Not all students can practice their movements because not all parents are responsive in sending their children's movement assignments through video assignments— [A4, T2]
5.	In your opinion, is an online platform used effectively?	Although the Measurement cannot be done with certainty, online learning is considered less effective for children with disabilities. This can be seen from the results of the students' scores being unsatisfactory. In addition to external factors in the form of internet access and facilities constraints, other obstacles arise from internal factors, namely students' motivation [A5, T1]	In terms of delivering online learning materials, it is less effective and has several limitations. APE learning requires practice and direct teacher guidance, so PJJ is difficult to achieve learning success.[A5, T2]
6.	What are the obstacles faced during the online/DL?	Online learning is not as ideal as conventional learning because students' motivation in learning is lower, the sense of responsibility is also reduced because the teacher is not physically present. [A6, T1]	There are several obstacles including internet access and quotas. Goal achievement learning and learning outcomes can be measured based on the presence of students, daily tests, and assignments. [A6, T2]

Students consider online learning to be more complicated than offline learning. These problems are becoming increasingly complex, coupled with the increase in the cost of the internet quotas used to access learning. Besides the increasing cost of buying quotas, students also have difficulty accessing the internet and learning facilities such as cellphones and laptops. As a result, students feel that understanding the learning material is relatively better when learning is carried out offline.

4 DISCUSSIONS

The COVID-19 pandemic is a situation that was not imagined by adaptive physical education (APE) teachers. Instead, distance learning (DL) is generally a planning concept implemented during pandemics. Learning theories in online learning/DL include behaviorism, cognitivism, and constructivism (Schunk, 2014).

Behaviorism in online learning is the mind as the "main motor" in responding to stimulus during DL learning quantitatively. Cognitivism in online learning is a dynamic type of student learning with various online learning strategies during pandemics. Finally, constructivism in online learning emphasizes the learning situation that affects the way students contextualize the teaching materials received.

The study tried to uncover the phenomenon of APE during the pandemic through interviews with two teachers from different regions. This differences refer to culture, conditions, types, and so on. Therefore, differences in conditions and characteristics are expected to reveal some of the latest images of APE learning during the COVID-19 pandemic in Indonesia.

The profile of APE learning changes during the Covid-19 pandemic in this study was pared down through influencing indicators. As a result, there are six pared indicators in the APE learning change profile during a pandemic. Further explanations related to the profile of changes in APE learning during the pandemic are shown in Figure 2.

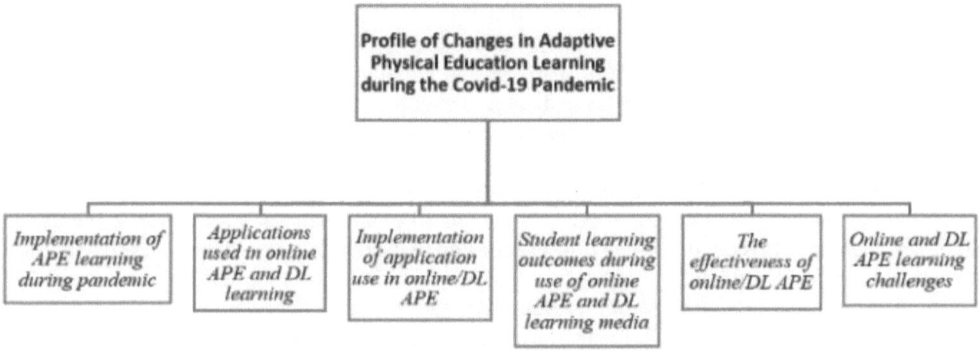

Figure 2. APE learning change profile during the Covid-19 pandemic.

Implementation of APE learning during the pandemic has undergone a change that was initially face-to-face into DL/online. The following are the results of the interview [A1, T1]:

"Before the COVID-19 pandemic, APE learning was done offline, namely face-to-face. However, nowadays conditions are changing with all learning including APE learning being forced by conditions to be done online." [A1,T1].

The situation of changes in the implementation of APE learning began since the Indonesian state declared large-scale social restrictions in March and still tends to be strict until September 2021. This is following the results of the interview [A1, T2].

"The COVID-19 pandemic caused learning in schools to be carried out online from March 2020 until now. So that APE learning is also done online. and DL". [A1,T2].

Switching from learning offline to online is an essential form of rapid response in APE teacher learning strategies for learning during pandemics. The health hazards of the COVID-19 virus will spread in crowded activities in face-to-face study groups. This is the reason behind the change in APE learning to online. The above is also relevant to research from Finlnd (Europe) and China (Asia), where APE's implementation has also changed. The change that occurred in both countries is the switch from the learning of APE from face-to-face (offline) to DL (online) (Ng, 2020). Applications used in online APE and DL learning vary. According to the interviews [A2, T1] and [A2, T2]:

"The platforms used during distance learning are Zoom meeting, Google Classroom, and WhatsApp (WA)." [A2, T1].

"The applications used during DL APE learning are WA and occasionally phone." [A2, T2]

The applications or platforms used by APE teachers during learning vary greatly. Several factors cause this to vary. Internal factors are interests, motivations, health conditions, and other aspects of the individual APE students. External factors are the availability of laptops/gadgets, family economic conditions, signal conditions, and other aspects unrelated to the self/individual APE students. The above is also relevant to research in Finland (Europe), China (Asia), and South Korea (Asia) where the applications are used in online APE and DL learning vary. Applications that are often used in Finland and China include the Zoom meeting (Ng, 2020), and in South Korea Zoom meeting and google classroom are used (Jeong & So, 2020). Implementation of application use in online/DL APE is the best solution available today during a pandemic. The following are the results of the interview [A3, T1]:

"APE learning during a pandemic that requires us to keep our distance. So DL learning using online platforms/applications is the best way today to adapt to various conditions that exist." [A3, T1]

Online platforms/apps may be one of the best options right now. However, it is not always adequate to do this. Sometimes implementation through assignments is delivered through short phone messages. The following are the results of the interview [A3, T2]:

"Measurement and evaluation of APE cannot be done with certainty because online learning is less effective for students with disabilities. It is seen that the student's grades have not met the minimum standards. The existence of internal factors such as motivation and external factors such as internet access is a common cause of platforms/applications. Face-to-face learning is still not allowed to this day. Learning through tasks delivered over the phone to parents. The role of parents is very high in success, but the reality is that parents do not always play an active role in fostering their children's learning." [A3, T2]

Implementation in online applications/platforms is indeed an alternative when face-to-face can not be done. However, there are times when there are other alternatives, such as assignments via phone text message. This condition can differ according to the conditions and the role and condition of each parent of APE students.

The above follows previous research in South Korea and Indonesia related to the implementation of various online learning platforms during the pandemic. In South Korea, learning during pandemics is mainly done with online platforms/applications (Jeong and So, 2020). Furthermore, in Indonesia, physical education learning in non-disability students is also applied in various ways during pandemics. In Indonesia, learning is also done online/DL and adjusts existing conditions, like in other countries (Hambali et al., 2021). The assignment is one of the alternatives applied when the application/platform can not be made.

Student learning outcomes during the use of online APE and DL learning media differ according to what students with disabilities experience. The following is the result of the interview [A4, T1]:

"Learning outcomes are sometimes seen by direct practice through observation of motion in zoom or motion video being sent." [A4, T1]

Learning outcomes can not always be seen in the development during online/DL. This is because online platforms are not always effective in learning outcomes. The following is the result of the interview [A4, T2]:

"Learning outcomes are ineffective on online platforms. Not all students can practice their movements, because not all parents are responsive in sending their child's movement tasks through the task video." [A4, T2]

Learning outcomes in APE vary according to the learning conditions received by students in the country. Students who have good learning outcomes may be able to carry out online learning/DL smoothly. However, I also still found learning results that were not as expected. Therefore, barriers to acceptance and delivery of tasks become one of the factors in learning success.

The above follows previous research in the United States, South Korea, Colombia, and Indonesia relating to student learning outcomes during pandemic learning. Research in the United States,

South Korea, and Colombia showed that online platforms' learning outcomes are pretty effective (Zapata-Cuervo et al., 2021). This condition may be the same as other countries, but there can also be different learning outcomes due to the differences in the characteristics of other countries. In Indonesia, the learning outcomes of physical education in non-disability students do not always follow what is expected. During pandemics, learning in Indonesia is not always effective because various input problems impact learning output (Amran, Suherman, & Asmudin, 2021).

The effectiveness of online/DL APE tends to be less effective. Factors that are generally the cause of less effectiveness are internet access and student motivation. The following is the result of the interview [A5, T1]:

"Measurement and evaluation of APE cannot be done with certainty because online learning is less effective for students with disabilities. Nevertheless, it is seen that the student's grades have not met the minimum standards. The existence of internal factors such as motivation and external factors such as internet access is a common cause." [A5, T1].

Other factors also influence effectiveness. Some other factors include the importance of students getting direct guidance or face-to-face contact. The following is the result of the interview [A5, T2]:

"The delivery of learning materials online, I think, is less effective and has some limitations. APE learning requires practice and direct guidance from me as a teacher. This DL condition will be difficult to achieve success in APE learning." [A5, T2].

Teachers feel that online/DL is less effective, especially for APE learning in students with disabilities. APE learning requires optimal movement training. Online learning requires students to practice movements via online video (via zoom) or offline (videos sent to WhatsApp). However, this cannot be done with various obstacles experienced by students. Students experience network/internet access constraints and internet quotas. While another obstacle faced by students is the lack of motivation that comes from themselves to follow the learning. Students sometimes only attend and afterward do not do activities related to learning activities, and at the most extreme do not follow APE learning.

The above follows previous research in the United Kingdom (UK) and Indonesia relating to the effectiveness of student learning during pandemic learning. The research in the UK showed that there is ineffective student learning because of motivation and problems on online learning platforms (Goodyear et al., 2021). This condition may be the same or different from the differences in characteristics of other countries. In Indonesia, the effectiveness of physical education learning in non-disabilities is not always what is expected. Teachers feel that the dominant thing in physical education is psychomotor or motion learning. However, motion limitations become massive during online learning/DL (Amran, Suherman, & Asmudin, 2021). Therefore, students cannot get direct guidance as in offline learning before the pandemic.

Online and DL APE learning challenges. This system is not as idealized as offline learning before the pandemic. The online learning system /DL raises internal student problems. The following is the result of the interview [A6, T1]:

"The online learning I do is not as ideal as conventional learning. This is because the student's motivation in learning is lower, the sense of responsibility is also reduced because the teacher is not physically present"—[A6, T1].

"The challenge of online/DL APE learning is further reviewed from the external factors of the student. External factors are related to outside factors that affect the online learning/DL of students with disabilities. This is following the results of the interview"—[A6, T2].

"I found several obstacles become a challenge, namely internet access and quotas. Furthermore, the achievement of learning goals and learning outcomes can be measured based on student attendance, daily repeats, and assignments. External factors affect the student's learning achievement"—[A6, T2].

Online APE/DL learning in practice does not escape the challenges in its implementation. The condition of this learning system is not always ideal compared to conventional learning. However, conventional defense is not recommended during pandemics. The implementation of learning is also constrained by internet access. In online learning/DL, uneven communication causes students difficulty in understanding the material, especially APE learning, which is dominant in psychomotor skills.

Behind the problem of APE online/DL, there is the potential to optimize it. Strategy in optimization is through synergy between components, including teacher knowledge, parent role, student motivation, availability of supporting facilities (HP/laptop, internet quota, and network). Based on these components, the role of parents becomes dominant in online APE learning/DL. This dominant role is because learning is carried out 50% or more at home. Parents who are also affected by economic problems in their families will affect the smooth learning process of APE.

The above is in line with previous research in South Korea and Poland regarding challenges in implementing online APE/DL learning during pandemics. In South Korea, learning during a pandemic has not escaped various challenges. Online learning/DL influences students' interests and motivations (Jeong & So, 2020). There is a tendency for both of these things to decrease with the boredom of learning that is always at home. Furthermore, the challenge of APE learning in Poland is in the internal and external factors of students. Internal factors include student anxiety and external factors include the role of parents during pandemic learning (Parczewska, 2020). Anxiety and boredom are challenges to be overcome by APE teachers. In addition, there is a role for parents who actively influence the success of students' learning.

5 CONCLUSIONS

Based on the study results, it can be concluded that learning has changed from before the pandemic, namely offline to online, which is considered the best solution at this time. However, various online learning problems arise. They are considered less than ideal because there are several shortcomings, especially in APE learning, requiring movement practice and direct motion evaluation. Student learning outcomes tend to be unsatisfactory. Various obstacles have caused the APE online/DL during the COVID-19 pandemic to be less than optimal. Therefore, it takes a variety of ways and to achieve success in APE online/DL.

Therefore, this study has implications for teachers' knowledge of implementation and strategies in achieving APE success on the DL during pandemic times. The author recommends further research into the implementation and learning strategies of adaptive physical education during the COVID-19 pandemic on a wide scale. This study will contribute to the development of adaptive physical education in the future.

REFERENCES

Adedoyin, O. B. and Soykan, E. (2020) 'Covid-19 pandemic and online learning: the challenges and opportunities', *Interactive Learning Environments*, 0(0), pp. 1–13. doi: 10.1080/10494820.2020.1813180.

Alrashed, F. A. *et al.* (2021) 'Implications of the COVID-19 pandemic on the mental health and professional psychomotor skills of dental students', *Psychology Research and Behavior Management*, 14(1), pp. 675–685. doi: 10.2147/PRBM.S317641.

Amran, Suherman, W. S. and Asmudin (2021) 'Physical Education Online Learning During the Covid-19 Pandemic?: Effectiveness , Motivation , and Learning Outcomes', *The International Journal of Social Sciences World*, 3(1), pp. 123–137.

Ayre, C. and Scally, A. J. (2014) 'Critical values for Lawshe's content validity ratio: Revisiting the original methods of calculation', *Measurement and Evaluation in Counseling and Development*, 47(1), pp. 79–86. doi: 10.1177/0748175613513808.

Bland, A. R. *et al.* (2021) 'The impact of COVID-19 social isolation on aspects of emotional and social cognition', *Cognition and Emotion*, 0(0), pp. 1–10. doi: 10.1080/02699931.2021.1892593.

Brom, C. *et al.* (2020) 'Mandatory Home Education during the COVID-19 Lockdown in the Czech Republic?: A Rapid Survey of 1 st -9 th Graders' Parents', *Frontiers in Education*, 5(103), pp. 1–11. doi: 10.31234/osf.io/fbhn3.

Burhaein, E. (2020) 'Bagaimana Pendidikan Jasmani Adaptif di Era New Normal', in *Bunga Rampai Strategi, Proses, Evaluasi, dan Model Pembelajaran Pendidikan Jasmani, Olahraga, dan Kesehatan (PJOK) pada Era Pandemi Covid-19*. Surabaya: UNESA University Press.

Burhaein, E. (2021) 'Aktifitas Fisik, Pendidikan Jasmani Adaptif, dan COVID-19', in *Akselerasi Berpikir Ekstraordinari Merdeka Belajar Pendidikan Jasmani dan Olahraga era Pandemi COVID-19*. 1st edn. Tulungagung: Akademia Pustaka.

Burhaein, E., Phytanza, D. T. P. and Demirci, N. (2020) 'The development and validation of a revised Friendship Activity Scale and Adjective Checklist for use in the Indonesian Unified Sports program', *International Sports Studies*, 42(e), pp. 18–28. doi: 10.30819/iss.42-e.03.

Burhaein, E., Tarigan, B. and Phytanza, D. T. P. (2020) 'The experience and understanding of the K-13 curriculum implementation of Indonesian teachers of Adapted Physical Education (APE)', *International Sports Studies*, 42(e), pp. 29–42. doi: 10.30819/iss.42-e.04.

Castañeda-Babarro, A. *et al.* (2020) 'Physical activity change during COVID-19 confinement', *International Journal of Environmental Research and Public Health*, 17(18), pp. 1–10. doi: 10.3390/ijerph17186878.

Chen, N. *et al.* (2020) 'Epidemiological and clinical characteristics of 99 cases of 2019 novel coronavirus pneumonia in Wuhan, China: a descriptive study', *The Lancet*, 395(10223), pp. 507–513. doi: 10.1016/S0140-6736(20)30211-7.

Direktorat Pembinaan PK-LK Republik Indonesia (2013) *Pedoman Penyelenggaraan Program Penjas Adaptif bagi Peserta Didik Berkebutuhan Khusus di Sekolah Inklusi*. Jakarta: Kementerian Pendidikan dan Kebudayaan.

Fiorenzato, E. *et al.* (2021) 'Cognitive and mental health changes and their vulnerability factors related to COVID-19 lockdown in Italy', *PLoS ONE*, 16(1 January), pp. 1–25. doi: 10.1371/journal.pone.0246204.

Goodyear, V. A. *et al.* (2021) 'Social media use informing behaviours related to physical activity, diet and quality of life during COVID-19: a mixed methods study', *BMC Public Health*, 21(1), pp. 1–14. doi: 10.1186/s12889-021-11398-0.

Hambali, S. *et al.* (2021) 'The effectiveness learning of physical education on pandemic covid-19', *International Journal of Human Movement and Sports Sciences*, 9(2), pp. 219–223. doi: 10.13189/saj.2021.090208.

Jeong, H. C. and So, W. Y. (2020) 'Difficulties of online physical education classes in middle and high school and an efficient operation plan to address them', *International Journal of Environmental Research and Public Health*, 17(19), pp. 1–13. doi: 10.3390/ijerph17197279.

Karina, P. M., Maryanti, R. and Wulandary, V. (2021) 'Affective and Psychomotor Aspects of 5th-Grade homepage: against Students with Students in Online Learning', *Jurnal Asesmen Dan Intervensi Anak Berkebutuhan Khusus*, 1(1), pp. 45–60.

Ko, N. Y. *et al.* (2020) 'Cognitive, affective, and behavioral constructs of COVID-19 health beliefs: A comparison between sexual minority and heterosexual individuals in Taiwan', *International Journal of Environmental Research and Public Health*, 17(12), pp. 1–10. doi: 10.3390/ijerph17124282.

Lawshe, C. H. (1975) 'A quantitative approach to content validity', *Personnel Psychology*, 28(4), pp. 563–575. doi: 10.1111/j.1744-6570.1975.tb01393.x.

Lawshe, C. H. (1985) 'Inferences From Personnel Tests and Their Validity', *Journal of Applied Psychology*, 70(1), pp. 237–238. doi: 10.1037/0021-9010.70.1.237.

Li, G. *et al.* (2020) 'Coronavirus Infections and Immune Responses Coronavirus infections and immune responses', *Journal of Medical Virology*, 92(4), pp. 1–9. doi: 10.1002/jmv.25685.

Mahmood, S. (2021) 'Instructional Strategies for Online Teaching in COVID-19 Pandemic', *Human Behavior and Emerging Technologies*, 3(1), pp. 199–203. doi: 10.1002/hbe2.218.

Miles, M. B., Huberman, A. M. and Saldana, J. (2014) *Qualitative data analysis: A methods sourcebook (3rd ed.)*. Thousand Oaks, CA: Sage.

Miles, M. B., Huberman, A. M. and Saldaña, J. (1994) *Qualitative Data Analysis*.

Miners, S., Kehoe, P. G. and Love, S. (2020) 'Cognitive impact of COVID-19: looking beyond the short term', *Alzheimer's Research and Therapy*, 12(1), pp. 1–16. doi: 10.1186/s13195-020-00744-w.

Ng, K. (2020) 'Adapted physical activity through COVID-19', *European Journal of Adapted Physical Activity*, 13(1), pp. 1–3. doi: 10.5507/EUJ.2020.003.

Nurulfa, R. *et al.* (2021) 'Physical Education Survey during the COVID-19 Pandemic in Eastern Indonesia', *International Journal of Human Movement and Sports Sciences*, 9(4), pp. 668–675. doi: 10.13189/saj.2021.090410.

Parczewska, T. (2020) 'Difficult situations and ways of coping with them in the experiences of parents homeschooling their children during the COVID-19 pandemic in Poland', *Education 3-13*, 0(0), pp. 1–12. doi: 10.1080/03004279.2020.1812689.

Phytanza, D. T. P., Burhaein, E. and Pavlovic, R. (2021) 'Gross Motor Skills Levels in Children with Autism Spectrum Disorder during the COVID-19 Pandemic', *International Journal of Human Movement and Sports Sciences*, 9(4), pp. 738–745. doi: 10.13189/saj.2021.090418.
Purwanto, P., Lumintuarso, R. and Burhaein, E. (2021) 'Impact of Running Techniques through the Sprint Ability in Athletes during the COVID-19 Pandemic', *International Journal of Human Movement and Sports Sciences*, 9(4), pp. 717–724. doi: 10.13189/saj.2021.090416.
Putra, A. N. (2020) *Implementasi Pembelajaran Pjok Pada Masa Pandemi Covid-19 Di Sd Negeri Se-Kecamatan Lendah Kabupaten Kulon Progo daerah Istimewa Yogyakarta*. Universitas Negeri Yogyakarta.
Putra, Y. M., Purwanto, S. and Burhaein, E. (2021) 'Effect of Limb Muscle Power Training with Leaps on Athlete's Speed during the COVID-19 Pandemic', *International Journal of Human Movement and Sports Sciences*, 9(3), pp. 461–465. doi: 10.13189/saj.2021.090310.
Schunk, D. H. (2014) *Learning theories: an educational perspective*. London: Pearson.
Simbolon, M. E. M. *et al.* (2021) 'DAMPAK COVID-19 PADA PEMBELAJARAN PENDIDIKAN JASMANI OLAHRAGA DAN KESEHATAN SMP DI PROVINSI KEPULAUAN BANGKA BELITUNG', *Riyadhoh: Jurnal Pendidikan Olahraga*, 4(1), pp. 9–16. doi: 10.31602/rjpo.v4i1.4226.
Simpson, R. J. *et al.* (2020) 'Can exercise affect immune function to increase susceptibility to infection??', *Exercise immunology review*, 26(1), pp. 8–22.
Sulaksono, G. and Wibowo, A. (2021) 'Studi Deskriptif Pelaksanaan Pendidikan Jasmani Olahraga dan Kesehatan pada Tingkat SMA Kelas XI Selama Masa Pandemi Covid-19', *Jurnal Pendidikan Kesehatan Rekreasi*, 7(1), pp. 82–90. doi: 10.5281/zenodo.4420514.
UNESCO (2020) *Covid-19 Impact on Education Data: COVID-19 Education Disruption and Response, United Nations Educational, Scientific and Cultural Organization.*
Wei, X. S. *et al.* (2020) 'A cluster of health care workers with COVID-19 pneumonia caused by SARS-CoV-2', *Journal of Microbiology, Immunology, and Infection*, (XXXX). DOI: 10.1016/j.jmii.2020.04.013.
Woods, J. *et al.* (2020) 'The COVID-19 pandemic and physical activity, *Sports Medicine and Health Science*, 2(2), pp. 55–64. DOI: 10.1016/j.smhs.2020.05.006.
Wu, J. and Zha, P. (2020) 'Preventive, Mitigating, and Treatment Strategies for COVÝD-19', *SSRN*, 2020(February 25), pp. 1–38. doi: https://dx.doi.org/10.2139/ssrn.3544428.
Yunitasari, R. and Hanifah, U. (2020) 'Pengaruh Pembelajaran Daring terhadap Minat Belajar Siswa pada Masa COVID 19', *Edukatif: Jurnal Ilmu Pendidikan*, 2(3), pp. 236–240.
Yunus, N. R. and Rezki, A. (2020) 'Kebijakan Pemberlakuan Lock Down Sebagai Antisipasi Penyebaran Corona Virus Covid-19', *SALAM: Jurnal Sosial dan Budaya Syar-i*, 7(3). doi: 10.15408/sjsbs.v7i3.15083.
Zapata-Cuervo, N. *et al.* (2021) 'Students' Psychological Perceptions Toward Online Learning Engagement and Outcomes during the COVID-19 Pandemic: A Comparative Analysis of Students in Three Different Countries', *Journal of Hospitality and Tourism Education*, 00(00), pp. 1–15. doi: 10.1080/10963758.2021.1907195.

Innovation on Education and Social Sciences – Maureen et al (Eds)

The effectiveness of online lectures on art students

D.N. Mukti, D. Djatiprambudi & W.H.N. Ruci
Universitas Negeri Surabaya, Surabaya, East Java, Indonesia

ABSTRACT: The purpose of this research is to identify the effectiveness and enthusiasm of art students in online learning. Further discussion of this research will be carried out by linking the data findings with relevant references. Sampling was done randomly by distributing survey questionnaires about the effectiveness and enthusiasm of art students through social media. The results obtained are that more than half of the respondents want to return to the offline learning pattern, almost 40% prefer to do the hybrid pattern, and the rest are comfortable with being fully online. The autonomy of selecting the relevant method refers to the basic need to self-regulate and regulate behavior according to one's self-awareness; competence means the need to develop personal abilities and skills and interact effectively with the environment; connectivity is the need to feel socially connected in a learning environment that occurs both online and offline.

1 INTRODUCTION

The future of education can be read in extreme terms as a future that has no space. Space in this article is interpreted as a physical building. When the education system is assumed to be in a more traditional understanding, which is always associated with buildings, future education may no longer be relevant. It can be seen how rapidly the development of education and the world of technology is. Space can be created in cyberspace, online classes with invisible boundaries but a function similar to classrooms with physical buildings.

In 2020 a pandemic swept across the world. As a result, many sectors have suffered, including the education sector. In Indonesian education, there are many policies made by the government to deal with the pandemic. At the beginning of the pandemic, all schools were closed without any teaching and learning process. Yet several months later, an online learning method was determined to be carried out. It does not go smoothly, but of course, there are many pros and cons because when the government decides to conduct online learning, not all elements of education are ready to do it. But ready or not, this pattern must continue. In their research, Duncan and Howell (2010) suggested that face-to-face training did not significantly affect teaching beliefs and teaching innovation. It has also not produced a long-term and lasting impact on pedagogy (Duncan & Howell, 2010). It reinforces the reason for the relevance of the change from face-to-face methods to online methods.

When initially implemented, there were many problems. Teachers who usually teach face-to-face in class must master various tools to support online teaching and learning. As a result, there is much confusion, and the learning process is less effective. Younger teachers are a transitional generation from analog to digital, and they still have high adaptive power because these devices are not foreign to them. However, for more senior teachers, a generation accustomed to working analogously, it is undoubtedly challenging to use digital technology for the online teaching and learning process.

In different cases, online learning is considered a meaningful way to develop teachers' thinking skills and promote their professional development (Quinn et al., 2019). Therefore, encouraging changes in learning methods to online learning, at the same time, will improve the skills of the

DOI 10.1201/9781003265061-4

teaching staff. Other studies have stated that in any format, online learning has advantages over face-to-face learning, such as balancing educational and development resources, reducing costs, and promoting educational equity (Johnson & Aragon, 2003).

Higher education in the arts or at the art campus is also inseparable from changes in teaching patterns, all of which are directed to conduct online teaching and learning activities. In art lectures, in general, there are two characteristics of the material given. The first is the theoretical aspect, and the second is practical lectures. The theoretical aspect usually revolves around the history of art, aesthetics, and today's art studies. In contrast, the material for practice for fine arts includes painting, sculpture, drawings of shapes, choreography (dance subject), motion, the practice of stringed instruments (music subject), wind instruments, swipe, and so on. Quite a lot of courses or subjects require practice because it relates to the skills of each individual in a particular technique. With online lectures being implemented, it will be a different problem to maintain students' enthusiasm in taking classes and practicing. Indeed, many variables affect how enthusiastic students are in attending lectures. However, in this study, we tried to discuss the effectiveness of art students in qualitative online lecture methods.

In this study, the research would be directed to find out the enthusiasm and effectiveness of online lectures from the students' point of view. An art course with two lecture characters, namely theory and practice, has different challenges from other lectures. Of course, when talking about art, there are aesthetic variables that need to be considered in making works in practical classes. This aesthetic variable can not only emerge by studying the technique. Furthermore, a specific learning pattern must be continuously carried out and cultivate skills, but taste is essential. With the challenges faced, it is crucial to conduct further research on the enthusiasm and effectiveness of online lectures for art students.

2 METHODS

The descriptive survey method used in this study is also known as the normative survey method. It is based on the assumption that certain phenomena usually follow a general pattern or a specific pattern. This is closely related to the purpose of the research, which is to identify the effectiveness and enthusiasm of art students when studying online. Further discussion of this research will be carried out by linking the data findings with relevant references. Sampling was done randomly by distributing survey questionnaires about the effectiveness and enthusiasm of art students through social media. A questionnaire is a form that is filled out online. The provided form is divided into four types of questions, the first is a general question related to the identity of the questionnaire filler and general inquiries related to online lectures, the second question is specific about terrorism courses, the third question deals with courses practice, and the fourth is an additional question.

3 RESULTS AND DISCUSSION

The distribution of the questionnaire was carried out from May 27, 2021 to May 30, 2021. Forty-three respondents have filled the form, but there was one data that the researcher discarded because the major of the respondent was not relevant to the research being conducted. Thus the final number of respondents was 43 people. In the questionnaire distributed by students from several campuses, there were 30 respondents from the State University of Surabaya, five respondents from the State University of Malang, three respondents from PGRI Adibuana University Surabaya, three respondents from the Indonesian Institute of the Arts Yogyakarta, and one respondent from the Universitas Sebelas Maret, Surakarta. There were 23 respondents from the department of fine arts, 12 respondents from the department of fine arts education, one respondent from the drama department, and five respondents from the visual communication design department.

3.1 *Online lecture pattern*

In the first and second questions, three answer choices are presented about online lecture patterns. This type of learning can be in a synchronous or asynchronous format. The first requires learning to take place at a fixed time, similar to traditional offline learning in a fixed location; the latter is relatively flexible in time, enabling self-directed learning and the exchange of resources through communication networks (Khan, 1998). The first pattern is a real-time online lecture pattern using zoom (or a video call platform with many other people) and having discussions. This pattern is commonly used; even a new term, the 'zoomer' generation, has appeared due to its popularity. Using a video conferencing platform makes it possible to hold classes face-to-face without having to be in the same room.

There are many advantages to using this pattern. In addition to being able to conduct lectures in real-time, the host or co-host can also present material. In addition, online learning can promote the sharing and co-construction of educational experiences through effective social interactions (Kent, Laslo, & Rafaeli, 2016). But of course, in addition to the many disadvantages, one of which is the unstable connection, some places do have good signal quality or WiFi networks. Still, many locations also have bad connectivity. Doing lectures in real-time using the platform video conferencing requires a good and stable connection so that users can display their video and sound well. If the signal is terrible, the video can freeze and the sound will be distorted, making it difficult to listen to get around this. Students usually turn off the video so that they can minimize data consumption. But this is usually also a strategy for students when attending online classes. In reality, they do not only attend classes but also do other things like eating, drinking, even sleeping. So the material given during online lectures becomes less effective. Researchers believe that strong desires can help people stay motivated and direct and predict their performance (Jeno, Adachi, Grytnes, Vandvik, & Deci, 2019). It can also guide their engagement and learning behavior (Reeve, 2006).

The second pattern is an online lecture using WhatsApp group and discussion via chat. In this condition, students mostly had no difficulties with internet connection. When learning using the WhatsApp group platform, the teaching process would be well-conducted because this pattern does not require a good and stable connection. It can be done with a mediocre connection. This pattern is relatively easy to do, but once the lecturers find it difficult to monitor students in real-time because if it is done in a chat, the lecturer cannot see what the students are doing.

The third pattern is that the material was provided, and the lecturer asked the students to learn it or do the task independently. This pattern can usually be done for quite some time depending on the deadline of assignments. However, if students only need to study the material, then there will be no deadline and the study duration can be relatively long. The obstacle that is often faced from this pattern is that students are reluctant to learn the material provided by the lecturer. This is very dependent on the enthusiasm that students have for the material, if it is felt that the material is less interesting and enjoyable. It is possible for the material not to be read at all.

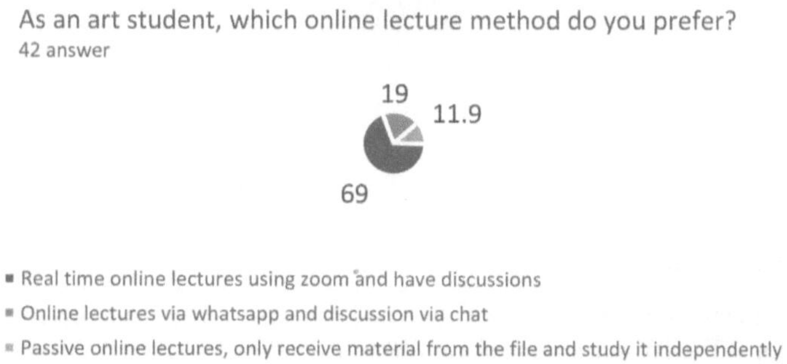

From the chart above, it showed that 29 people prefer real-time online lectures by using video conferencing platforms and conducting discussions. Then online lectures using WhatsApp and discussing it through chat were chosen by eight people. Five people prefer passive online lectures, which involve only receiving material or tasks and studying them independently. From this question, it can be seen that 69% of respondents have high enthusiasm to learn in real-time and have discussions.

In the second question, the respondent was asked to argue which pattern is more effective for online lectures.

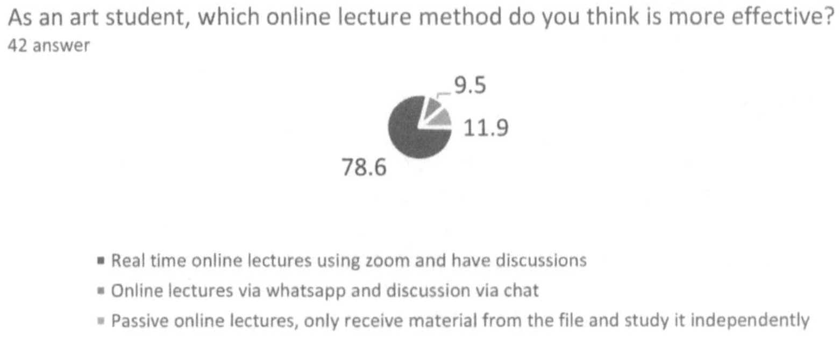

When talking about effectiveness, online lectures in real-time using a video conference platform and being able to discuss directly are considered to be most effective. This is proven by 33 respondents or 78.6% choosing this pattern. Then the third option, namely passive online lectures chosen by five respondents, and also studying through WhatsApp, which was chosen by four respondents. This evidence shows that most respondents think that online lectures using a video conferencing platform are an effective method.

The third question was related to how students responded when online lectures took place. The question asked was 'do you (students) do the material provided by the lecturer?'. This question requires students to find out their intensity in listening to the course. However, by asking this question, it is intended as an initial identification to find out how students respond to the material given by the lecturer during online lectures. From the answers provided by 42 respondents, 42% or a total of 30 respondents chose to record the material provided by the lecturer as much as 21.4% with a total of 9 people taking notes during online lectures, then around 7.1% with a total of 3 respondents not taking notes during online lectures.

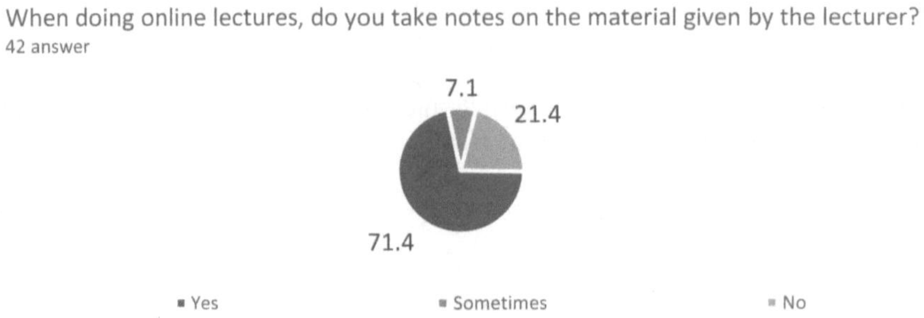

3.2 *Theory courses conducted online*

In looking at the effectiveness of the online theoretical courses, the researchers asked the respondents three questions. The first question asked 'Is the provision of art theory through online lectures effective?'. This question is intended to find out how the respondents see the process of online lectures. From 42 respondents, 18 people answered 'enough'. Enough in this question is interpreted as an answer that states the attitude is quite effective and ineffective. Sixteen respondents chose a value of 4 to express the ineffectiveness of the online lecture pattern on theoretical material. The majority of respondents chose sufficient and less effective, indicating that they prefer the teaching pattern in theory courses to be done physically. This is related to the previous lecture pattern, where all lectures were carried out physically. There is only one person or 2.4% of people who think that theoretical lectures are effective online. This answer comes up because the respondent is interested in learning patterns that are suitable for them to take for the theory courses.

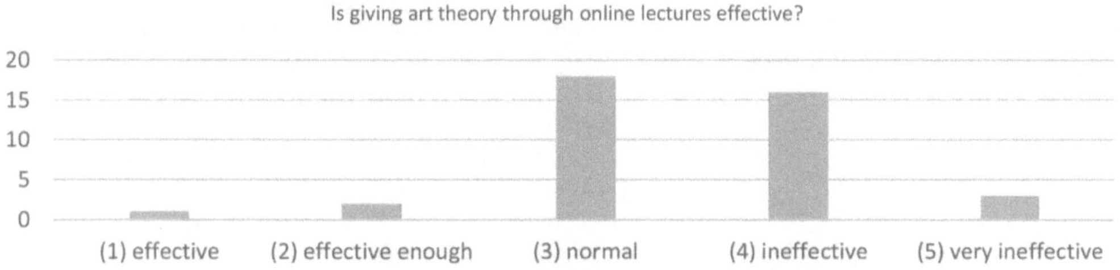

The second question is 'do you prefer to get theoretical material by.' The choices given in this question are online, offline, or hybrid (online and offline). 54.8% of respondents prefer to do it offline, 38.1% of respondents choose to make a hybrid lecture pattern (online and offline), and 7.1% with three respondents doing online or online lectures.

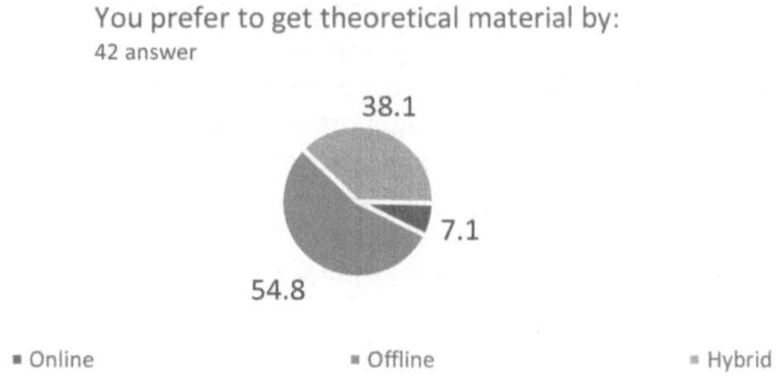

The third question asks the motivational variables for online lecture participants, and the third question is 'how is your general motivation when doing theory lectures independently.'

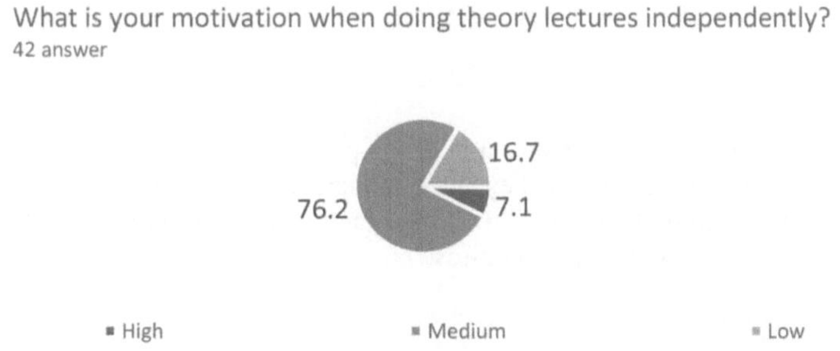

The majority of respondents answered moderate with 76.2%, low with 16.7%, and the ones who responded high were only 7.1%. The answer to this question is related to the previous question. From the responses, it can be concluded that why they prefer to do theory lectures physically is because when they study theory independently (by themselves) they are less motivated. This is possible because when learning in class and together, many variables exist and can make students more motivated in doing lectures.

3.3 *Practical courses conducted online*

Practical courses are subjects that are always available in art colleges because in art college, students are required to understand not only theoretical material, students are also required to have creativity related to a work of art.

The first question posed to the respondent was 'is doing practical courses through online tutoring effective?'. The answer to this question is answered using a scale with one representing effective and five representing ineffective. The results of this question show a graph that increases gradually from numbers 1 to 5. With the most answers being number five with 17 respondents, they think that practical lectures are ineffective online.

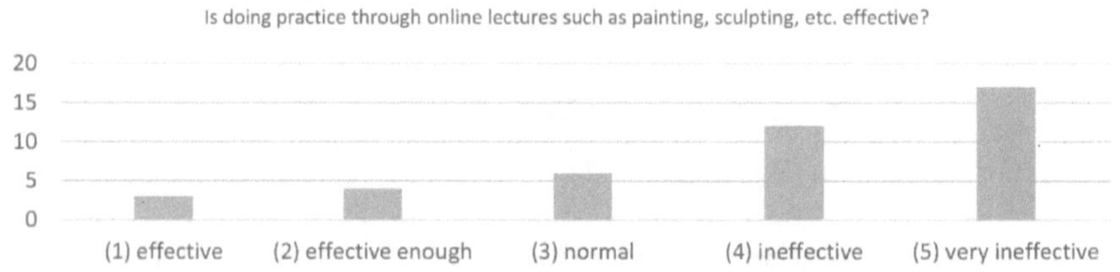

The process of creating works for each individual is different, some prefer to do it in confusion, but there are also those who prefer to do it in groups. The biggest obstacle in practice is the availability of space to be able to practice. It suggests that the environment also affects the psychological mechanism of student participation in the learning process. In the end, mental involvement affects learning behavior and involvement in the learning process (Wang, Wang, Cui, & Zhang, 2021). In practical courses, apart from the need to take advantage of existing facilities on campus, it can also be read as a psychological need to proceed in a supportive environment.

The second question is 'do you prefer to get practical material in', the answers given are online, offline, and hybrid (online and offline). The response from the majority of respondents is that they choose to do practical lectures offline with a percentage of 78.6%, then eight respondents choose to do practical lectures in a hybrid way, and only one respondent chooses to do practical lectures online.

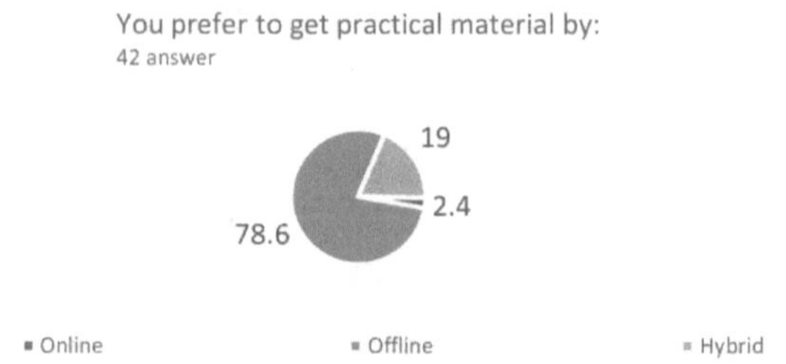

The next question is respondents' motivation in doing practical courses online. The answers from the majority of respondents stated that they were moderate with a percentage of 71.4%, 26.2% indicated that they had high motivation when doing practical courses online. Only one person had low motivation in doing online practical lectures.

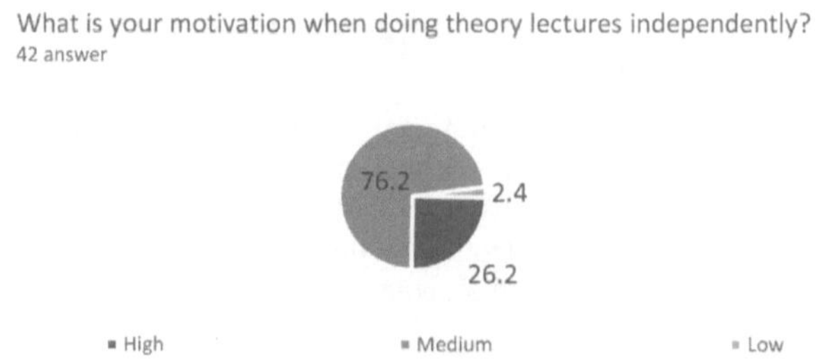

Emerging data shows that more students choose to take practical courses offline. This is confirmed by the next question that asks about motivation, where the majority of respondents stated that the level of motivation was moderate.

3.4 *Expected lecture pattern*

The final question posed in this questionnaire is to combine opinions while studying on the subject and which theory is the pattern chosen by the respondent, return to the pre-pandemic pattern where everything is done offline, full of online lectures, or by conducting hybrid lectures (online) and offline).

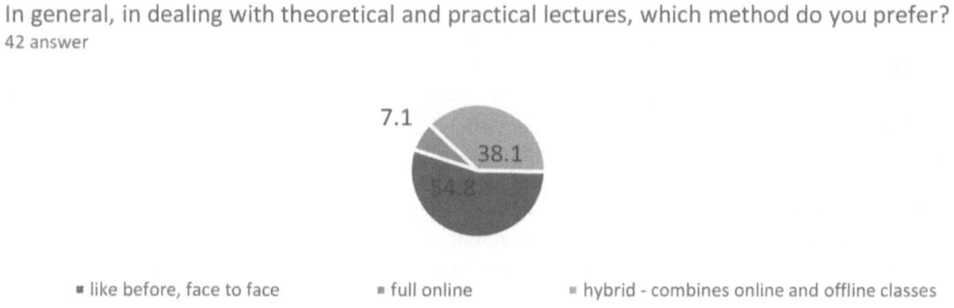

Of the 42 respondents, 54.8% chose to return to the offline pattern before the pandemic, 38.1% chose the hybrid pattern, and only 7.1% or three respondents decided to continue studying with the online pattern.

The data that appears implies that the pattern of art lectures in the future is a more hybrid lecture pattern that connects offline and online work. It is because by doing a mixed lecture pattern, more possibilities will occur and the art lecture pattern will be more efficient. According to Garn's (2019) description, autonomy related to the preferred lecture method refers to the basic need to self-regulate and regulate behavior, one's self-awareness; competence (developing personal abilities and skills and interact effectively with the environment); connectedness (the need to feel socially connected in a learning environment) (Garn, Morin, & Lonsdale, 2019). Students who have been in a physical teaching and learning situation, and today they are in a condition with online methods. They can decide which method is more comfortable for them based on their experience.

4 CONCLUSION

With the shift in teaching and learning methods where previously the teaching and learning process was dominated by physical learning patterns, all students were required to be in one room and follow the process. However, because of the pandemic, the education system is encouraged to change teaching and learning patterns to turn to online methods. Teachers and students are not required to be in a physical room, but all of them conducted learning in an imaginary space using a video conference application. In the transition, of course, there are many pros and cons because something new will appear in the education system. However, when adaptation can be applied effectively, the changes that occur will be very profitable for the world of education in the future.

Likewise, in the arts education community, this research is focused on identifying the effectiveness and enthusiasm of art students when studying online. The results of this study were that more than half of the respondents chose to return to the offline learning pattern, and almost 40% chose to do a hybrid teaching and learning pattern, and the rest chose to carry out the online teaching and learning process fully.

Returning the learning environment back to the offline pattern is a logical choice. This study revealed that the respondents subjectively explained that they are used to the teaching and learning process conducted in an offline pattern. Thus it is certainly not easy to suddenly change into an online teaching and learning pattern. However, gradual adaptation will occur. This can be seen in how almost 40% of respondents chose to carry out the teaching and learning process in a hybrid way. The vital thing to note is how these learning patterns can build student motivation. Motivation is a variable that cannot be separated, and its role is very dominant. Therefore, whatever the pattern is to pay attention to how to maintain student motivation, it must still be built and maintained so that the data processing runs well and smoothly.

REFERENCES

Duncan-Howell, J. (2010). Teachers making connections: Online communities as a source of professional learning. *British Journal of Educational Technology*, 41(2), 324–340. https://doi.org/10.1111/j.1467-8535.2009.00953.x

Garn, A. C., Morin, A. J. S., & Lonsdale, C. (2019). Basic psychological need satisfaction toward learning: A longitudinal test of mediation using bifactor exploratory structural equation modeling. *Journal of Educational Psychology*, 111(2), 354–372. https://doi.org/10.1037/edu0000283

Jeno, L. M., Adachi, P. J. C., Grytnes, J. A., Vandvik, V., & Deci, E. L. (2019). The effects of m-learning on motivation, achievement and well-being: A Self-Determination Theory approach. *British Journal of Educational Technology*, 50(2), 669–683. https://doi.org/10.1111/bjet.12657

Johnson, S. D., & Aragon, S. R. (2003). An instructional strategy framework for online learning environments. *New Directions for Adult and Continuing Education*, 2003(100), 31–43. https://doi.org/10.1002/ace.117

Kent, C., Laslo, E., & Rafaeli, S. (2016). Interactivity in online discussions and learning outcomes. *Computers and Education*, 97, 116–128. https://doi.org/10.1016/j.compedu.2016.03.002

Khan, B. H. (1998). Web-based instruction (wbi)?: An introduction. *International Journal of Phytoremediation*, 21(1), 63–71. https://doi.org/10.1080/0952398980350202

Quinn, F., Charteris, J., Adlington, R., Rizk, N., Fletcher, P., Reyes, V., & Parkes, M. (2019). Developing, situating and evaluating effective online professional learning and development: A review of some theoretical and policy frameworks. *Australian Educational Researcher*, 46(3), 405–424. https://doi.org/10.1007/s13384-018-00297-w

Reeve, J. (2006). Teachers as facilitators: What autonomy-supportive teachers do and why their students benefit. *Elementary School Journal*. https://doi.org/10.1086/501484

Wang, M., Wang, M., Cui, Y., & Zhang, H. (2021). Art teachers' attitudes toward online learning: An empirical study using self determination theory. *Frontiers in Psychology*, 12. https://doi.org/10.3389/fpsyg.2021.627095

Innovation on Education and Social Sciences – Maureen et al (Eds)

The effect of extroversion/introversion on English proficiency

K. Kusumarasdyati
Universitas Negeri Surabaya, Surabaya, East Java, Indonesia

ABSTRACT: The role of the learners' personality in learning English as a Foreign Language (EFL) has drawn the interest of researchers and educators. The present ex-post-facto study added empirical evidence of this issue by investigating whether extroversion and introversion affected the learners' English proficiency. Seventy-five undergraduates majoring in English participated in this study and completed a questionnaire to find out whether they were extroverts, introverts, or ambiverts. They also took the Test of English Proficiency (TEP) to measure their language proficiency. The results indicated the English proficiency of the extrovert learners was not significantly different from that of the introvert or ambivert learners. On the basis of the findings, implications for the teaching of EFL and suggestions for further studies were presented.

1 INTRODUCTION

The classification of personality into extravert (or extrovert) and introvert was proposed at the beginning of the 20th century by Jung (1971), who observed how individuals differed with respect to the source and direction of expressing their energy. Whereas extroverts turn them outward, introverts tend to do the same inward. Extroverts regain energy by socializing and engaging in conversations with others. They think talking to people is more stimulating, hence the impression that they speak a lot. In contrast, introverts find conversing with a group of individuals draining their energy. They can refuel by withdrawing from others and reflecting in solitude.

These differences prompted some research that associate extroversion/introversion and language learning. For instance, Nussbaum (2002) found that extroverts and introverts approached small group discussion differently. The extroverted learners tended to express contradictions and counterexamples with their extroverted peers, while the introverted learners collaborated with the other introverted members of the group to find solutions. Another study demonstrated that the extroverts tended to begin the discussion, introduce new topics and restate ideas in language classrooms, whereas the introverts generally asked questions (Abali, 2006).

The contrast between these two personality types was not confined to the learners' behaviors only, but it also occurred in the use of language learning strategies (Chen & Hung, 2012). The introverted language learners used cognitive and metacognitive strategies more than the extroverted learners (Kayaoğlu, 2013), except for metacognitive strategies that promote opportunities for communication. Similarly, Wakamoto (2000) found a significant correlation between extroversion and functional practice strategies and social-affective strategies, but introversion was not associated with any specific strategies.

Although the above studies provided empirical evidence of the role that extroversion and introversion played in language learning, Brown (2000) cautioned against stereotyping the extroverts as better language learners because they seemed to speak more than their introverted peers in the classroom. Contrary to popular belief, learners were not categorized as extroverts simply because they expressed their ideas orally more frequently than others, and the introverted learners were not always the quiet, passive ones in class discussions either. Both the extroverts and the introverts might talk a lot during the lesson, and they differed in how they got the sense of wholeness:

DOI 10.1201/9781003265061-5

the former obtained it from others, while the latter needed to distance themselves from others to contemplate.

Despite the dissimilar traits, research indicated that the English proficiency of the extroverts was not significantly different from that of the introverts (Busch, 1982). The present study replicated the previous work to obtain more empirical evidence about this issue but one more personality type was added namely ambiversion. Coined by Edmund Smith Conklin a few years after Jung revealed extroversion and introversion, the term ambiversion refers to the characteristics of extroversion and introversion (Davidson, 2017). An ambiverted person is mainly extroverted in a particular condition and mainly introverted in another to adapt to the surrounding, and this is considered more realistic as no person is purely extroverted or introverted in real life.

This study aimed to find out whether the learners' extroversion, introversion and ambiversion significantly affected their English proficiency. The procedures of conducting this study will be taken up in the next section. Then the findings will be reported, followed by the implications from these findings.

2 METHODS

This research was conducted by using ex-post facto study as the research design because the independent variable, i.e. the learners' extroversion and introversion, occurred naturally and could not be manipulated (Ary, Jacobs, & Sorensen, 2010; Cohen, Manion, & Morrison, 2005; Gall, Gall, & Borg, 2003). The population of the study consisted of 90 third-year students at the English Department of a state university in the outskirts of Surabaya, Indonesia. All of them were approached to participate in this study, but only 75 undergraduates responded so they became the sample.

Two instruments were administered in this study: a questionnaire and a test. The questionnaire was employed to classify the participants into introverts, extroverts or ambiverts. It was devised by Cain (2011), consisting of twenty statements that the participants had to identify as true or false. Some examples of the statements were "I prefer one-on-one conversations to group activities" and "I enjoy solitude." If the answer to more than 10 items was "true," the participants tended to be introverts. On the other hand, they were categorized as extroverts if they answered "false" in more than ten items. Equal number of "true" and "false" (that is, 10 of each) meant that the participants could be classified as ambiverts. The second instrument was the Test of English Proficiency (henceforth, TEP), a standardized test which was developed by the Language Center of the university. As the name suggests, the test estimated the participants' English proficiency with respect to listening comprehension, reading comprehension and sentence structures.

To collect data from the participants, the researchers asked them to answer the questionnaire, resulting in information about their personality—extrovert, introvert or ambivert. They were assigned into three groups on the basis of their personality type. Afterwards, they submitted the score of TEP. Because they had to take TEP every semester, each participant had at least four scores. They disclosed only the latest score for this study. Data collection resulted in three sets of TEP scores from the three groups. These data were analyzed by means of ANOVA to test the difference between these groups with respect to their English proficiency. One way ANOVA was utilized because there was only one independent variable, namely, the participants' extroversion/introversion.

3 RESULTS AND DISCUSSION

Out of 75 participants in this study, the majority could be classified as introverts (N = 52), some of them belonged to the category of extroverts (N = 11) and the rest were ambiverts (N = 6). The proportion of the participants is shown in Figure 1.

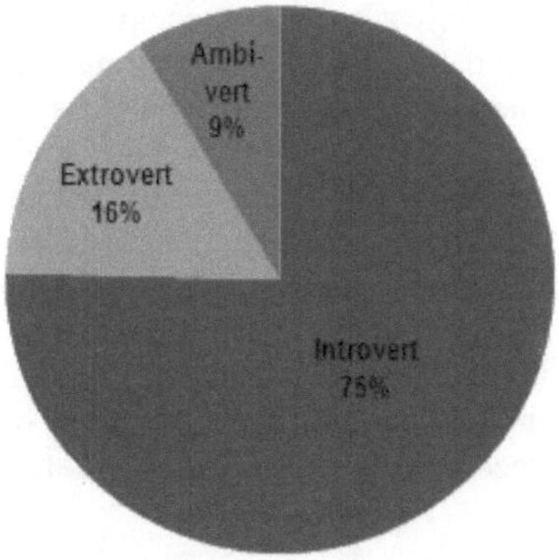

Figure 1. The percentage of introverted, extroverted and ambiverted undergraduates.

The results of the computation indicated that there was no significant difference in English proficiency between the extroverts, the introverts and the ambiverts. Table 1 reports the results of the computation.

Table 1. The results of ANOVA computation.

Groups	N	$\underline{X}$	F	P-value	F_{crit}
Extroverts	11	478.36			
Introverts	52	465.08	1.26	0.29	3.14
Ambiverts	6	454			

The means of TEP scores obtained from these three groups were only slightly different, suggesting that their English proficiency tended to be similar. The extroverted undergraduates had the highest mean ($\underline{X}_{extro} = 478.36$), the introverted ones had a slightly lower mean ($\underline{X}_{intro} = 465.08$) and the ambiverts generally scored the lowest ($\underline{X}_{ambi} = 454$). Although these groups appeared to be different, statistical tests using one-way ANOVA showed the difference in English proficiency between the extroverts, the introverts and the ambiverts was not significant ($F = 1.26$, $P = 0.29$).

This was consistent with the finding of the previous studies which demonstrated the extroverted learners did not have better English proficiency than the introverted ones (Busch, 1982; Chen, Jiang, & Mu, 2015). Regardless of the learners' place in a continuum of extroversion and introversion, they had an equal chance to be successful in learning a foreign language. Apparently, their personality types played an essential role in the *process* of language learning, but they did not affect the *outcome*. It explained why the extroverted learners expressed their ideas in a different manner from their introverted peers in classroom interactions (Abali, 2006; Nussbaum, 2002). The former also understandably put more emphasis on the use of social strategies to make language learning more effective for them, whereas the latter tended to make good use of cognitive and metacognitive strategies (Kayaoğlu, 2013; Liyanage & Bartlett, 2013; ;xford, 2003; Oxford, Hollaway, & Horton-Murillo., 1992; Wakamoto, 2000). While attempting to improve their English proficiency, the introverts needed to reflect and reason to perform their best so they utilized language learning strategies that facilitated reflection and reasoning. Consequently, what they articulated was mostly the results of these two cognitive behaviors. The extroverts, on the other hand, needed communication with others in thought processes, hence employing mostly social strategies in language learning. The ambiverts had the flexibility to adapt to any learning situations so they might reflect in solitude or engage in exchange with peers to optimize learning. The extroverts, the introverts and the ambiverts who selected the language learning strategies and activities that matched their type of personality well achieved the most and succeeded in improving their language proficiency.

This finding had an important implication in the teaching of English as a foreign language. English teachers should understand the personality type of the learners and assist them to orchestrate the appropriate learning strategies. Dedicating some class time for strategy instruction and raising the learners' awareness of extroversion, introversion and ambiversion is recommended whenever feasible. Additionally, teachers need to expose the learners to various learning activities to accommodate different personality types and respect these differences. As suggested by Brown (2000) earlier in this paper, the teachers should avoid considering the extroverted learners as superior or more desirable in the classroom simply because these learners seem to speak more than their peers. As a matter of fact, the teachers may lead the way in abolishing such stereotypes and embrace learners regardless of their personality types because—as this study has found—all of them have equal potential to succeed in learning English as a foreign language.

4 CONCLUSION

The study was an attempt to find out whether there was a significant difference in English proficiency between the extroverts, the introverts and the ambiverts. The results showed that such a difference did not exist among the undergraduates. It means the learners' personality types did not affect the outcome of learning, although they might determine the strategies and activities that the learners used in the learning process. Based on this finding, some implications for the teaching of English were suggested.

Further studies need to be conducted with larger samples to obtain more empirical evidence about this issue. Moreover, the construct of English proficiency in this study was only limited to listening comprehension, reading comprehension and grammar so further research should measure this construct more comprehensively by including other language skills in the test. Another potential phenomenon to investigate related to this issue is how the ambiverts mobilize language learning strategies to adapt to different learning environments. The strategies used by the introverted and extroverted learners have been well documented but similar research on the ambiverted language learners has been scant. In-depth studies to explore the language learning strategies utilized by the ambiverts will be an important piece of puzzle that enables educators to get a comprehensive insight into how learners with different personalities can be equally successful in learning a foreign language.

REFERENCES

Abali, F. (2006). *The effect of personality traits extroversion/introversion on verbal and interactive behaviors of learners.* Ankara: Bilkent University.

Ary, D., Jacobs, L. C., & Sorensen, C. K. (2010). *Introduction to research in education*. Belmont: Wadsworth.

Brown, H. D. (2000). *Principles of language learning and teaching*. New York, NY: Longman.

Busch, D. (1982). Introversion-Extraversion and the EFL Proficiency of Japanese Students. *Language Learning, 32*(1), 109–132. DOI: https://doi.org/10.1111/j.1467-1770.1982.tb00521.x

Cain, S. (2011). Quiz: Are you an Introvert or an Extrovert? (And Why It Matters). *Psychology Today*. Retrieved from: https://www.psychologytoday.com/us/blog/quiet-the-power-introverts/201103/quiz-are-you-introvert-or-extrovert-and-why-it-matters

Chen, M. -L., & Hung, L. -M. (2012). Personality type, perceptual style preferences, and strategies for learning English as a foreign language. *Social Behavior and Personality: An International Journal, 40*(9), 1501–1510. DOI: http://dx.doi.org/10.2224/sbp.2012.40.9.1501

Chen, Y., Jiang, Y., & Mu, Z. (2015). Survey study: The correlation between introversion/extroversion and oral English learning outcome. *Journal of Language Teaching and Research, 6*(3), 581-587.

Cohen, L., Manion, L., & Morrison, K. (2005). *Research methods in education*. London: Routledge Falmer.

Davidson, I. J. (2017). The ambivert: A failed attempt at a normal personality. *Journal of the History of the Behavioral Sciences, 53*(4), 1–19. DOI: 10.1002/jhbs.21868

Gall, M. D., Gall, J. P., & Borg, W. R. (2003). *Educational research: An introduction*. Boston, MA: Pearson Education.

Jung, C. G. (1971). *Psychological types*. Princeton, NJ: Princeton University Press.
Kayaoğlu, M. N. (2013). Impact of extroversion and introversion on language-learning behaviors. *Social Behavior and Personality: An International Journal, 41*(5), 819–825. DOI: 10.2224/sbp.2013.41.5.819
Liyanage, I., & Bartlett, B. (2013). Personality types and languages learning strategies: Chameleons changing colours. *System, 41*(3), 598–608. DOI: 10.1016/j.system.2013.07.011
Nussbaum, E. M. (2002). How introverts versus extroverts approach small-group argumentative discussions. *The Elementary School Journal, 102*(3), 183–197. DOI: 10.1086/499699
Oxford, R. L. (2003). Language learning styles and strategies: Concepts and relationships. *International Review of Applied Linguistics in Language Teaching, 41*, 271–278. DOI: 10.1515/IRAL.2003.012
Oxford, R. L., Hollaway, M. E., & Horton-Murillo, D. (1992). Language learning styles: research and practical considerations for teaching in the multicultural tertiary ESL/EFL classroom. *System, 20*(4), 439–456. DOI: 10.1016/0346-251X(92)90057-A
Wakamoto, N. (2000). Language learning strategy and personality variables: Focusing on extroversion and introversion. *IRAL – International Review of Applied Linguistics in Language Teaching, 38*(1), 71–81. DOI: 10.1515/iral.2000.38.1.71.

Innovation on Education and Social Sciences – Maureen et al (Eds)

The use of Xreading books & audios for extensive reading program

W.E. Savitri & A. Munir
Universitas Negeri Surabaya, Surabaya, East Java, Indonesia

ABSTRACT: In doing the Extensive Reading (ER), teachers are suggested to use graded readers materials which can also be found online. Xreading is one platform that provides it. It is an online library that provides its readers with graded readers' books and their audio versions. The audios are available for most of the books. The availability of the audio enables students to do Reading While Listening (RWL) instead of Reading Only (RO). Nevertheless, there has been inadequate information on how students utilize the books and audios in Xreading. Therefore, this study aims to reveal how the books and audio in Xreading are used in an ER program, students' tendencies to do RWL or RO, and the correlation between reading time and listening time. The participants of this study are 154 freshmen university students in Indonesia who belong to the ER program using Xreading for one semester. The data of this study are digital records in Xreading and students' responses to a questionnaire. The data were analyzed quantitatively. The research found that Xreading reading books and audiobooks have been used by the participants. However, audiobooks have not been used maximally for various reasons. Personal preference and different needs have been the main reasons for the underuse of audiobooks.

1 INTRODUCTION

Extensive reading (ER) is a reading activity in which students do not 'read to learn', but they 'learn to read'. In ER, students practice their reading skills in a free-formulated atmosphere. They develop their reading fluency (ERF, 2016). Day and Bamford (Day & Bamford, 2002) put forward 10 principles for ER. Prowse and Macalister support these 10 principles (Macalister, 2015; Prowse, 2002); However, Robb argues that the ten principles are too stiff and some are not doable. He reminds that ER should be flexible and adjusted with the students' context (Robb, 2002). To respond to it, Jeon and Day justify the effectiveness of core ER principles and reveal that in actual practice, not all ten principles are applied by ER practitioners (R. Day, 2015; Jeon & Day, 2015). Principle #5 of ER mentions that reading should be for pleasure. In line with it, despite the experts' different points of view, they all agree that ER should be enjoyable and make the students feel good in doing it.

To make ER a pleasurable program, the books used in it should be easy for the students. Easy means the learners know at least 98% of the words without the help of dictionaries and other assistance (Hu & Nation, 2000). It is relevant with ER principle #1, which clearly states that the reading material is accessible (ERF, 2016). Too hard materials will cause stressful experiences. To create an enjoyable atmosphere, teachers who organize the ER program are suggested to use graded readers materials (R. R. Day, 2012). Graded readers are books specially prepared to develop reading speed and fluency and facilitate reading for pleasure (ERF, 2016).

Graded readers cover fiction and non-fiction. The term 'graded' indicates that the books have graded difficulty levels (Brierley, Nakamura, & Niimura, 2020; Ramonda, 2020).

 DOI 10.1201/9781003265061-6

Graded readers cover fiction and non-fiction. The term 'graded' indicates that the books have graded difficulty levels (Brierley, Nakamura, & Niimura, 2020; Ramonda, 2020).

Graded readers can be in the form of prints or online. Printed graded readers have been published for several years. Recently, online graded readers have become more favored due to the advance of mobile technology. These online graded readers are helpful and easy to access as long as the students have an internet connection. Some online platforms provide graded readers (Al Damen, 2018; Milliner, 2017; Milliner & Cote, 2015). Not all the platforms, however, offer free service. Xreading is one of them. To access it, an institution (or teachers & students) should pay some amount of annual charge. Subscribing this platform enables the users to get additional benefits because Xreading provides graded readers books and their audio versions. This is one strength of Xreading as an online library.

Previous studies found that offline and online ER, which are implemented worldwide, really help students improve their English ability, especially their mastery of vocabulary and grammar (Cote & Milliner, 2014; Nakanishi, 2015; Shih, Chern, & Reynold, 2018;. ER also helps to increase the reading rate (Huffman, 2014). Yet, this strategy also faces challenges (Van Amelsvoort, 2016). Teachers and students who implement both modes of ER have various perceptions on online and offline ER [20] (Collett, 2018; Howarth & Bollen, 2019). However, ER's audiobooks are still very limited, although the audios are available, especially for ER with online platforms.

In Xreading, the audios are available for most of the books. The audio availability allows the students to do Reading While Listening (RWL) instead of Reading Only (RO). While reading and listening can be done when somebody wants to improve their reading skill, listening skill, or both. Previous studies proved that students who did RWL outperformed their peers who did not do it (A. C.-S. Chang, 2009; C. Chang & Chang, 2011; Milliner, 2019;. It also helps to reduce anxiety (Ranto Rozak, Saleh, Linggar Bharati, & Sutopo, 2019) and establish students' self-regulated learning strategies (Wang, 2004). However, there has been inadequate information on how students utilize the books and audios in Xreading. Therefore, this study aims to reveal how the books and the audios in Xreading are used by the students in an ER program and students' preference in using them. Specifically, it aims to:

1. Depict how the books and audios available in Xreading are used by the students in the ER program.
2. Determine students' tendency to use the books and audios for conducting either RWL or RO.
3. Reveal whether there is a significant correlation between students reading time and listening time.

2 METHODS

The participants of this study are 154 freshmen university students in Indonesia who took an ER program using Xreading. The students were obliged to do an extensive reading for the whole semester. In order to be able to access Xreading, they had to subscribe to it. This subscription gave them opportunities to access hundreds of books in Xreading and all features embedded there, including the audios.

There are two kinds of data utilized in this study. The primary data of this study are digital records in the Xreading platform containing students' ER activities for the whole semester. The records reveal the number of books the students target and the ones they read, the books' levels, their reading speed, their reading time, and their listening time. The data have been analyzed quantitatively.

In addition, students' responses to a questionnaire that has been designed in line with the students' digital records in the ER program are analyzed as supporting data. The questionnaire mainly asks students' reasons behind Xreading books and audio during the ER program. These data are also analyzed quantitatively.

3 RESULTS AND DISCUSSION

3.1 *Depiction of how the books and audios available in Xreading are used by the students in ER program*

In general, the study found that the books of Xreading were used more frequently than the audios. This information is inferred from the marked difference between reading time and listening time. Details of the use of Xreading by the participants are presented in Table 1.

Table 1. Average use of Xreading books and audios.

	Average
Books added	27.08
Books passed	22.73
Words read	58,826.77
Book level	4.61
Read speed (word/minute)	116.2831
Reading time (minutes)	476.79
Listening time (minutes)	44.36

Table 1 shows that the number of books the students passed is less than the number they planned to read. It indicates that many students failed to meet their reading targets. Table 1 also shows that most students did not meet the institution's expectation of reading 60,000 words minimally for the whole semester.

Related to the book level, the digital record shows that it turns out to be level 5 out of 11. It is quite normal that, on average, the students' level is still below level 5 because the participants of this study are freshmen. As new graduates of senior high school, they are commonly not familiar with books in English because it is not common for traditional Indonesian schools to use books written in English as learning materials. Therefore, they lack experience and exposure with English used in books beyond English lesson books.

It is also revealed that the students' average reading speed is only about 116 words per minute only. It is below the normal speed of 200–400 words per minute for college students (Brysbaert, 2019; Rayner, Schotter, Masson, Potter, & Treiman, 2016). Again, it might occur because the students involved were freshmen who are non-native speakers of English.

As stated previously, Table 1 indicates that the audios in Xreading were less frequently used than the books. The discrepancy between the reading time and listening time is considerable. The table shows that the listening time is only one-tenth of the reading time. This suggests that the audio is underused.

3.2 *Students' tendency in using the books and audios for conducting either RWL or RO*

It is found that most students tended to implement RWL than RO. Besides, the students tried to read books that met their reading level. As many as 81% of the students read books that are at their reading level. Only less than 20% of students reported reading books that were not suitable for their reading level. It can be inferred that the students felt secure when they read books within their reading level. They did not dare themselves too much to read beyond their level.

Another finding is that most students know that Xreading has audio facilities and that the audios' speed can be adjusted. However, their familiarity with the Xreading audios did not automatically make them use the audiobooks or implement RWL.

There are 60 out of 154 students who did RO, without using audiobooks. Meanwhile, 94 out of 154 students read books in Xreading using RWL. Thus, the students know that there are audiobooks in Xreading, and more than half of them use the audiobooks. However, their use of audiobooks is limited to only 10% of their reading time. The students' familiarity with the audiobooks is shown by 98% of the students.

Next, the students' reasons for conducting either RWL or RO is elaborated. Students' responses to the questionnaire reveal that the students had some reasons for doing RO. Their reasons are as follows.

1) Inability to do multitasks

This is related to a problem with concentration and focus. The students get confused about the tasks. They are confused about whether they should focus on reading or on listening. For these students, the audios easily distract their reading focus so that in the end, both tasks do not run well. It is a fact that not everyone can multitask, especially in learning a foreign language.

2) RWL slows down their reading pace

In the students' opinion, the audiobook's speed is slower than their reading speed. Thus, they felt that the audio did not fulfill their need to read at their own reading pace. The students felt that RWL slowed down their reading speed. This opinion initially seems questionable because Xreading audiobook speed is known to be adjustable. The data from the questionnaire is proof that the students know that the speed can be adjusted. Most of the students (68%) know it. However, when the audiobooks are rechecked, many of them have no speed adjusting button. It needs more study to check whether the students who felt that RWL slows down their reading pace belong to those who admit that they do not know that the audiobooks' speed is adjustable. It is necessary to check more whether the cause is basically because of the non-existence of the speed adjusting button.

3) Preference and comfort

This reason is related to the students' personal preference and their feeling of comfort. They stated that in ER, what they want to do is read, not listen. Therefore, they did not feel it necessary to listen to the available audiobooks or do RWL, although they know that the Xreading books they read have their audio version.

Related to comfort, the students felt a lot more comfortable focusing on their reading without distraction from any audio except particular music they found. This group of students says that they try to focus on achieving their reading target, and they believe that reading without listening is their comfortable way to reach their goal. They believe that they can already understand the books without the help of the audio so that they do not need the audio. Some of them mentioned that they tried to use the audio earlier, but then they changed and set aside the audio because it turned to distract rather than assist with the reading itself.

4) RWL is not interesting

The students think that RWL is not attractive. They mentioned that their purpose is to read; therefore, they do not listen to audiobooks. Moreover, the audiobooks are not attractive enough for them to trigger them to do RWL. The students think that the narrator's voice in Xreading audio books is too robotic and lacks expression, so it is unpleasant to hear. For them, it would be more interesting if the sound

The responses to the questionnaire also reveal students' reasons for doing RWL. Amongst the reasons are as follows.

1) Learning pronunciation

The students chose to do RWL because they wanted to know the acceptable pronunciation of the words found in the books they read. They believe that Xreading audio provides good examples of how to pronounce correctly. During RWL, students could learn to pronounce and check the spelling of the words in the virtual book. RWL strengthens the students' pronunciation accuracy and, at the same time, improves the students' vocabulary mastery.

2) Helping to be more focus on their reading

For some students, the availability of audiobooks in Xreading assisted them to be more focused on their reading. For these students, it felt like being guided to read carefully and meaningfully when listening to the audio and reading the book simultaneously. This answer leads to the suspicion that these students belong to an auditory type of learning style. Unfortunately, there was no data from both digital records and questionnaires to confirm or to refuse it.

3) Helping to understand the book

Most students (two-third of the student number) believe that reading while listening helps them understand the book better. These students felt that they benefited from the availability of audiobooks. Since RWL helped them focus more on the reading, their understanding of the text improved simultaneously. Again, this is what the students think and feel of their ability. Thus, it is not automatically proof of their natural ability. To know their actual ability, it will be necessary to conduct a thorough reading assessment. Yet, this is in line with a previous research finding that integrates both digital books and their audios help to understand the books and get engaged to the reading (Larson, 2015)

4) Reducing fatigue and eyesore

Long hour screen time during pandemic Covid-19 makes students' eyes exhausted. Thus, it is logical that students mentioned it as their reason. By listening to the audiobooks, the students might rest their eyes here and there during RWL whenever they feel tired. There is a possibility that in the end there were students who did not even read the book but simply listened to the audiobook. If it happened, it might be called extensive listening rather than extensive reading. The questionnaire, however, did not provide information about it. It simply caught the students' statement that using audiobooks in RWL helped lessen their fatigue and eyesore. This also helps students who got a problem with their eyes as well as eyes disorder like low vision or those who have reading disabilities (Esteves & Whitten, 2011)

5) Improving storytelling skill

Although the students who chose not to do RWL said that the narrator's voice is robotic, the students who chose to do RWL stated the opposite. They stated that the narrator's way of telling a story could be an excellent example and establish their storytelling skill. The students learned how to tell a story interactively, emphasize essential details, produce different voices to adjust the characters, etc.

6) Improving listening

This reason is predictable. RWL unconsciously trains the student to listen to aural materials. Hence, their ears are familiar with English sounds. They also learned linguistic features of spoken utterances and learned to derive meanings from the sentences they heard. Since the books also accompanied the sounds, it was easier for the students to check complex or unfamiliar words when they missed the words in the audio.

In general, the findings related to students' tendencies unexpectedly reveal that the students have a significantly different opinion on RWL and then trigger curiosity that different learning styles might cause this different tendency. Despite the dispute on the actual existence of learning styles, it is widely believed that learning styles contribute to the students learning (Miqawati & Sulistyo, 2014; Göktürk & Altay, 2015; Grossmann, 2011; Tempelaar, Rienties, & Giesbers, 2011;. Visual students might belong to those who preferred RO because they prefer to focus on the text. For this kind of student, audio might reduce their reading focus. On the contrary, auditory students might belong to the ones who did RWL. For this sort of student, audio helps them to increase their focus on the text they read. Unfortunately, this study did not investigate the students' learning styles. Therefore, further study can be done on this basis.

3.3 *Correlation between students' reading time & listening time*

Further analysis on the Reading time and Listening time reveals how reading time and listening time using Pearson correlation shows there is significant positive moderate correlation ($r = 0.543$, $p = 0.000$). This indicates that the longer the students read, the longer they listen. As the listening time is only one tenth of the reading time, those who read less might not listen to the audiobook. This might suggest that those who read less consider listening to audiobooks might interfere with their reading, i.e. reducing their comprehension of the text.

On the contrary, those who read more might have used the audiobooks longer. In other words, these students have made use of advanced reading strategies, not solely depending on their reading the print. This kind of student seems to take advantage of the mobility offered by RWL activity in which they do not have to rely on their eyes, but more on their ears. Some scholars call this type of reading activity Reading by Listening (Tattersall Wallin, 2021; Tattersall Wallin & Nolin, 2020).

4 CONCLUSION

Many participants have used Xreading reading books and audiobooks; however, the audiobooks have not been used maximally for various reasons. Personal preference and needs for reading strategies have been the main reasons for the underuse of audiobooks. The different ways of using xreading books and audios are also thought to impact different learning styles.

In response to the findings, it is suggested that teachers do not force students to do either RO or RWL in implementing ER using Xreading. It would be more convenient if teachers repeatedly show and explain how to use the books and the audiobooks effectively and efficiently to accommodate different preferences and learning styles. Since there has been no information on the correlation between students' proficiency, students' actual practice of ER, and their reading strategies in doing ER in this study (whether they conduct RO or RWL), it is suggested that a further study be carried out in this particular topic.

REFERENCES

Al Damen, T. M. (2018). The effectiveness of M-reader in promoting extensive reading among Arab EFL learners. *Arab World English Journal*, (1), 3–23. https://doi.org/10.24093/awej/mec1.1

Brierley, M., Nakamura, R., & Niimura, M. (2020). *Difficulty Levels of Graded Readers*. 219–228.

Brysbaert, M. (2019). How many words do we read per minute? A review and meta-analysis of reading rate. *Journal of Memory and Language*, *109*(July), 104047. https://doi.org/10.1016/j.jml.2019.104047

Chang, A. C.-S. (2009). Gains to L2 listeners from reading while listening vs. listening only in comprehending short stories. *System*, *37*(4), 652–663. https://doi.org/10.1016/j.system.2009.09.009

Chang, C., & Chang, A. C.-S. (2011). The effect of reading while listening to audiobooks: Listening fluency and vocabulary gain. *Asian Journal of English Language Teaching*, *21*(2), 43–64. Retrieved from http://search.proquest.com/docview/919962561?accountid=8330%5Cnhttp://library.anu.edu.au:4550/resserv?genre=article&issn=10262652&title=Asian+Journal+of+English+Language+Teaching&volume=21&issue=&date=2011-01-01&atitle=The+Effect+of+Reading+While+Listenin

Collett, P. (2018). A comparison of two online systems for extensive reading. *Journal of Extensive Reading*, *6*(3). Retrieved from http://jalt-publications.org/content/index.php/jer/

Cote, T., & Milliner, B. (2014). Extensive reading on mobile devices?: Is it a worthwhile strategy?? *Proceedings of the 12th Asia TEFL and 23rd MELTA International Conference*, (August), 979–990. Retrieved from www.xreading.com

Day, R. (2015). Extending extensive reading. *Reading in a Foreign Language*, *27*(2), 294–301.

Day, R. R. (2012). Graded readers. *The Encyclopedia of Applied Linguistics*. https://doi.org/10.1002/9781405198431.wbeal0459

Day, R., & Bamford, J. (2002). Top ten principles for teaching extensive reading. *Reading in a Foreign Language*, *14*(2), 136–141.

ERF. (2016). The extensive reading foundation's guide to extensive reading. In *ERF*. https://doi.org/10.1093/elt/cct010

Esteves, K. J., & Whitten, E. (2011). *Reading Horizons?: A Journal of Literacy and Language Arts Assisted Reading with Digital Audiobooks for Students with Reading Disabilities Assisted Reading with Digital Audiobooks for Students with Reading Disabilities*. *51*(1).

Göktürk, N., & Altay, A. (2015). The relationship between Efl learners ' learning styles and their scores in audio and video-mediated L2 listening tests. *Journal of Theory and Practice in Education*, *11*(3), 971–988.

Grossmann, D. (2011). A study of cognitive styles and strategy use by successful and unsuccessful adult learners in Switzerland. *Dissertation Master of Arts*.

Howarth, M., & Bollen, D. (2011). *Teacher Perceptions of an Online Extensive Reading Platform by*. 35–40.

Howarth, M., & Bollen, D. (2019). Student perception of online extensive reading platform. *Procedings of Sojo University Vol. 44*, 145–152.
Hu, H.-C., & Nation, P. (2000). Unknown vocabulary density and reading comprehension. *Reading in a Foreign Language, 13*(April), 403–30.
Huffman, J. (2014). Reading rate gains during a one-semester extensive reading course. *Reading in a Foreign Language, 26*(2), 17–33.
Jeon, E.-Y., & Day, R. R. (2015). The effectiveness of core ER principles. *Reading in a Foreign Language, 27*(2), 302–307. Retrieved from http://nflrc.hawaii.edu/rfl
Larson, L. C. (2015). E-books and audiobooks: Extending the digital reading experience. *Reading Teacher, 69*(2), 169–177. https://doi.org/10.1002/trtr.1371
Macalister, J. (2015). Guidelines or commandments? Reconsidering core principles in extensive reading. *Reading in a Foreign Language, 27*(1), 122–128.
Milliner, B. (2017). One year of extensive reading on smartphones: A report. *JALT CALL Journal, 13*(1), 49–58. https://doi.org/10.29140/jaltcall.v13n1.211
Milliner, B. (2019). Comparing extensive reading to extensive reading-while-listening on smartphones: Impacts on listening and reading performance for beginning students. *Reading Matrix: An International Online Journal, 19*(1), 1–19.
Milliner, B., & Cote, T. (2015). Mobile-based extensive reading. *International Journal of Computer-Assisted Language Learning and Teaching, 5*(4), 1–15. https://doi.org/10.4018/ijcallt.2015100101
Miqawati, A. H., & Sulistyo, G. H. (2014). The PQRST strategy, reading comprehension, and learning styles. *Indonesian Journal of Applied Linguistics, 4*(1), 123–139.
Nakanishi, T. (2015). A meta-analysis of extensive reading research. *TESOL Quarterly, 49*(1), 6–37. https://doi.org/10.1002/tesq.157
Prowse, P. (2002). Top ten principles for teaching extensive reading: A response. *Reading in a Foreign Language, 14*(2).
Ramonda, K. (2020). Extensive reading and class readers: The case for no choice. *ELT Journal, 74*(3), 277–286. https://doi.org/10.1093/elt/ccaa017
Ranto Rozak, R., Saleh, M., Linggar Bharati, D. A., & Sutopo, D. (2019). Reading While Listening (RWL) in an extensive listening course to reduce student teachers' foreign language listening anxiety (FLLA). *KnE Social Sciences, 3*(10), 349. https://doi.org/10.18502/kss.v3i10.3916
Rayner, K., Schotter, E. R., Masson, M. E. J., Potter, M. C., & Treiman, R. (2016). So much to read, so little time: How do we read, and can speed reading help? In *Psychological Science in the Public Interest, Supplement* (Vol. 17). https://doi.org/10.1177/1529100615623267
Robb. (2002). Extensive reading in an Asian Context – An alternative view. *Reading in a Foreign Language, 14*(2), 146–147. Retrieved from http://nflrc.hawaii.edu/rfl/October2002/discussion/robb.pdf
Shih, Y.-C., Chern, C., & Reynold, B. L. (2018). Bringing extensive reading and reading strategies into the Taiwanese junior college classroom. *Reading in a Foreign Language, 30*(1), 130–151.
Tattersall Wallin, E. (2021). Reading by listening: Conceptualising audiobook practices in the age of streaming subscription services. *Journal of Documentation, 77*(2), 432–448. https://doi.org/10.1108/JD-06-2020-0098
Tattersall Wallin, E., & Nolin, J. (2020). Time to read: Exploring the timespaces of subscription-based audiobooks. *New Media and Society, 22*(3), 470–488. https://doi.org/10.1177/1461444819864691
Tempelaar, D. T., Rienties, B., & Giesbers, B. (2011). Student learning preferences in a blended learning environment: Investigating the relationship between tool use and learning approaches. In *Building Learning Experiences in a Changing World*. https://doi.org/10.1007/978-94-007-0802-0_12
Van Amelsvoort, M. (2016). Extensive reading onboarding: challenges and responses in an optional program key words. *Juntendo Journal of Global Studies, 1*, 95–106.
Wang, C. (2004). Self-regulated learning strategies and self-efficacy beliefs of children learning English as a second language. *ProQuest Dissertations and Theses*.

Innovation on Education and Social Sciences – Maureen et al (Eds)

Variative etude as skill-forming on the Youth Level Viola major course at music study program

H.Y. Karyawanto, M. Sarjoko & V.E. Dewi
Universitas Negeri Surabaya, Surabaya, East Java, Indonesia

ABSTRACT: Music practice courses require assistance in the learning process. Hence, innovation in textbook materials development must be adapted to the instrument characteristics. This study discusses the significance of etude variations in the development of Youth Level Viola learning books as a trigger for students' skills. This study uses a mixed method with the Research and Development method developed by ADDIE models. The development of Youth Level Viola learning books is empirically able to show significant results in the learning process. The product trial was conducted at the Music Study Program, Universitas Negeri Surabaya with purposive sampling. Data analysis of product validity results is arranged in a questionnaire scale. The aspect of product validity test is appropriate with the Indonesian National Education Standards Agency criteria. As a result of the effectiveness, the percentage of 90% indicates that the Youth Level Viola learning book can be a guide for students to practice.

1 INTRODUCTION

The domain of music education and the music industry moves dynamically in an increasingly competitive and promising direction (Ferdian, Putra, & Yuda, 2020). Other than that, the improvement in the realm of education is currently experiencing some changes in rapid ways. Electronic media that is growing rapidly affects the way of thinking and the pattern of life, especially in education. Innovation and creativity are buzzwords in society. Innovation as a concept refers to the use of inventive ideas and novel ways to act and create solutions. Creativity as a part of innovation is considered essential for economic growth and as an everyday life skill. Innovation is seen as a precondition for general success and well-being. On the basis of innovation strategies, governments around the world pursue innovations by establishing new innovation centers (Lindfors & Hilmola, 2016). The Innovation process is a simple way to teach students creative skills (Thorsteinsson & Denton, 2003). The old methods have been left behind and replaced with the latest findings and innovations, and those left behind are starting to adapt to changes, which is globally known as the 4.0 industrial revolution era and 5.0 society. For an individual, a nation, and humankind to survive and progress, innovation and evolution are essential. Innovations in education are of particular importance because education plays a crucial role in creating a sustainable future (Serdyukov, 2017). By seeing the polemic, it is needed to improve better learning that can also improve students' understanding, especially in the field of performing arts. Innovation is based on creative emphasis in both teaching and coursework (Thorsteinsson, 2002). Performing art education becomes an identity for a cultured nation. Universities that have Music Study Programs have innovations in developing their learning to improve the quality of their education, especially in the main subject in performing art, namely musical practice.

Art practice teaching in colleges and universities refers to the professional art students' activities through various art forms, the class teacher's teaching knowledge and skills combined with practical and flexible use to realize a kind of knowledge to deepen the teaching activities (Zhou, 2018). In line with that statement, practical courses have different characteristics from other courses, which

DOI 10.1201/9781003265061-7

require time and intensive assistance. Through this learning material book, it is hoped that it will provide solutions for a more effective learning process. The learning material book can guide students in learning and can also be a guide in understanding a basic concept of the knowledge being studied. Students in the learning process need references, but sometimes it is constrained by existing material books. By the observations, the following results were found: (1) the lack of material books on the viola instrument in Indonesia, (2) students need books with Indonesian translations since most viola books are still in foreign languages, and (3) the importance of strengthening techniques in playing the viola in youth level. The technique can be improved through practicing etude. By the results of that study, it is crucial to develop this viola learning book to solve the needs of the lectures process at the Sendratasik Department, Faculty of Languages and Arts, Universitas Negeri Surabaya.

2 METHODS

This research focuses on developing the Youth Level Viola learning book by using a mixed-method (Research and Development method developed by ADDIE models). There is a learning design based on a systems approach. The components contained in the ADDIE Model, among others: Analyze, Design, Develop, Implementation, and Evaluation. The results of testing the model's effectiveness show that the learning model of the viola instrument course can improve students' skills in understanding the practical material of the instrument, as proved by the increase in learning outcomes achieved. This study pays attention to three characteristics. First, the learning model purpose is focused on independent learning efforts centered on students as learning subjects (student-centered). Second, the role of the lecturer is as a facilitator who assists and motivates students in developing integrity learning. In an innovative work, the individual and his ideas unite since students work with something in their environment that is their own need. This personal identification is an essential motivational factor (Lindfors & Hilmola, 2016). Third, learning is still implemented in the classroom by combining classical meetings with e-learning (blended learning). Using the internet as a learning resource, the internet network connection is an essential requirement. Music educators need to stay abreast of developments in technologies and their use by students (Southcott & Crawford, 2011). Thus, internet connection facilities on campus must be adequate to support the lectures process during the current pandemic.

3 RESULTS AND DISCUSSION

In the beginning, it was explained that the analysis of student needs in this study was based on the students' need for learning materials for the Youth Level Viola major course. Viola itself is one of the bowed string instruments (Zhang & Ji, 2016). As an advanced level from the *Indria* level, the Youth level will focus more on training the etude material whose complexity is gradual. In detail, at the Youth level, students will be able to get focused on how to achieve the principle of viola voice, which are: (1) audible sound body (driver), connect body, is the natural musical instrument sound body responsible for voice device, (2) the friction between the connecting body is connected to the sound body and resonance body part, responsible for the voice adjustment and resonate with the interactive relationship between them, and (3) the resonator is to speak when the body to generate sound, let the sound resonance vibration in a fixed space, generally refers to the particular structure of the instrument (Zhao, 2015). In the process of determining the etude material, it is determined from several book materials, which aim to provide a learning book for students that are practical, affordable, and easy to use. Learning materials in the context of learning are one of the components that must exist since learning materials must be studied, observed, studied and used as material to be mastered by students and can provide guidelines for studying them. Without learning materials, learning will not produce anything (Hernawan, Permasih, & Dewi, 2012).

Johnson and Onwuegbuzie (2007) state that the validation process for mixed methods (quantitative and qualitative) in research development is called legitimacy. The legitimacy process of data validity is carried out by several parties who have competent competence in the realm of music. Therefore, several components become the main points in the process of testing the validity of the Youth Level Viola learning book, including:

3.1 *Eligibility content component*

- The material used in the Youth Level Viola learning book includes variative etude material completed by accompaniment music.
- The material in the learning book can be applied independently by students (student-centered).
- Reference material used in the Youth Level Viola learning book can be applied up to date.

3.2 *Presentation eligibility component*

- The title displayed in the Youth Level Viola learning book is in accordance with the overall etude material concept.
- The material is presented in a coherent, gradual manner and can direct students' understanding into the process of learning the material.
- The Youth Level Viola learning book is thoroughly presented, starting from the introduction, table of contents, appropriate text size, and standards.

3.3 *Language feasibility component*

- Students easily understand the grammar used.
- The grammar used is under the General Guidelines for Indonesian Spelling.
- Systematic preparation of the Youth Level Viola learning book is in accordance with the rules of writing learning material modules for the university level.

3.4 *Component of eligibility in graphics*

- The readability aspect (the size and type of letters) used can be read easily.
- Regarding the learning book's visual quality, it contains image sharpness.

The product effectiveness test of the Youth Level Viola learning book was conducted in two meeting phases. The instruments used in the effectiveness test include the following:

1) Structured interviews with music practitioners, lecturers of music study programs, and students are conducted. This instrument is used to obtain information about responses from music practitioners, music study program lecturers, and students about the preparation of the Youth Level Viola learning book and their responses during the learning process using the learning book.
2) Student activity observation sheet during the learning process is also used in this study. A student activity observation sheet is used to obtain information about the stages of learning undertaken by students during the viola learning process using the Youth Level Viola learning book. This sheet is filled out by a Music Study Program lecturer whose job is to observe and give assessments during the implementation of viola learning using the Youth Level Viola learning book.

In the process of the development, product analysis is arranged in the form of a scale questionnaire. The type of scale used is the Likert Scale. Riduwan (2012) said that the Likert scale is used to measure the attitudes, opinions, and perceptions of a person or group of people about social events or symptoms. In the research process, this social phenomenon has been specified in detail and specifically by the researcher, further referred to as the research variable. Every answer or statement describing aspects of the validity test of product development, according to the Indonesian National Education Standards Agency, will be linked to the form of a statement in the following words:

- Scale 1 = Very Poor / Very Poor / Very Disagree.
- Scale 2 = Poor / Poor / Disagree.

- Scale 3 = Fair / Moderate / Neutral.
- Scale 4 = Good / Agree.
- Scale 5 = Very Good / Very Agree.

This questionnaire was handed over to the validator team. Below are the examples of applying a Likert scale (with checklist form) for the validity test of the learning book quality related to the eligibility content component (Table 1).

Table 1. Analysis of score on validity test of eligibility content component.

Assessment point	Score				
	5	4	3	2	1
The material used in the Youth Level Viola learning book includes variative etude material completed by accompaniment music.	√				
The material in the learning book can be applied independently by students (student-centered).		√			
Reference material used in the Youth Level Viola learning book can be applied up to date.		√			

Based on the validation results from the validator team determined, the formula for calculating the score used in the Likert Scale is by calculating the percentage of each item. For example, there are three validators to test the eligibility content component, then the description of the score calculation is as follows:

Answering 5: 2 people; Answering 4: 1 person; Answering 3: 0; Answering 2: 0; and Answering 1: 0.

Calculate scores by:

$$
\begin{aligned}
&\text{The number of scores for 2 people answered} && 5:2\times 5=10\\
&\text{The number of scores for 1 person answered} && 4:1\times 4=4\\
&\text{The number of scores for 0 people answered} && 3:0\times 3=0\\
&\text{The number of scores for 0 people answered} && 2:0\times 2=0\\
&\text{The number of scores for 0 people answered} && \underline{1:0\times 1=0}\\
& && \text{Total }=14
\end{aligned}
$$

The ideal score for item number 1 (highest score) = 5 x 3 people = 15

The lowest scores for item number 1 = 1 x 3 people = 3

Explanation of percentage interpretation of scores for Likert Scale (Hernawan et al., 2012):

Figures 0% - 20% = Very Less.
Figures 21% - 40% = Less.
Figures 41% - 60% = Enough.
Figures 61% - 80% = Good.
Figures 81% - 100% = Very Good.

Based on the scores obtained from several validators, the material aspects used in the Youth Level Viola learning book have a percentage: 14/15 x 100% = 93.33%. Therefore, it is classified as very good. Achievement targets will be produced in the form of a learning book that contains guidelines, viola playing theory, and practice materials for the Youth level. In the long term, the results of this research are expected to produce a learning book that is innovative, practical, and effective (Tables 2 and 3).

Table 2. Assessment test score of the Youth Level Viola learning book phase I.

Assessment point	Score				
	5	4	3	2	1
Every stage in learning activities can be carried out coherently.		√			
Students can use the learning book.			√		
Students can get to know and recall the etudes in the learning book.		√			
Students are able to understand and be able to apply the basic concepts and techniques of viola playing.			√		

First trial score recapitulation (highest score amount: 5 x 4 items = 20):

No item gets a score of	$5 : 5 \times 0 = 0$
There are 2 items that have a score of	$4 : 4 \times 2 = 8$
There are 2 items that get a score of	$3 : 3 \times 2 = 6$
No item gets a score of	$2 : 2 \times 0 = 0$
No item gets a score of	$1 : 1 \times 0 = 0$
	Total : 14

$$\text{Percentage results} = \frac{\text{Trial score results} \times 100\%}{\text{Highest score amount}}$$

$$= 14/20 \times 100\%$$

$$= 70\% \text{ (valid, quality, can be used with minor improvements).}$$

Table 3. Assessment test score of the Youth Level Viola learning book phase II.

Assessment point	Score				
	5	4	3	2	1
Every stage in learning activities can be carried out coherently.		√			
Students can use the learning book.		√			
Students can get to know and recall the etudes in the learning book.	√				
Students are able to understand and be able to apply the basic concepts and techniques of Viola playing.	√				

Summary of second phase score (highest score number: 5 x 4 items = 20):

There are 2 items that get a score of	$5 : 5 \times 2 = 10$
There are 2 items that have a score of	$4 : 4 \times 2 = 8$
No item gets a score of	$3 : 3 \times 0 = 0$
No item gets a score of	$2 : 2 \times 0 = 0$
No item gets a score of	$1 : 1 \times 0 = 0$
	Total : 18

$$\text{Percentage results} = \frac{\text{Score results} \times 100\%}{\text{Highest score amount}}$$
$$= 18/20 \times 100\%$$
$$= 90\% \text{ (very good, very valid).}$$

Regarding the percentage classification in the two phases process of the Youth Level Viola learning book, it can be described in Table 4.

Table 4. Score percentage classification for effectiveness test of the Youth Level Viola learning book.

Criteria for achieving	Validity level value
76 % – 100 %	Very valid, very high quality, can be used without improvement.
51 % – 75 %	Valid, quality, can be used with minor improvements.
26 % – 50 %	Invalid, not qualified, cannot be used.
1 % – 25 %	Very invalid, very not qualified, cannot be used.

The effectiveness of using the Youth Level Viola learning book can be measured from every activity that students do while using learning books. It is shown from the percentage of observations of student activities in the learning process of viola course that showed a score of 90% (very good, very valid), which means that the students are able to process the etude they have read from the learning book into the act of playing the viola instrument, that has been able to be done well by students. Thus, this means that the aim of increasing the psychomotor domain of students to the level of P5 (Skilled Movements) can be achieved.

Furthermore, based on Bloom's taxonomy analysis along with the percentage of observations of student activities during the learning activity, the effectiveness of the use of the Youth Level Viola learning book can also be shown from the opinions of students and lecturers of Music Study Programs, during the learning book being used on two phases. Based on the interview results of Music Study Program lecturers, it is found that the procurement of the Youth Level Viola learning book was very beneficial for increasing students' skills. In this case, music playing skills were expected to be a solution for students to accelerate playing viola instruments.

The effectiveness of the use of the Youth Level Viola learning book also can be seen in Figure 1.

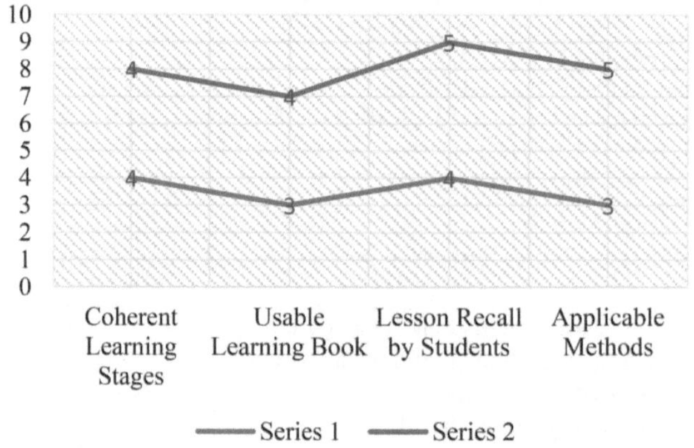

Figure 1. Score of assessment test of the Youth Level Viola learning book.

4 CONCLUSION

The development of the Youth Level Viola learning book empirically is able to show significant results in the learning process of viola instruments. It is shown in the results of the effectiveness of the Youth Level Viola learning book, which presents a percentage of up to 90%, where students are able to process and implement every etude on the learning book.

Referring to Bloom's taxonomy stage, the Youth Level Viola learning book is included in the P5 Psychomotor realm (Skilled Movement). This achievement shows that learning books for Youth Level Viola subjects using etude variations can improve skills and facilitate students in practicing.

ACKNOWLEDGMENTS

This research is fully supported by Universitas Negeri Surabaya Research Grant.

REFERENCES

Ferdian, R., Putra, A. D., & Yuda, F. (2020). Preparation of learning materials for basic flute instrument based on locality and ABRSM curriculum. *Proceedings of the 1st International Conference on Lifelong Learning and Education for Sustainability (ICLLES 2019)*, pp. 145–150. https://doi.org/https://doi.org/10.2991/assehr.k.200217.030

Hernawan, A. H., Permasih, H., & Dewi, L. (2012). Pengembangan bahan ajar. *Direktorat UPI, Bandung*, *4*(11).

Johnson, R. B., & Onwuegbuzie, A. J. (2007). Toward a definition of mixed methods research. *Journal of Mixed Methods Research*, *1*(2), 112–133. https://doi.org/10.1177/1558689806298224

Lindfors, E., & Hilmola, A. (2016). Innovation learning in comprehensive education? *International Journal of Technology and Design Education*, *26*(3), 373–389. https://doi.org/10.1007/s10798-015-9311-6

Riduwan, M. B. A. D. (2012). *Skala pengukuran variabel-variabel penelitian*. Bandung: Alfabeta.

Serdyukov, P. (2017). Innovation in education: What works, what doesn't, and what to do about it? *Journal of Research in Innovative Teaching & Learning*. https://doi.org/10.1108/JRIT-10-2016-0007

Southcott, J., & Crawford, R. (2011). The intersections of curriculum development: Music, ICT and Australian music education. *Australasian Journal of Educational Technology*, *27*(1).

Thorsteinsson, G. (2002). *Innovation and practical use of knowledge*.

Thorsteinsson, G., & Denton, H. G. (2003). The development of innovation education in Iceland: A pathway to modern pedagogy and potential value in the UK. *Journal of Design and Technology Education*, 8(3), 172–179.

Zhang, R., & Ji, H. H. (2016). Study and analysis of various violin playing schools BT, *Proceedings of the 2016 International Conference on Economy, Management and Education Technology*. 1595–1597. https://doi.org/https://doi.org/10.2991/icemet-16.2016.368

Zhao, G. (2015). The study on the performance characteristics of the violin tone in the computer music BT, *Proceedings of the 2015 International Conference on Education Technology, Management and Humanities Science* (pp. 44–47). https://doi.org/https://doi.org/10.2991/etmhs-15.2015.11

Zhou, Y. (2018). Analysis on teaching path of art practice in colleges and universities-taking violin as an example BT, *Proceedings of the 8th International Conference on Education, Management, Information and Management Society (EMIM 2018)*. 709–713. https://doi.org/https://doi.org/10.2991/emim-18.2018.142

Innovation on Education and Social Sciences – Maureen et al (Eds)

Folklore for developing plurilingual and pluricultural competence (PCC) in foreign language classroom

P.V. Asteria, A. Yuniatin & E.K. Dewi
Universitas Negeri Surabaya, Surabaya, East Java, Indonesia

ABSTRACT: This research aims to produce an intermediate-level BIPA module design to develop plurilingual and pluricultural (PCC) competencies based on folklore. The research model used is the adaptation of the ADDIE model. The developed module prototype with the module format includes the initial part (reflection, video observation of folklore, disclosure of video content, reading folklore, synopsis, motivation), the core part (reflection, synopsis discussion, discussion of folklore values with the relevance of people's lives, comparing values in stories the people towards the values of the country of origin), and the final part (acting out the dialogue, uploading to social media, providing captions based on the content of folklore). The module is presented with a flipped learning (FL) strategy with stages before class (reflection, observing folklore videos, disclosing video content, reading folklore, making synopsis, motivation) and in-class (reflection, synopsis discussion, discussion of values, i.e., folklore with relevance to people's lives, compare the values in folklore to the values of the country of origin), and after class (acting in dialogues, uploading to social media, providing captions based on folklore content).

1 INTRODUCTION

Foreign language education cannot be separated from cross-language and cross-cultural interactions (Kramsch, 1998). Language differences occur because of differences in thinking between individuals, which become their identity as part of a particular society (Kramsch, 1998). Cross-language and cross-cultural differences that occur would require a 'third space' to provide understanding for someone in interpret cultural differences from various perspectives that can be accepted by individuals who have different cultures (Bhabha, 1994). This will create tolerance for language and cultural differences inherent in other people. There is a hope that it will be easier to learn foreign languages, which is one of the demands of people globally.

In foreign language classes, students need to be aware of thinking and be open so that it is easy to accept language and cultural differences. Foreign language learners, usually adult learners, have knowledge and experience of their native language and culture. This knowledge and experience of the target language should not only be considered an obstacle in learning a foreign language but can also be used to accelerate language acquisition and cultural understanding of the target language. Therefore, teachers should be positive with the belief that the plurilingual and pluricultural competence (PCC) of students in foreign language classes needs to be developed.

Plurilingual and pluricultural competence (PCC) is a person's ability to use his language and culture to learn foreign languages. PCC is essential because it helps students to better understand the similarities and differences between the language and culture of origin and the foreign language and culture that is the target to be mastered. By having good PCC, students can communicate appropriately when interacting across different cultures to achieve communicative competence in understanding knowledge and applying attitudes and skills when interacting with heterogeneous cultures (Deardorff, 2019).

 DOI 10.1201/9781003265061-8

Interlanguage learning constantly interacts with different languages spoken by several people who have different cultural backgrounds; it needs to be developed according to the various circumstances during the interaction process (Coste, Moore, & Zarate, 2009). Learning a language involves a process that each learner feels differently when interacting with other people who use a heterogeneous language (Franceschini, 2005). Linguistic experiences inherent in students based on cultural differences are involved when students study multilingualism to understand speech in several contexts such as educational or professional (Thamin & Simon, 2010).

Furthermore, the purpose of interlingual and intercultural education needs to be taught to students to foster understanding in carrying out their role as part of a heterogeneous society (Council of Europe, 2020). The diversity of text types can support education and shape understanding of diverse cultures (Beacco et al., 2016). Language and cultural education is the right combination and has a significant impact on increasing understanding of humanity and globally that everyone has differences that can be overcome by generating feelings of empathy through education (Feng, Byram, & Fleming, 2009). Language and cultural education are components that are bound to provide an understanding of language and cultural diversity that encourages the creation of a prosperous life. Communication is needed to support the role of students in interpreting the cultural differences that occur around them (Catana, 2014).

The selection of non-fiction/fiction texts can be used as material to teach interlingual and intercultural competence accompanied by learning activities that support intercultural education (Byram, Gribkova, & Starkey, 2002). Readers' responses to the texts they read reference the different ways of thinking due to the influences of different cultures, thus giving rise to different responses to texts (Hayik, 2011). Various books containing intercultural topics such as literature, history, and fiction/non-fiction stories are used to increase sensitivity and awareness of cultural differences that should not be used as reasons for discrimination.

Based on these considerations, the purpose of this study is to produce an Indonesian language module design for intermediate-level foreign speakers (BIPA) to develop plurilingual and pluricultural (PCC) competencies based on folklore. It sets to design competency development materials in foreign language classes by utilizing folklore. Folklore is a culture that is spread and passed down from generation to generation (Fitrianita, 2018). In folklore, some values can be considered a representation of the values in a particular society. Thus, this module is expected to be a language and cultural stimulus to BIPA students because it provides an opportunity to identify similarities and differences in the values of the country of origin and Indonesia.

2 METHODS

This type of research is R&D with the adaptation of ADDIE model (analysis, design, development, implementation, and evaluation). In this study, the stages of the ADDIE model were carried out only until the development stage, namely the development of teaching module products.

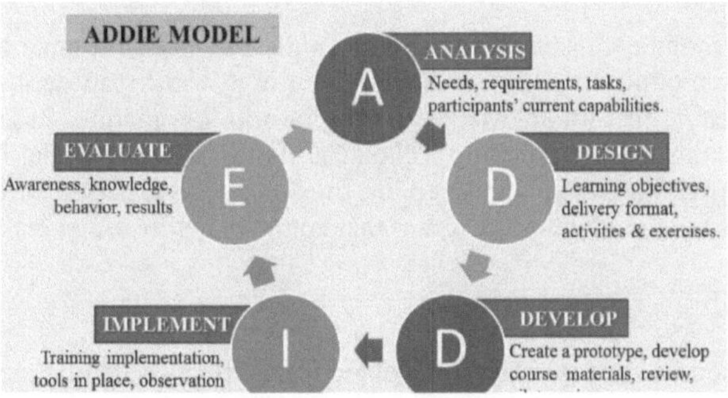

Figure 1. ADDIE development model (Sugihartini, 2018).

The stages of the applied ADDIE Model were Analysis, Design, and Develop, which are described as follows:

2.1 *Analysis*

At this stage, the activities carried out are needs, requirements, tasks, and current capabilities. In need, the writer analyzes the needs of BIPA students in learning foreign languages. There is a requirement that BIPA students who are allowed to use this module must have an intermediate level. In the task, there are language and cultural tasks. In participants' current capabilities, BIPA students must be at the intermediate level.

2.2 *Design*

At the design stage, the things that must be done are learning objectives, delivery formats, activities, and exercises. In learning objectives, appropriate learning objectives are arranged according to the analysis stage. In the delivery format, the presentation of the module includes cover, module identity, reflection, reading and watching videos, compiling a synopsis, an overview of the area of origin of folklore, synopsis discussion, vocabulary review, discussion of values, discussion of the relevance of values to Indonesian people's lives, comparing Indonesian values with country values, arranging group dialogues, practicing role-playing, recording role-playing activities, and uploading videos to social media. In activities and exercises, there are activities and exercises for BIPA students to test understanding.

2.3 *Develop*

At the development stage, the activities carried out were to create a prototype, develop course materials, and review. In the create prototype, develop the module according to the design that has been designed, such as using A4 paper, and so on according to the design at the design stage. In developing course materials, developing teaching materials have been designed at the analysis and design stages. In the review, the author re-examines the finished mid-level BIPA module product whether it is appropriate or not.

3 RESULTS AND DISCUSSION

The resulting intermediate-level Indonesian Language for Foreign Speakers (BIPA) module develops plurilingual and pluricultural competence (PCC) based on folklore. The developed module design is as follows.

3.1 *Beginnings*

The first part of the module consists of a cover, module identity, and table of contents. On the cover of the teaching module for BIPA students, there is a world map depicting BIPA students from various countries in the middle. Moreover, several folklore pictures in Indonesia are on the cover, illustrating that this teaching module will discuss folklore in Indonesia. The module identity section contains information on the author, editor, title, and several pages. Next, on the command instructions, the use of the module is a diagram that contains before class, class, and after class.

3.2 *Core part*

The core part of the module consists of learning objectives, reflection, reading and watching videos, compiling a synopsis, an overview of the area of origin of folklore, synopsis discussion, vocabulary review, discussion of values, discussion of the relevance of values to the lives of Indonesian people, comparing Indonesian values with the values of the country of origin, arranging group dialogues, practicing role-playing, recording role-playing activities, and uploading videos to social media.

There are three learning objectives for this module: being able to identify similarities and differences in people's behavior based on culture, identifying possible misunderstandings in cross-cultural communication from Indonesia, and being able to communicate according to the role they have. In the reflection section, there are introductory sentences that reflect the experiences and feelings of students in understanding certain materials. Next, in the reflection section of the introduction one can reflect on the material in this module. Furthermore, in the reading and watching videos section, there are ten folklore texts and also barcodes that can be scanned by BIPA students containing folklore videos.

Furthermore, in the section compiling the synopsis, there is a task to compile an overview of the folklore that has been read and watched. There is a collection link to make it easier for students to collect their assignments at this stage. Furthermore, in the overview of the area of origin of the folklore, there is an overview of the area of origin in the folklore by including a map of the area of origin, description of the area, dances, food, and tours. This module is also equipped with a barcode containing a video of the area of origin that BIPA students can scan.

Furthermore, in the synopsis discussion, there are instructions to discuss the synopsis because they have been able to pass the stage of compiling a synopsis. Furthermore, in the vocabulary review section, there is a chart for writing complex vocabulary according to BIPA learners. Furthermore, in the discussion of values, there are instructions for BIPA students to discuss with their classmates the values that have been captured in folk tales that have been read and watched. There are synopses compiled and there is a chart containing the contents of 3 columns, namely the title, value, and reason. Furthermore, in the discussion section on the relevance of values to the life of Indonesian people, there is a column chart used by BIPA students to determine the relevance of values in folklore to the lives of Indonesian people.

Furthermore, in the section comparing Indonesia's values with the values of their home country, there is a chart. Using that chart BIPA students can try to compare Indonesian values and those of their home country. Furthermore, in the section on compiling group dialogues, there are instructions for compiling group dialogues. The example synopsis is converted into dialogues so that BIPA students have an idea to arrange their dialogues according to their respective folklore. In this module, there is also a link for collecting results from the dialogues that have been made to make it easier for students to collect assignments. Furthermore, there are instructions for BIPA learners to practice role-playing according to their respective folklore in the practicing role-playing section. There are instructions for BIPA students to record their role-playing activities.

3.3 *Final part*

Reference, Author Profile, and the back cover. In writing articles and making intermediate level BIPA modules, they are compiled with references and the authors' profile includes 3 authors.

This module has advantages because of integrated PCC development. The following are the advantages of this module:

3.4 *Assignments*

There are individual assignments of compiling a synopsis based on folk tales that have been read to train BIPA students' sensitivity to intercultural and interlingual differences. Second, the group assignment trains students' collaborative work skills; even though they come from different cultural and linguistic backgrounds, they can work together to achieve the goals they want to achieve together (Reich & Reich, 2006), namely composing a dialogue in groups and then playing a game. The roles are based on dialogues that that have been arranged so that the students can interact with other students with different cultural and language backgrounds to discuss and role-play to relate the culture contained in folklore with the culture of Indonesian society today (Figure 2).

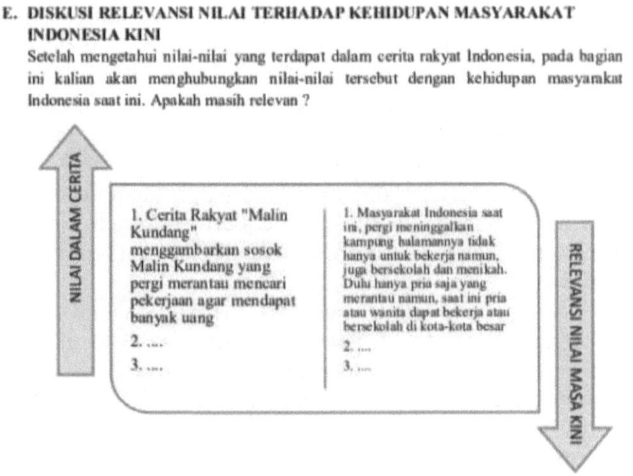

E. DISKUSI RELEVANSI NILAI TERHADAP KEHIDUPAN MASYARAKAT INDONESIA KINI

Setelah mengetahui nilai-nilai yang terdapat dalam cerita rakyat Indonesia, pada bagian ini kalian akan menghubungkan nilai-nilai tersebut dengan kehidupan masyarakat Indonesia saat ini. Apakah masih relevan ?

Figure 2. The relevance of folklore's value to today's life.

Communication between students to express ideas and emotions can increase the vocabulary mastered (Jones, 2001). Role-playing is a means to develop vocabulary mastery significantly, accompanied by preparation of the proper language structure (Livingstone, 1998). The following is one of the tasks in identifying the relevance of the values in folklore to the life of Indonesian people.

3.5 *Illustrations*

The visualization supports the material's content that describes the different cultures in each folklore that is used as material (Figure 3).

MALIN KUNDANG (SUMATERA BARAT)

Dulu, terdapat keluarga miskin yang terdiri dari seorang ibu dan anaknya yang bernama Malin Kundang. Karena ayahnya telah meninggalkannya, ibunya pun harus bekerja keras sendiri untuk bisa menghidupi keluarganya. Malin adalah anak yang pintar tapi sedikit nakal. Ketika dia beranjak dewasa, Malin merasa kasihan pada ibunya yang dari dulu bekerja keras menghidupinya. Kemudian Malin meminta izin untuk merantau mencari pekerjaan di kota besar."Bu, saya ingin pergi ke kota. Saya ingin kerja untuk bisa bantu ibu di sini." pinta Malin. "Jangan tinggalkan ibu sendiri, nak. Ibu hanya punya kamu di sini." kata sang ibu menolak. "Izinkan saya pergi, bu. Saya kasihan melihat ibu terus bekerja sampai sekarang." kata Malin. "Baiklah nak, tapi ingat jangan lupakan ibu dan desa ini ketika kamu sukses di sana" kata ibu berlinang air mata.

Keesokan harinya Malin pergi ke kota besar dengan menggunakan sebuah kapal. Setelah beberapa tahun bekerja keras, dia berhasil di kota rantauannya. Malin sekarang menjadi orang kaya yang bahkan mempunyai banyak kapal dagang. Dan Malin pun sudah menikah dengan wanita cantik di sana. Berita tentang Malin yang menjadi orang kaya sampai ke ibunya. ibunya sangat senang mendengarnya. Dia selalu menunggu di pantai setiap hari, berharap anak satu-satunya kembali dan mengangkat drajat ibunya. Tetapi Malin tak pernah datang. Suatu hari istiri Malin bertanya tntang ibu Malin dan ingin bertemu dengan nya. Malin pun tidak bisa menolak keinginan istri yang sangat dicintainya itu. Malin menyiapkan perjalanannya menuju desanya menggunakan sebuah kapal pribadinya yang besar dan cantik. Akhirnya Malin pun datang ke desanya beserta istri dan anak buahnya.

Figure 3. The images support material's content.

Illustrations take the form of figures based on roles in folklore. Their different clothes and ways of life illustrate the existence of a heterogeneous society that is able to show a great sense of tolerance and live side by side in peace. Visual media in the form of images, objects, models, or tools makes the students get the learning experience. Real vision can clarify the delivery of the material presented so that students can understand less explicit material and improve affective functions by encouraging students to carry out more learning activities (Sudjana & Rivai, 2003).

3.6 *Motivation*

Motivation is applied in BIPA learning to support the performance of BIPA students by training the sensitivity of BIPA students as part of a very varied global society. Therefore studying interlingual and intercultural education is very important to increase the role of BIPA students as members of the community and to ask them to respect differences (Figure 4).

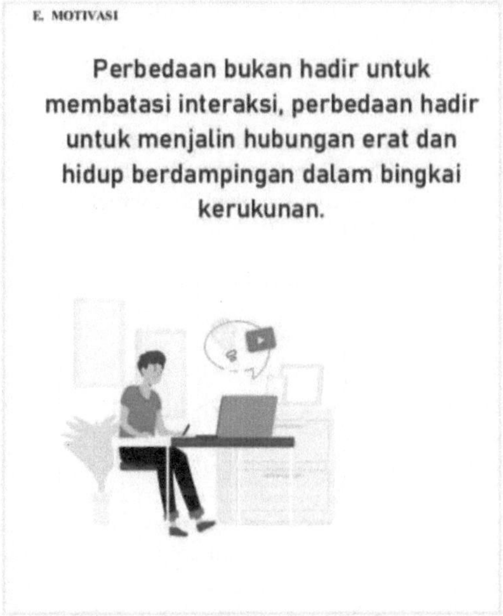

Figure 4. Motivation: Respect the Differences.

Motivation has a relationship with developing students' thinking and attitudes that seek to achieve learning goals (Ryan & Dechi, 2000). Maintaining motivation needs to be done by designing learning that can meet the needs of students by implementing learning strategies that can attract students' focus and involve them in learning interactions and learning tasks (Ryan & Dechi, 2000).

3.7 *Flipped learning strategy*

Flipped learning allows students to get information before studying in class. Thus, students can actively engage in cross-cultural interactions in class discussions. Flipping in the classroom, students study the material, such as watching a video presented before getting information in the form of content in class (Cheng, Ritzhaupt, & Antonenko, 2018) (Figure 5).

Figure 5. Flipped Learning Strategy.

The learner benefits through flipped learning strategies because students can control videos on folklore barcodes that are available independently. Therefore students can adjust the volume and frequency of video playback or repeat video playback, which can reduce cognitive stress and demands (Clark, Nguyen, & Sweller, 2011).

4 CONCLUSION

The design of the intermediate level BIPA module produced in this study aims to develop plurilingual and pluricultural competence (PCC) based on folklore. The module format includes the initial part (reflection, video observation of folklore, disclosure of video content, reading folklore, synopsis, motivation), the core part (reflection, synopsis discussion, discussion of folklore values with the relevance of people's life, comparing values in the people's story towards the value of the home country), and the final part (acting the dialogue, uploading to social media, providing captions based on the contents of folklore). This module has advantages in the aspect of tasks, and flipped strategy illustrations. Thus, BIPA teachers are advised to use this module in Intermediate level BIPA classes. In addition, other researchers are expected to continue the development of PPC-based teaching materials by utilizing other aspects.

REFERENCES

Beacco, J.-C., Byram, M., Cavalli, M., Coste, D., Cuenat, M.E., Goullieret, F., et al. (2016). Guide for the development and implementation of curricula for plurilingual and intercultural education. Retrieved from https://rm.coe.int/CoERMPublicCommonSearchServices/DisplayDCTMContent?documentId=09000016806ae621.

Bhabha, H.K. (1994). The location of culture. London: Routledge.

Byram, M., Gribkova, B., & Starkey, H. (2002). Developing the intercultural dimension in language teaching. A practical introduction for teachers. Retrieved from https://www.coe.int/t/dg4/linguistic/Source/Guide_dimintercult_EN.pdf.

Catana, S.E. (2014). Teaching cross-cultural communication issues – a way of successfully integrating into the multicultural knowledge society. Procedia - Social and Behavioral Sciences, 128, 343–348.

Cheng, L., Ritzhaupt, A.D., & Antonenko, P. (2018). Effects of the flipped classroom instructional strategy on students' learning outcomes: A meta-analysis. Educational Technology Research and Development. https://doi.org/10.1007/s11423-018-9633- 7.
Clark, R.C., Nguyen, F., & Sweller, J. (2011). Efficiency in learning: Evidence-based guidelines to manage cognitive load. John Wiley & Sons.
Coste, D., Moore, D., & Zarate, G. (2009). Plurilingual and Pluricultural Competence. Strasbourg: Council of Europe.
Council of Europe. (2020). Plurilingual and intercultural education: Definition and founding principles. Retrieved from https://www.coe.int/en/web/platform plurilingual-intercultural-language-education/the-founding-principles-of-plurilingual-and-intercultural-education.
Deardorff, D.K. (2019). Manual for developing intercultural competencies: Story circles. https://doi.org/10. 4324/9780429244612.
Feng, A., Byram, M., and Fleming, M. (2009). Becoming Interculturally Competent through Education and Training. Clevedon: Multilingual Matters.
Fitrianita, E. (2018). Membangun etos dan kearifan lokal melalui foklor : studi kasus foklor di tembalang Semarang. Semarang.https://ejournal.undip.ac.id/index.php/endogami/article/download/21306/14278 Diakses tanggal 17 Juli 2021.
Franceschini, R. (2005). Sprachbiographien: das Basel-Prag-Projekt (BPP) und einige mögliche Generalisierungen bezüglich Emotion und Sprachwerb. In R. Franceschini & J. Miecznikowski (Eds.), Leben mit mehren Sprachen. Vivre avec plusiers langues. Sprachbiographien. Biographies langagières (pp. 121–145). Berne: Peter Lang.
Hayik, R. (2011). Critical visual analysis of multicultural sketches. English teaching, 10(1), 95–118.
Jones, P. (2001). Fun Class Activities: Games and Activities with Laughter. London: The Penguin Group.
Kramsch, C. (1998). Language and culture. Oxford, England UK: Oxford University Press.
Livingstone, C. (1998). Role-play in Language Learning. Cambridge: Cambridge University Press.
Reich, S.M., & Reich, J.A. (2006). Cultural competence in interdisciplinary collaborations: A method for respecting diversity in research partnerships (pp. 51–62). 10.1007/s10464-006-9064-1.
Ryan, R.M., & Deci, E.L. (2000). Intrinsic and extrinsic motivations: Classic definitions and new directions. Contemporary educational psychology, 25(1), 54–67.
Sudjana, N. & Rivai, A. (2003). Teknologi Pengajaran. Bandung: Sinar Baru Algesindo.
Sugihartini. (2018). Addie sebagai model pengembangan media intruksional edukatif (Mie) mata kuliah kurikulum dan pengajaran. Bali.https://ejournal.undiksha.ac.id/index.php/JPTK/article/view/14892 Diakses tanggal 17 Juli 2021.
Thamin, N., & Simon, D.-L. (2010). Biographies langagières, compétences plurilingues et affiliation sociale: Enquête auprès de cadres étrangers en entreprise internationale de la région Rhône-Alpes. Cahiers de l'ACEDLE, 7(1), 30.

Trends and issues in education

Innovation on Education and Social Sciences – Maureen et al (Eds)

Building evidence-informed innovation. The case of 21st-century skills

D.D. Curtis
Associate Professor, College of Education Psychology and Social Work, Flinders University, Australia Director, Transforming Education, Australia

ABSTRACT: I review the need for innovation in education and outline critical elements of the innovation cycle. In doing this, I refer to resources that are valuable to educators and researchers. I illustrate parts of this innovation cycle with an example from recent and current projects. I conclude by advocating for two critical areas of research that can enhance educational practice—the application of well-known and highly effective but under-used research-based knowledge and assessment. I sound cautious about investing in some claims that are published in the education research literature.

1 INNOVATION IN EDUCATION

Innovation is a process that seeks opportunities, mobilizes resources, generates a product or service, and engages stakeholders. The innovation lies in the process and not always in the product. We need to innovate in education to improve outcomes, enhance equitable access, and adapt to society's changing structures and requirements. This is true for the U.S. (Serdyukov, 2017), for Australia (Thomson, De Bortoli, Underwood, & Schmid, 2019), for Indonesia (McKinsey Global Institute, 2012), and, doubtless, many other countries.

Beyond education, we have seen innovation on a large scale. New products or services are brought to consumers, and those innovations are often highly disruptive. Innovation has occurred in education, e.g., MOOCs. To date, even though enrolments in them have significantly increased, MOOCs have not been disruptive (Kent & Bennett, 2017). They do disintermediate the relationship between teachers and learners. Learners do not have to enrol at a university nor pay a fee for that enrolment, although they have the option of taking the course for credit and having assignments graded. But MOOCs have not created the disruption that we have seen in the commercial world. Indeed many institutions have used them as advertising for their courses. Their MOOCs become 'tasters' designed to entice students to enrol in paid courses.

The COVID pandemic has posed challenges for education—for schools and universities. School education involves relationships between teachers and learners, but critical learning involves a set of peer relationships. The move to online learning has removed the immediate one-to-one teacher-student interaction and has used forms of technology, print or communications technologies, to mediate that relationship. In most instances, the peer networking relationship has been lost. Of course, there have been many other consequences of school closures and the use of technologies to facilitate learning. Most schools were unprepared for the benefit of technology, even for those schools and students who had access to the devices and infrastructure. In Australia, we found little impact on their learning for some students, those with ready access to technology.

On the other hand, students from disadvantaged backgrounds lacked access to devices, e.g., the entire family shared a single smartphone or tablet and had poor network access. Learning for those students has been adversely impacted. Is this an opportunity for innovation?

DOI 10.1201/9781003265061-9

2 THE INNOVATION CYCLE

Innovation is a cyclic process: it is never complete. It may lead to satisfactory outcomes, but there is always scope to push to the next level.

2.1 *Exploration: Problem or aspiration and baseline data*

The process of innovation begins with a problem or an aspiration in a context. The context must be explored, and the Problem or aspiration unpacked—two approaches to exploring and understanding the problem appeal to me. I like Bacchi's (2009) approach in which she poses a series of questions designed to elicit a deep understanding of the Problem. She begins by asking, 'What's the problem represented to be?' and 'What is left unproblematic about it?'

An alternative is the 'Five Whys' in which teachers or leaders state the Problem then ask 'why.', E.g., the literacy achievement of boys is much lower than that of girls; why? This is a problem statement and a first attempt to understand it. If the answer is boys spend less time reading, or boys are reluctant readers, the next 'why' seeks a deeper understanding. Conventional wisdom suggests that it takes five iterations to reach a sufficiently deep understanding of the Problem.

Whichever approach is taken, it is important to generate a complete understanding of the Problem so that the chosen solutions do indeed address the core problem and not simply a superficial symptom of a fundamental problem.

You may not have a problem; rather, you may have an aspiration for higher levels of achievement or participation. However, an aspiration must be understood in relation to the current context; therefore, the analytical approaches suggested above remain appropriate.

Understanding the context is more than identifying a problem or aspiration. It is important to be aware of any potential barriers to the implementation of an innovation. For this, the Hexagon Tool (Metz & Louison, 2018) is valuable for evaluating the site's readiness for innovation.

Critically, it is necessary to state precisely what outcomes are intended and how they will be measured in advance of the subsequent stages of the innovation cycle. If a change in student achievement is envisaged, what is the magnitude of that change? For whom is it intended (e.g., all students or a subset of them)? How will you measure progress? This final question requires you to generate baseline data using the same tools you will use when you later measure the outcome. The difference between the pre-and post-intervention assessments reveals the effectiveness of the innovation in your context.

2.2 *Investigation: Evidence of efficacy*

Once educators are clear about the Problem or aspiration, their task is to find a solution—an innovation—that is fit for purpose and for which there is substantial evidence for the effectiveness of the innovation. This phase of innovation is vital. A poorly chosen intervention will not lead to the desired improvements, and it will waste time and resources that could have been employed in implementing an effective solution.

The education literature is a vast collection of potential innovations. Some are well-established and have a strong evidence base. Sadly, however, the education research literature is replete with 'plausible ideas' that are utterly ineffective. Learning styles is one of these. So, where can educators find initiatives that are worthy of consideration? I recommend three sources.

Visible learning is my first 'go-to source.' Based on more than 800 meta-analyses, Hattie (2009) listed about 150 interventions along with their effect sizes.

My second source is the Education Endowment Foundation. Their work is school-focused. They trial promising interventions in real schools (in the U.K.) and report on the cost of implementation, the strength of the evidence, and on change in learning outcomes (measured in months of learning gain).

I also use the What Works Clearinghouse. Their methods are rigorous. They place a high value on randomized control trials (RCTs). Not all interventions are amenable to this type of investigation, so relying only on this source would not reveal the full range of potentially valuable innovations. Further, RCTs tend to be based on narrowly prescribed settings and may be very context-dependent and therefore may not translate well to your context.

The investigations that I suggest are designed to identify feasible innovations—those for which a strong evidence base exists. This stage of the innovation process should lead to a shortlist of potentially valuable interventions. The Hexagon Tool (Metz & Louison, 2018) is a useful device for documenting your exploration of potentially valuable interventions and reviewing readiness. This stage culminates in the selection of one of the preferred interventions for implementation.

2.3 *Implementation: Adopt or adapt?*

Implementation must be planned. Through your exploration, you will have identified a promising innovation for which you have evidence of efficacy. Before rushing to implement it, check that your site is ready for it. The Hexagon Tool is again helpful as it provides a structure that enables you to judge the need for the innovation, its fit with your site, and the capacity of your site to manage the implementation. Suppose your site is not ready for the innovation, e.g., because your stakeholders are not sufficiently engaged or because other initiatives are being introduced. In that case, it is time to pause to do any necessary preparatory work.

Having selected an intervention that appears to hold the greatest promise of success, it must be implemented. Here, two competing challenges must be faced. Conventionally, the fidelity of implementation is the default position. It is desirable to implement the innovation as it has been applied in those studies on which the evidence of its effectiveness is based. On the other hand, your context may be rather different from the one(s) in which the innovation was trialed. You may need, therefore, to adapt aspects of the innovation. Any changes, and the reasons for them, must be documented. This documentation will be informative in subsequent phases of the innovation cycle.

Time is a critical factor in achieving desired educational outcomes. We need to know how long it is likely to take before we can reasonably expect an observable change to occur. Short-duration interventions rarely produce large effects, so all participants must be patient. In some interventions, especially those that involve changes in teachers' practices and students' learning behaviors, we may see an initial decline in performance before gains are realized. Brown, Roediger, and McDaniel (2014, pp. 51–53) cited such a case in implementing the interleaved practice, a highly effective learning strategy. When it was first introduced, the pace of learning slowed, but on later application tests, those who used this method substantially outperformed those who used mass-practice techniques.

This phase of the innovation cycle must include measuring expected outcomes so they can be compared with the baseline data collected prior to implementation.

2.4 *Evaluation: Did it work?*

Evaluation minimally seeks to answer the question, 'Did it work?' Or perhaps, 'How well did it work?' Evaluation, while it must do this, can do much more. It may help you to find out how the innovation worked, why it worked for some participants but not all, and what other outcomes were observed.

In education, we are often concerned with achievement, and we expect interventions to lead to better learning and higher levels of achievement. These situations are amenable to quantitative methods. We compare students' achievements before and after the intervention. We find out whether there is a difference, how large that difference is, whether it is statistically significant and whether the effect size is of sufficient practical importance to justify the effort invested in its implementation.

Education involves students' and teachers' attitudes and dispositions. They can be assessed using either open-ended or semi-structured interviews or through the use of reliable instruments. They can be found by searching the literature, but excellent and brief survey instruments are included in the background questionnaires for teachers and students in international studies such as PISA, TIMSS, and PIRLS. I favor the brief attitude scales used in those studies because, while evaluation must be robust, it should not consume more than 10% of the resources allocated to the innovation, nor should it be onerous on participants.

A combination of quantitative and qualitative data is highly enlightening. Interviews with participants, teachers, and students, can be very revealing. In two recent evaluations, teachers and school leaders were very informative about what worked and why it worked, and in some sites, why the same innovation did not work by revealing specific local challenges. In one project, we found that at least four different but related initiatives were being implemented in a group of schools. The workload on teachers was unreasonable, and they were annoyed by the demands on their time. It was impossible to know which of the initiatives led to the outcomes that were achieved. This situation was unhelpful for any of the projects that were being pursued. In another project, we found that some leaders were using the professional learning opportunities to engage teachers who were reluctant to commit to the school's programs. These teachers were unlikely to participate meaningfully in the project and therefore undermined the intentions of the initiative. If we had not undertaken interviews with stakeholders, this information would not have been revealed. These cases highlighted situations in which qualitative information supplemented and contextualized the information we gleaned from the analysis of quantitative achievement data.

An evaluation must take into account the multilevel structure of educational settings. Education is a complex, multi-factorial, and multilevel enterprise. Student outcomes result from both school and non-school influences. In some countries, where school education is highly regulated, non-school forces—mainly family factors—account for about 80% of the variance in student achievement. This has several implications for project expectations and evaluations. It places limits on the magnitude of effects that can be expected to arise from innovation. It requires project managers to consider all possible influences on outcomes and attempt to capture the most salient ones in the design of the evaluation.

2.5 *Consolidation*

Having evaluated the innovation, you will know whether or to what extent it achieved its intended outcomes. A thorough evaluation will also inform you about whether its outcomes were uniform across all participants. Qualitative data will have enabled you to determine any impediments to implementation and any side-effects, which may be positive or negative.

Reflection on the findings of the evaluation is an essential element of the consolidation phase. Knowing what worked, what did not, and for whom will enable you to reject, accept, or revise the innovation. Reflection must cover both the innovation itself and features of the context, as innovations in education are context-dependent.

2.6 *Generalization*

If the evaluation provided evidence of the effectiveness of the innovation, it is desirable to consider whether the data enable a case for the broader implementation of the intervention. This may include applying the innovation more broadly in the current site, e.g., in different subjects or at other year levels, or recommending it for other locations.

Typically, an initial implementation of a novel idea may lead to evidence that the solution is viable and worthy of further development or more rigorous evaluation. The Education Endowment Foundation identifies some innovations as 'promising practices,' which may be subject to further trials, including case-control experimental studies. Such problems could lead to evidence that the What Works Clearinghouse requires before it fully endorses an innovation.

Innovations that are taken to a comparative trial stage should be reported, whether or not they were found to be beneficial. It is customary to publish reports of successful interventions but less familiar to share negative or inconclusive findings. However, these can be valuable as they counter the positive bias of current published literature, and they serve a cautionary note to others who might be considering the intervention.

3 A CASE STUDY: 21ST CENTURY SKILLS

C21 skills or capabilities subsume the knowledge, skills, and dispositions individuals will increasingly require to navigate their changing personal, social, and occupational contexts. Capabilities, for which many labels have been used, are not new, but an attempt to implement capabilities systems remains a challenge and qualifies as an innovation. Curtis and McKenzie (2001) identified four pairs of issues that must be addressed for the successful implementation of 21st-century skills. The pairs of issues are their: selection and definition, dissemination and implementation, assessment and reporting, and certification and recognition. These issues reflect phases of the innovation cycle. Implementation of capabilities is a difficult problem, for which failure has been as frequent as success (Curtis, 2019). Here I focus on definition and assessment, as they are two key problems that have prevented the effective implementation of capabilities.

3.1 *Defining capabilities*

Any capability, e.g., teamwork or Problem solving, is a complex construct. It must first be selected as worthy of inclusion in a set of capabilities that are right for the context and then be understood, elaborated, and operationalized. This process begins with a systematic literature search. The most productive part of this search is the location of one or more theoretical accounts of it. In the case of problem-solving, four theoretical models were found (Curtis, 2010), and one was selected as the basis of an elaboration of the construct. The most useful models enable researchers to develop a description that guides the construction of a framework that is teachable and learnable and that leads to an assessment tool. The framework includes the critical dimensions of the construct, avoiding the key threats to validity—construct under-representation and construct-irrelevant variance—identified by Messick (1995, 1998). Each dimension of the construct must be reflected in several indicators, and for each indicator, a limited number of readily discriminable performance levels must be described. That is, the framework leads directly to the development of a generalizable assessment rubric. Validation of the framework occurs through the application of the rubric that is based on the framework.

3.2 *Assessing capabilities*

Assessment drives learning, and effective assessment leads to high-quality learning, so it is worth investing in assessing capabilities if we value them. In some systems, specific tests of capabilities have been developed, and these have been quite separate from other learning that students do (e.g., Australian Council for Educational Research, 2001). An alternative, shown to be effective, is to embed the assessment of capabilities within existing assessment arrangements (Curtis, 2010; Curtis & Denton, 2003). This has the advantage that students perceive the assessment to be authentic as it is embedded in the discipline-specific learning that they do in their courses.

The method involves asking teachers and lecturers to review all assessment tasks and selecting those that allow students to develop and demonstrate a chosen capability. This process is repeated for all subjects that students take and all targeted abilities. The affordance for capabilities varies by both subject and assessment type. For example, in a current investigation in which experts review assessment types and criteria, some tasks are found to be highly suited to the target capability while others have little to offer. However, across all assessments in all subjects in a course, numerous opportunities are available. We have recommended that students undertake at least two and up to four assessments of each capability so that a capabilities profile can be reported with confidence on graduation. Encouraging students to develop and demonstrate their capabilities in multiple contexts enhances the encoding of the capability and enables its flexible application.

When an existing task has been selected for a capability assessment, the normal subject-specific assessment is undertaken and its grade recorded. Immediately following that assessment, the capability rubric is used to assess the student's demonstration of the capability. While this imposes an additional load on teachers and on students, this extra onus is modest, and by limiting the number of occasions on which capabilities are assessed during a course, the net burden is constrained. In the courses in which we took this dual assessment approach, we have required students to self-assess and point to evidence of the capability in their subject-specific task responses. This passes some responsibility to learners and limits the burden on the assessor. In a current project in which this approach is being implemented, we have noted the existence of a series of related formative assessment tasks that culminate in a summative task. By encouraging students to self-assess their capability development over the formative tasks, they develop the capability and then demonstrate it in the final assessment. The final capability assessment is logged in the student's academic record and reported on their final transcript.

In evaluations of the above approach to capabilities development, we validated the assessment rubric and demonstrated that it was sufficiently reliable and that students were able to describe their command of the capabilities and to illustrate it by reference to examples. Employers commented on students' ability to articulate their skills.

4 CONCLUSION

We do need to be prepared to innovate, but we must be conscious that a vast body of research has generated substantial evidence of effective practices that are not routinely implemented. Fundamental research on learning and recent research on assessment appear to be productive avenues for exploring potential initiatives. Brown et al. (2014) pointed out that applying basic learning principles, such as spaced, interleaved, and varied practice along with reflection and elaboration, leads to sustained improvements in learning. Their work revealed an exciting innovation. Rather than focus on education and how to improve it, they explored its converse—forgetting—and how to minimize it. This innovative reframing of the Problem helped to generate insights that led to novel applications of fundamental research.

Assessment is a productive arena for further basic research and application. In particular, the evidence for peer-assessment is compelling (Double, McGrane, & Hopfenbeck, 2019; Li, Liu, & Steckelberg, 2010), although some contextual barriers must be overcome, e.g., having peers assign grades is counter-productive, and student acceptance of peer rather than teacher judgment must be developed (Liu & Carless, 2006).

Innovation is required as education systems, and goals adapt to changing social and labor market contexts and requirements; as most countries must look outwards to find their places in an increasingly connected and interdependent world, the goals of education change. How can individuals navigate their way through the education system to develop the personal competencies they need, how can they develop the capabilities that will enable them to contribute to the emerging social structures, and how can they develop the knowledge, skills, and dispositions that will be required in the labor market? While much effort has been invested in addressing these questions since at least the early 1990s (e.g., The Secretary's Commission on Achieving Necessary Skills, 1991), these initiatives have largely failed (Curtis, 2019), and that failure can be attributed in large measure to an unwillingness to develop effective assessment regimes by applying what is known about effective assessment. The losses can also be attributed to a reluctance to apply a robust innovation cycle. There is a clear need to apply what is known about effective assessment to the broad set of capabilities that transcend disciplinary silos so that students graduate from our schools, vocational education providers, and universities with the necessary capabilities to flourish in emerging social and work contexts.

The implementation of any innovation must be carefully planned, seeking evidence of the efficacy of the proposed innovation and understanding the context into which the innovation is to be applied. Critically, a strategic evaluation must be built into the implementation.

It is appropriate to end with a note of caution. Not all innovations will yield the intended outcomes. A cautious and skeptical approach is warranted as too many initiatives involving much effort have not led to the claimed benefits.

REFERENCES

Australian Council for Educational Research. (2001). *Graduate Skills Assessment. Summary Report. GSA Exit 2000.*

Bacchi, C. (2009). *Analysing policy. What's the Problem represented to be?* Pearson.

Brown, P. C., Roediger, H. L., & McDaniel, M. A. (2014). *Make it stick. The science of successful learning.* Harvard University Press.

Curtis, D. D. (2010). *Defining, assessing and measuring generic competences* Flinders University]. Adelaide.

Curtis, D. D. (2019). Developing Problem solving and other general capabilities: What will it take and how can we be sure of success? In H. Askell-Williams & J. Orrell (Eds.), *Problem solving for teaching and learning. A festschrift for Emeritus Professor Mike Lawson*. Routledge.

Curtis, D. D., & Denton, R. (2003). *The authentic performance-based assessment of problem-solving*. NCVER.

Curtis, D. D., & McKenzie, P. (2001). *Employability skills for Australian industry. Literature review and framework development*. Australia: Australian Council for Educational Research.

Double, K. S., McGrane, J. A., & Hopfenbeck, T. N. (2019). The impact of peer assessment on academic performance: a meta-analysis of control group studies. *Educational Psychology Review*. https://doi.org/https://doi.org/10.1007/s10648-019-09510-3

Hattie, J. (2009). *Visible learning. A synthesis of over 800 meta-analyses relating to achievement*. London: Routledge.

Kent, M., & Bennett, R. (2017). What was all that about? Peak MOOC hype and post-MOOC legacies. In R. Bennett & M. Kent (Eds.), *Massive Open Online Courses and Higher Education : What Went Right, What Went Wrong and Where to Next?* (pp. 1–8). London: Routledge.

Li, L., Liu, X., & Steckelberg, A. L. (2010). Assessor or assessee: How student learning improves by giving and receiving peer feedback. *British Journal of Educational Technology, 41*(3), 525–536. https://doi.org/10.1111/j.1467-8535.2009.00968.x

Liu, N.-F., & Carless, D. (2006). Peer feedback: the learning element of peer assessment. *Teaching in Higher Education, 11*, 279290.

McKinsey Global Institute. (2012). *The archipelago economy: Unleashing Indonesia's potential*. http://www.mckinsey.com/~/media/McKinsey/dotcom/Insights%20and%20pubs/MGI/Research/Productivity%20Competitiveness%20and%20Growth/The%20archipelago%20economy/MGI_Unleashing_Indonesia_potential_Full_report.ashx

Messick, S. J. (1995). Validity of psychological assessment. Validation of inferences from persons' responses and performances as scientific inquiry into score meaning. *American Psychologist, 50*(9), 741–749.

Messick, S. J. (1998). Alternative modes of assessment, uniform standards of validity. In M. D. Hakel (Ed.), *Beyond multiple-choice: Evaluating alternatives to traditional testing for selection* (pp. 59–74). Lawrence Erlbaum.

Metz, A., & Louison, I. (2018). *The Hexagon: an exploration tool*. National Implementation Research Network.

Serdyukov, P. (2017). Innovation in education: what works, what doesn't, and what to do about it? *Journal of Research in Innovative Teaching & Learning, 10*(1), 4–33.

The Secretary's Commission on Achieving Necessary Skills. (1991). *What work requires of schools. A SCANS report for America 2000*. U.S. Department of Labor.

Thomson, S., De Bortoli, L., Underwood, C., & Schmid, M. (2019). *PISA 2018. Reporting Australia's results, volume I. Student performance*. ACER.

Innovation on Education and Social Sciences – Maureen et al (Eds)

The impact of principal transformational leadership and teacher commitment to organizational change in elementary schools

Windasari, E. Roesminingsih & S. Trihantoyo
Educational Management, Faculty of Education, Universitas Negeri Surabaya, Surabaya, Indonesia

ABSTRACT: Organizational change is a fundamental strategy to ensure that public organizations remain relevant in the changing environment. Schools today are required to be able to achieve sustainable competitive advantage and make changes in order to adapt to existing changes. This study aims to determine the impact of transformational leadership and teacher commitment principles on organizational change in elementary schools. Multiple linear regression was used to measure the impact of the principal's transformational leadership variable and teachers' organizational commitment on organizational change. This study was of teachers of elementary schools in East Java with 84 respondents using a simple random sampling method. The results show that transformational leadership and teacher commitment simultaneously impact organizational change in elementary schools. Furthermore, both variables have a significant impact through school organizational change of elementary school in East Java.

1 INTRODUCTION

1.1 *Research background*

The current socio-cultural, economic, political, and technological developments are increasingly complex and require management approaches and organizational changes. Organizational change is a fundamental strategy for ensuring that public organizations remain relevant in a changing environment. Anderson and Anderson (2001) stated that organizations could manage change and maintain their productivity and relevance. Conversely, an organization that cannot manage change will impact organizational losses from the financial aspect, opportunities, productivity, and reputation (Anderson & Anderson, 2001).

School change and development are urgently needed for the school's sustainability as an organization and achieving progressive social change in society (Rosenblatt, 2004). Schools today are required to be able to achieve sustainable competitive advantage and make changes in order to adapt to existing changes. In order to achieve sustainable competitive advantage, schools need to develop organizational change strategies, management approaches, and leadership styles. As with other organizations, organizational change in schools is defined as any change, improvement, restructuring, or adjustment in the educational process in schools (Dimmock, 1996).

An effective leader is a leader who does not use power to resist change but who can accelerate change with personal solid qualities, inspire subordinates and realize the organization's vision. Based on various studies that have been conducted, transformational leadership has been recognized as one of the best leadership styles to be adopted in managing challenges and school restructuring (Barnett, McCormick, & Conners, 2001). Leithwood and Jantzi (2000) stated that transformational leadership could increase the level of commitment among teachers. Walumbwa, Lawler, Avolio, Wang, and Shi (2005) stated that transformational leaders succeeded in increasing their staff's level of commitment and satisfaction. Principal leadership has an impact on the behavior of staff and

DOI 10.1201/9781003265061-10

teachers in schools. Teachers' level of commitment in schools is also one of the supporting factors in creating organizational change in schools.

Currently, the data shows that many elementary schools in East Java cannot adapt to the stakeholder needs and global developments, and the total number of elementary schools in East Java has decreased in the last three years (Figure 1).

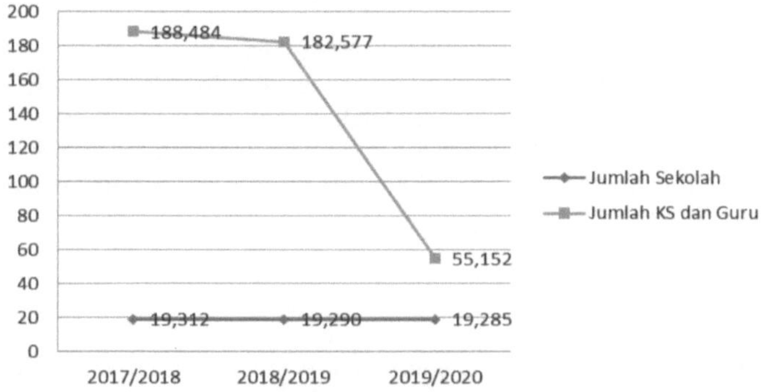

Figure 1. Graph of the number of schools, principals, and primary school teachers in East Java in 2017–2020. **Source*: http://statistik.data.kemdikbud.go.id/index.php/page/sd.

The graph above shows the decreasing number of schools in East Java in the last three years. The decline in the number of schools is in line with the significant decrease in the number of principals and teachers. Schools are required to survive and be competitive so that their existence can be accepted by the community continuously. Schools that can survive are schools that are able to adapt and make changes so that they are still relevant to the needs of today's society.

1.2 *Research problem*

Based on the problems above, the research problem in this study is as follows: do transformational leadership and teacher commitment impact the organizational change of elementary schools in East Java?

2 LITERATURE REVIEW

2.1 *Transformational leadership*

Transformational leadership is a new concept used in educational leadership. Transformational leadership must transform schools into practical, innovative schools and produce high-quality teachers (Gkolia & Belias, 2014). Kuhnert and Lewis (1987) defined transformational leadership as an activity process involving other people to create good relationships and increase employee motivation.

Transformational leadership helps employees internalize the organization's values, thereby encouraging employees to develop their abilities and contribute to organizational achievement (Kanungo, 2001). On the other hand, transformational leadership impacts admiration, trust, loyalty, and respect for the leader. As a result, employees will be motivated to work better than expected, show a high sense of responsibility, and have a suitable emotional attachment to work (Bass & Riggio, 2006).

Transformational leadership is identified as consisting of four components: charisma, inspiration, intellectual stimulation, and individual attention (Northouse, 2007). Charisma: A transformational leader is a figure whose behavior becomes a role model for his employees. Leaders who have charisma are shown by the strength of their vision and mission, demonstrate attitudes with high moral standards, create respect and trust from employees, and become employees' inspiration. Inspirational: Transformational leaders can communicate organizational goals by creating morale, enthusiasm, and motivating employees. Transformational leaders generate new ideas and encourage employees to initiate change in the organization. Intellectual stimulation: Transformational leaders always motivate employees to be innovative, analytical, and creative. Leaders always encourage employees to be able to solve problems through creative solutions. This intellectual stimulation stimulates awareness of the problem and the leader's way of thinking and stimulates employee creativity. Individual attention: Transformational leaders always act as mentors and mentors within the team. Leaders provide ongoing direction and feedback to help employees succeed and develop their skills. The principal's transformational leadership can motivate teachers to change attitudes and values by being committed to the mission and vision of education. Transformational leadership can move the organization towards a clear vision, mission, and organizational goals (Leithwood & Jantzi, 2000).

2.2 *Teacher commitment*

Elizur and Koslowsky (2001) argue that commitment is something related to personal values to mark the existence of individuals in the organization. Yuet (2017) has a commitment view from emotional attachment between employees and the organization. Teacher commitment has been identified as a critical aspect to reform school capacity.

The capacity of schools to make sustainable depends on the commitment of teachers to change. According to Meyer, Allen, and Smith (1993), organizational commitment is divided into three main characteristics: (1) Affective commitment, is defined as the desire of employees to be emotionally committed to the organization. Employees will remain actively involved in the organization based on their own will and desire. (2) Continuing commitment refers to the commitment of employees to stay in the organization by weighing the costs incurred by leaving the organization against the benefits of staying in the organization. (3) Normative commitment, the last dimension, refers to the type of mandatory commitment in which employees remain in the organization because of a sense of security and responsibility. This commitment involves an employee's sense of obligation to remain with the organization.

Commitment to school organization is defined as the commitment of teachers to adopt and achieve school goals and values ??and maintain the desire to remain in school. Committed teachers are always responsible for the actions taken, carry out their duties as well as possible, and improve the education system's efficiency, effectiveness, and productivity.

2.3 *Organizational change*

Change is a complicated process and quite hard for the organization. Armstrong (1992) defined change management as a process that ensures an organization can make significant changes in its culture, policies, structures, and systems. Organizational change is an attempt by the organization to move from current conditions to expected future conditions to increase its effectiveness (Lunenburg, 2010). Meanwhile, according to Jones (2013), it is a process in which an organization optimizes its performance in order to achieve ideal conditions.

Schools, generally accepted as open systems, have structures that are more susceptible to internal and external pressures of change (Beycioglu & Aslan, 2010). Like other organizations, organizational change in schools is any change, improvement, restructuring, or adjustment in the educational process in schools (Dimmock, 1996).

Hargreaves (2005) highlights that the process and initiation of educational change are intended to change aspects of learning and teaching in schools. However, according to Hoy and Sweetland (2001) changing school organization is not intended to eliminate the problems schools face but to innovate by considering internal and external environmental aspects. Therefore, principals are faced with the best choices to respond to changes that occur in schools. In general, the targets for organizational change include vision, strategy, culture, structure, systems, production technology, and leadership style (Marcus, van Dam, Marcus, & van Dam, 2020).

3 METHOD

3.1 *Instrument*

This research questionnaire consists of two parts. Part I was developed to identify the condition of the respondents' demographic data. Containing information on gender, age, and length of employment, part II is developed from a literature review and conceptual framework on transformational leadership, teacher organizational commitment, and organizational change. The measurement scale uses a 5-point Likert scale. 1=strongly disagree, 2=disagree, 3=neutral, 4=agree, and 5=strongly agree. Instrument validation uses content validity from a team of experts in education management who will provide input and criticism of the draft instrument that has been compiled. The expert team's input and feedback were used to refine the draft of the instrument and test the validity of the content of the instrument that had been prepared. The content validity test is intended to determine whether the factors that have been developed are appropriate. Pilot tests were conducted in 10 elementary schools in Surabaya to achieve the reliability of instruments. The data collected in this research used primary data. In this study, the data collection method used a questionnaire. Respondents were given an online questionnaire consisting of open and closed questions. The prepared questionnaire was distributed via the Google Form link to teachers in all elementary schools in East Java.

3.2 *Population and sample*

The population in this study is elementary school teachers in East Java. The sampling method in this research is through a simple random sampling technique. The questionnaire was distributed to all teachers of the elementary schools in East Java. Therefore, the sample in this study was 84 teachers of an elementary school. The demographic profile of the respondents is given in Table 1.

Table 1. Respondent demographic.

Criteria	Total Respondents
Gender	
Male	21
Female	63
Age	
20–30 years old	37
31–40 years old	36
years old	11

3.3 *Data analysis method*

This study uses a quantitative approach using multiple linear regression analysis techniques. Multiple linear regression was chosen to measure the impact of the principal's transformational leadership variable and teacher organizational commitment on organizational change. This study uses two independent variables and one dependent variable. The independent variables in this study are the principal's transformational leadership and teacher organizational commitment, while the dependent variable is organizational change.

3.4 *Research hypothesis*

Ho: There is no positive and significant impact of the principal's transformational leadership and teacher commitment on organizational change.
Ha: There is a positive and significant impact of the principal's transformational leadership and teacher commitment on organizational change.

4 RESULT AND DISCUSSION

4.1 *Result*

The results of data interpretation based on each variable are shown in Table 2.

Table 2. Descriptive result.

Variable	N	Mean	Standard Deviation	Category
Transformational leadership	84	68.02	8.81	High
Teacher commitment	84	44.93	5.76	Moderate
Organizational change	84	62.15	7.72	High

The transformational leadership variable and organizational change have an average score of high category, while the teacher commitment variable has a moderate average score (Table 3).

Table 3. Multiple regression results.

Hypothesis	R Square	Correlation coefficient (sig)	Level of significance	Decision
There is a positive and significant effect of the principal's transformational leadership and teacher commitment to organizational change	0,516	0,00	0,05	Accepted

Based on the results of simultaneous testing of leadership variables and teacher commitment to organizational change, the results obtained are sig 0.000 < p-value 0.05. It means Ha is accepted where transformational leadership and teacher commitment simultaneously have an impact on organizational change. R square is 0.516 or 51.6%, which means that transformational leadership and teacher commitment simultaneously affect organizational change by 51.6%

4.2 *Discussion*

Current developments require the education sector to make changes. Education expects high quality from a stakeholder point of view. Therefore, schools as a manifestation of an educational institution must make changes and develop to meet stakeholder expectations for a good education. According to Crandaleisman and Louis (1986), school-based change is divided into pedagogic change and organizational change. Organizational change is changes in the school's organizational structure and processes such as management patterns, the hierarchy of authority, communication networks, decision-making styles, and school climate.

The analysis of the data shows that transformational leadership and teacher commitment affect organizational change in schools. 51.6% of organizational change at school is affected by transformational leadership and teacher commitment. In addition, the principal's leadership has a positive impact on organizational change. Moreover, a high level of commitment from teachers is expected to make education more prepared and face all kinds of challenges and obstacles. Teachers also have organizational commitment, the degree to which a person works to identify with the organization and its goals, and the desire to maintain membership. In general, teacher organizational commitment is the emotional attachment of a teacher to the school, and they are willing to survive, identify themselves, and be involved in achieving school goals. With the involvement of teachers in achieving school goals, it is expected that they will make a significant contribution to the changes in schools.

The teacher's organizational commitment occurs from the beginning when the teacher joins the school organization. The organizational commitment can increase or decrease; the circumstances that affect it are either due to the teacher's internal or external factors. One of the external factors that influence teacher commitment is supportive leadership. Transformational leadership will encourage teachers to increase their commitment to the school, thus it could lead to higher productivity. The principal must be able to have advanced thinking and a strategic concept map in leading the school. A visionary mindset and empowering teachers in schools are essential points for carrying out transformational leadership. Teachers should be encouraged to be active and empower themselves to help organizational change in schools. Transformational leaders are leaders who adopt a democratic approach to their leadership style (Giltinane, 2013). The principal seeks to involve and provide opportunities for teachers to actively participate in sharing their ideas in support of school programs. As a result, when the principal implements a transformational leadership style well, it will potentially involve stakeholders to achieve educational goals (Bush, 2015).

School organizational change focuses on aspects of management change, school vision and mission, school structure, school system, and school leadership. Organizational changes in schools are expected to make schools to be of better quality and more adaptable. School organizational change requires a change agent who carries out a process of influencing all organization personnel to receive the program, creativity, and innovation. Although principals as leaders have an essential role in implementing organizational change in educational institutions, they must also function as agents of reformation and innovation.

5 CONCLUSION

Educational organizations are required to realize changes following future situations or conditions. Changes in educational organizations are influenced and affected by transformational leadership and teacher commitment. The general goal is to increase the effectiveness and health of the school organization. Transformational leadership and teacher commitment have an essential role simultaneously in implementing organizational change in elementary schools. The results of this study can be used as a foundation for academics and school organizations so that organizational change can be achieved effectively and adequately. The limitation of this study is that this research consists of two variables. For future research, another variable can be used that has a contribution through school organizational change, such as school culture.

REFERENCES

Anderson, D., & Anderson, L. A. (2001). Beyond Change Management: How to Achieve Breakthrough Results Through Conscious Change Leadership (2nd edition). Pfeiffer. https://doi.org/978-0787956455

Armstrong, M. (1992). Human Resource Management: Strategy & Action. Kogan Page.

Arnold, K. A., Turner, N., Barling, J., Kelloway, E. K., & McKee, M. C. (2007). Transformational Leadership and Psychological Well-Being: The Mediating Role of Meaningful Work. Journal of Occupational Health Psychology, 12(3), 193–203. https://doi.org/10.1037/1076-8998.12.3.193

Barnett, K., McCormick, J., & Conners, R. (2001). Transformational leadership in schools – Panacea, placebo or problem. Journal of Educational Administration, 39(02).

Bass, B. M., & Riggio, R. E. (2006). Transformational Leadership. Lawrence Erlbaum Associates, Publishers.

Beycioglu, K., & Aslan, M. (2010). Change and innovation as main dynamics in school development?: Administrators and teachers' roles. February 2016.

Dimmock, C. (1996). Dilemmas for School Leaders and Administrators in Restructuring. In International Handbook of Educational Leadership and Administration (pp. 135–170). https://doi.org/10.1007/978-94-009-1573-2_5

Elizur, D., & Koslowsky, M. (2001). Values and organizational commitment. International Journal of Manpower, 22(7), 593–599.

Gkolia, A., & Belias, D. (2014). The Impact of Principals' Transformational Leadership on Teachers' Satisfaction: Evidence From Greece. European Journal of Business and Social Sciences, 3(6), 69–80.

Hargreaves, A. (2005). Educational change takes ages: Life, career and generational factors in teachers' emotional responses to educational change. Teaching and Teacher Education, 21, 967–983.

Hoy, W., & Sweetland, S. (2001). Designing Better Schools: The Meaning and Measure of Enabling School Structures. Educational Administration Quarterly – EDUC ADMIN QUART, 37(08), 296–321.

Jones, G. R. (2013). Design, and Change global edition. In Pearson Education Limited: Vol. Seventh Ed.

Kuhnert, K. W., & Lewis, P. (1987). Transactional and Transformational Leadership: A Constructive/Developmental Analysis. The Academy of Management Review, 12(4), 648. https://doi.org/10.2307/258070

Leithwood, K., & Jantzi, D. (2000). The effects of transformational leadership on organizational conditions and student engagement with school. Journal of Educational Administration, 38(2), 112–129. https://doi.org/10.1108/09578230010320064

Lunenburg, F. C. (2010). Approaches to Managing Organizational Change Lewin's Three-Step Change Model. International Journal of Scholarly Academic Intellectual Diversity, 12(1), 1–10.

Marcus, J., van Dam, N., Marcus, J., & van Dam, N. (2020). Organisational Change and Development. Handbook Organisation and Management, April, 627–660. https://doi.org/10.4324/9781003022435-17

Meyer, J. P., Allen, N. J., & Smith, C. A. (1993). Commitment to Organizations and Occupations: Extension and Test of a Three-Component Conceptualization. Journal of Applied Psychology, 78(4), 538–551. https://doi.org/10.1037/0021-9010.78.4.538

Northouse, P. G. (2007). Leadership theory and practice. Sage Publications, Inc.

Walumbwa, F., Lawler, J., Avolio, B., Wang, P., & Shi, K. (2005). Transformational Leadership and Work-Related Attitudes: The Moderating Effects of Collective and Self-Efficacy Across Cultures. Journal of Leadership & Organizational Studies, 11(03).

Yuet, F. K. C. (2017). Development and Validation of Teachersâ Commitment to Change Scale. International Journal of Academic Research in Business and Social Sciences, 7(7), 164–178. https://doi.org/10.6007/ijarbss/v7-i7/3087.

Innovation on Education and Social Sciences – Maureen et al (Eds)

A mismatch between high-status profession and altruistic motives, and why it matters for the teaching profession

S.I. Savira, Mustaji & D. Nurwidawati
Faculty of Educational Science, Department of Psychology, Universitas Negeri Surabaya, Surabaya, Indonesia

Afsana Ayub
Bangladesh National Commission for UNESCO, Ministry of Education, Dhaka, Bangladesh

ABSTRACT: The purpose of this study is to evaluate how instructors perceived their profession's status based on the characteristics of the profession and the motivations for entering and remaining in the teaching profession. Ten public elementary schools are participating in the research. The Rasch Model is used to examine the outcome. Based on the findings, it was discovered that there was more agreement on the features of high social status professions, but less agreement on the teaching profession's suitability for those traits. At the same time, the most common reason for becoming a teacher is a personal belief in the profession's importance. Further consideration is given to the notion that altruistic motives originated as an adaptive mental strategy for retaining the teaching profession rather than the reality that it is less monetarily rewarding. Teachers' worth may be influenced by social and cultural variables.

1 INTRODUCTION

In Indonesia, the teaching profession has attracted attention in recent decades. It is commonly quoted as one of the primary causes of the so-called "poor quality" of national education. The impact of globalization on the profession (e.g., Menter & Flores, 2021), and the cause–effect relationship between teacher education, teacher competence, and teacher certification (e.g., Suratman, Wulandari, Nugraha, & Narmaditya, 2010) have all been studied in depth. Surprisingly, the general consensus in the discourses remains low. The study's goal is to revisit what instructors believe about their profession's nature. This subject is posed in light of the premise that teachers' professional attitudes are typically influenced by their cultural and historical backgrounds (Beijaard, Meijer, & Verloop, 2004; Bjork, 2004; Chandratilake, McAleer, & Gibson, 2012; Escarlos & Tan, 2017). Moreover, teachers' attitudes are frequently linked to their likelihood of remaining in the profession. As a result, in order to teach professionalism, it is necessary to understand these types of inquiries.

This section of the study calls for a more in-depth assessment of the traits, status, and motivations for entering and remaining in the teaching profession. It is suggested that how instructors evaluate the state of the teaching profession has an impact on many elements of their personal and professional lives. Understanding teachers' perceptions of their profession will aid in predicting teachers' motivations for continuing to teach, despite the high workload and expectations, because the public frequently blames educators for their poor performance and skill, which is blamed for the poor education provided (Sanaky, 2005; Sauri, 2010). The growing focus on teachers' professional competency is partly due to rising remuneration, which is frequently cited as the primary motivator for students to pursue teacher education (Mukminin, Kamil, Muazza, & Haryanto, 2017). Many scholars, however, have looked at the negative effects of "teacher professionalism" (Gu & Day,

DOI 10.1201/9781003265061-11

2013; Valli & Buese, 2007). According to Fransson and Frelin (2016), at least five elements influence teachers' commitment to remain in the teaching: personal, teaching, school setting, system context, and professional growth. Mukminin et al. (2017), on the other hand, found three key motivations for becoming a teacher. The altruistic motive, sometimes known as a noble or heroic purpose, is the first. The second category includes intrinsic motivations such as intellectual motivations and self-improvement goals. The third is the extrinsic motive, which is linked to flexible working hours and a good social standing. Other studies believe it is critical to investigate teachers' motivations since they may be linked to their commitment, retention, performance (Escarlos & Tan, 2017), resilience (Gu & Day, 2013), and professional orientations (Evans, 2008, 2011; Richardson & Watt, 2016).

Further investigation uncovers a link between occupational prestige and health and job happiness (Ma & MacMillan, 1999). The prestige of occupation is divided into two categories in this study: characteristics of high-status professions and characteristics of teaching professions. According to Hargreaves et al. (2006), at least four indicators should be considered to understand the status of the teaching profession: defining status, comparing teaching status to other occupations, teacher change over time, and variables that may influence teacher's status in the future. Gender bias, on the other hand, is supposed to be taken into consideration when interpreting employment characteristics (Hibbard & Pope, 1987; Mukminin et al., 2017).

2 METHOD

This study employs a survey approach that is especially beneficial when attempting to collect a huge amount of data. Teachers of primary education in public schools make up the majority of the population. Ten schools were chosen at random, with ten teachers representing each. From a total of 100 questionnaires distributed, 98 were returned and were examined. The majority of the instructors in this study (81 percent) are certified teachers, which means they have met the essential qualifications and standards to teach as competent and professional teachers, as defined by the National Law of Teachers number 14/2005. Professional instructors have four skills, according to the legislation, namely professional, social, personal, and academic. Around 92% of the teachers were also civil or public servants employed by the government.

Hargreaves et al. (2006) based the questionnaire on three dimensions of the teaching profession: teaching status, teaching profession, and motives in becoming teachers. The three primary aspects of the instrument covered in this work are: the characteristics of high-status professions, the characteristics of teaching professions, and the motivations for becoming a teacher. The analysis of each segment will be delivered, followed by a round of questions and answers. Teachers were given a list of criteria of a high-status career and asked to rank each of them according to their assessment in the first phase. The same list was displayed in the second section, but they were asked to prioritize the qualities of the teaching profession. The two components were then compared to form hypotheses about how instructors viewed their career. The final section contains lists of reasons for becoming teachers, including Hargreaves et al. (2006)'s three indicators: motives for becoming a teacher, reasons for continuing a teacher, and respect and responsibility (Table 1).

The data analysis was conducted by applying the Rasch model analysis. The Rasch model is a method that analyzes the probability of specified responses as a function of person and item measure. This method of analysis is chosen because it provides a detailed analysis of the person and item reliability. It also allows deeper analysis to differentiate between favorable and least favorable items. The acceptable Zstd values (white) fall between –2.0 and +2.0, with sample sizes between about 30 and 300 (Bond & Fox, 2015). The data were statistically analyzed using Winstep 3.73. Item fit was tested using the Rasch model with the provisions: Fit Indices for Item Fit Statistics Fit Indices Outfit mean square values (MNSQ) 0.50–1.50, Outfit z-standardized values (ZSTD). –2.00 to 2.00, and Point Measure Correlation (PTMEA-CORR) 0.40–0.85 (Boone, Yale, & Staver, 2014). The reliability coefficient uses Cronbach's Alpha. The suggested coefficient is higher than 0.8 (Kasik et al., 2018).

Table 1. Description of aspects and indicators that is elaborated in the questionnaire.

Aspects	Indicators
Status of the teaching profession	Defining status: high professional status and the teaching profession
	Teacher's status compared to other occupations.
	Teacher's status transformation from time to time
	Factors that may influence teacher's status in the future
Teaching professionalism	Teaching as constructive learning
	Teaching as a trusted profession
	Teaching as collaboration with parents
	Teaching as autonomous
	Teaching as delivering standards
Reasons for becoming a teacher	Reasons for becoming a teacher
	Reasons of remaining teacher
	Responsibility and respect

3 RESULTS AND DISCUSSION

The first section of the instrument examines the status of teaching from the perspective of teachers. Teachers were required to rate various traits associated with high social status in this section. They were once again asked to rank the attributes they believe are associated with teaching. The first section of the questionnaire's reliability is as follows, according to the Rasch Model (Table 2).

Table 2. Person, item, and Cronbach's alpha reliability analysis result of characteristics of the high-status profession ($N = 98$).

Person reliability	.84
Item reliability	.94
Cronbach alpha	.89

The initial section of this questionnaire has a high level of personal reliability, while an item from the data above has an even higher level. In comparison to the individual or people in this study, the items in the questionnaire appear to be more consistent. As a result, the person, not the products, is responsible for any changes or variations in the response. The finding is consistent with the Cronbach's alpha value of.89. Further statistical analysis revealed that three patients scored more than 2 S.D. (two standard deviations), resulting in a logit of 8.11. Outliers in the score can be discovered. Outliers are reactions that do not match the model that was provided. The descriptive statistic is a number that describes something (Table 3).

Table 3. Descriptive statistic of the characteristic of the high-status profession ($N = 98$).

SD person	1.96
SD item	.77
Mean (person)	3.16

*SD = standard deviation.

The high social status of the profession is analyzed as follows. The lower the logit, the more favorable the item is.

The least favorable item recommended as a characteristic of a high-status profession is that external restrictions dominate the profession, as seen in Table 4.

Table 4. The rank of characteristics of the high-status profession based on the Rasch model analysis.

Characteristic	Logit
Positive working environments	−1.10
Perceived as a credible profession	−1.06
Respected and honored by society	−.80
Has a prospective career in the future	−.17
Has a positive image in social/mass media	−.09
Considered important and valued by the government	.09
Received high appreciation from society and government	.12
Received various benefits in addition to salary (remuneration, health insurance, etc.)	.78
Satisfying financially	.78
External regulations rule the profession	1.44

The teachers ranked the characteristics of the teaching profession in the second section of the questionnaire. The second portion of the questionnaire has the same items as the first, but this time teachers were asked to select traits linked with the teaching profession. As a result, the two components can be compared to see if teachers think the same way about their career as they do about high-status professions (Table 5).

Table 5. Person, item, and Cronbach's Alpha reliability analysis result of characteristics of the teaching profession ($N = 98$).

Person reliability	.87
Item reliability	.95
Cronbach alpha	.97

Similar to the first part, items in the questionnaire have higher reliability than the person or subject in this study. Cronbach's alpha also demonstrates a high level of consistency. As a result, any variation in responses in this study is due to the person, not the items (Table 6).

Table 6. Descriptive analysis of characteristics of the teaching profession ($N = 98$).

SD person	2.31
SD item	.91
Mean (person)	3.42

Furthermore, based on descriptive statistics, it was discovered that three participants have scores that do not match the suggested model. The outliers score is greater than 2 standard deviations (two times the standard deviation). Outliers, as mentioned in the first section, may also explain why person dependability is lower than item reliability.

The following are the characteristics of the teaching profession as ranked by the teachers in this study.

Table 7. The rank of characteristics of the high-status profession.

Characteristic	Logit
Positive working environments	−1.38
Perceived as a credible profession	−1.04
Respected and honored by society	−.81
Has a prospective career in the future	−.34
Has a positive image in social/mass media	−.25
Considered important and valued by the government	−.17
Received high appreciation from society and government	.26
Received various benefits in addition to salary (remuneration, health insurance, etc.)	.89
Satisfying financially	1.25
External regulations rule the profession	1.59

It can be deduced from a comparison of Tables 4 and 7 that teachers are quite positive about the value they place on their career, and they see some similarities between high-status professions and teaching. Teachers, on the other hand, appear to value the traits of high-status professions more than the attributes of their own job. It can be deduced by comparing the first part's mean (M = 3.16) to the second part's mean (M = 3.42). Another example is the second part's two least favorable items, which have logit ratings of 1.59 and 1.25, respectively, in comparison to the first part of the questionnaire, which included one most unpopular item with a logit value greater than one (= 1.44). As a result, it may be stated that, in comparison to appreciating high-status jobs, teachers' replies vary more in determining the character of their career. The findings could indicate that teachers have differing perspectives on their profession, which could be due to (1) differing perspectives on the value of their profession's privileges; (2) subjective satisfaction with the teaching profession based on each individual's experience; and (3) government policies affecting the construction of teaching professionalism.

By comparing the teachers' replies to the least favorite item from the first and second halves, the latter premise can be proven. 'The profession is governed by external regulation,' says the item. The teacher refers to the item as the least typical of a high-status profession in the first portion, implying that a high-status profession should be self-contained. The second section concerns teachers' perceptions of their profession, implying that their position is unaffected by external regulation. The latter is debatable, given that the only standard utilized to determine a teacher's professionalism is one imposed by the government. Beck (2008) is concerned that government regulation of teacher professionalism is often overbearing. However, as previously stated, the majority of the teachers in this study are government employees who may have different interpretations of the term "external regulation." Teachers may not think of the government as a source of external power (Bjork, 2004).

The motive for becoming a teacher is the third section of the questionnaire. Table 8 shows a summary of the reliability analysis.

Table 8. Person, item, and Cronbach's alpha analysis result ($N = 98$).

Person reliability	.87
Item reliability	.98
Cronbach alpha	.92

It can be concluded from Table 8 that person and item reliability is considered high. Table 9 summarizes the descriptive statistic of the motives of becoming a teacher.

Table 9. Descriptive analysis of motives of becoming a teacher ($N = 98$).

S.D. person	.83
SD item	1.11
Mean (person)	1.44

Table 10 summarizes the most favorable items shown by the higher to the highest negative value. Table 11 summarizes the most unpopular led by logit beyond 1.00.

Table 10. Summary of the most popular reasons for becoming a teacher ($N = 98$).

Reasons for becoming a teacher	Logit
I became a teacher because I believe that teachers have a significant role in the promotion of a nation	−1.35
I want to help promote the education of the nation	−1.31
A teacher is a noble profession	−1.31
I enjoy teaching	−1.22
I believe that teaching is an important job to provide children with the best modality of their life	−1.20
For me, teaching is a calling	−1.33
I am happy to know that I can do something to help children progressing in their development	−1.11
I feel comfortable working as a teacher	−1.04

Table 11. Summary of the least popular reasons for becoming a teacher ($N = 98$).

Reasons for becoming a teacher	Logit
I became a teacher because this profession is financially satisfying	1.12
I became a teacher because this profession is high rewarded	1.14
I did not have another option but to be a teacher	1.30
I became a teacher because of my high social status in society	1.44
I became a teacher because my parents told me so.	1.73
Teaching is the profession of the family, so I never thought of anything else better to do	1.92
I became a teacher because since I could not get into my favorite university	2.28

Based on the findings, it can be stated that one of the most common motivations for becoming a teacher is a personal belief that teaching is an essential career in which they can make a big contribution to the country. External factors, on the other hand, were forcing some of the least favorable products into the profession. It's also worth noting that professors do not enter the teaching profession because of their social standing. This could indicate that (1) teaching is not a high-status job, or (2) teaching is a high-status profession, but it is not the primary reason they chose it. When the other portions of the questionnaire are considered, the latter may be a closer possibility of explanation. According to Mukminin et al. (2017), although teachers regard their work as one of the highest-status professions in society, this is not the major reason they selected it, but rather because of 'altruistic considerations.'

However, if all three portions of the questionnaire are compared, it may be concluded that teachers do not regard their work to be financially rewarding. They agree that a high-status profession should pay well (logit.78), but they do not believe that teaching is financially rewarding (logit 1.25 for the same item on part 2). As a result, when instructors say that they did not pick this career because it pays well (Part 3), they may be implying that they do not believe teaching pays well. They might only have 'altruistic motives' or 'intrinsic motives,' as a result. Further consideration could be given to the hypothesis that altruistic intentions arose as a coping mechanism.

4 CONCLUSIONS

Although teachers are generally enthusiastic about their work, there is no consensus on the characteristics of their vocation, according to this survey. This may have an impact on how individuals define teaching, as well as their attitudes and performance in the classroom. The most beneficial motives for entering and remaining in the teaching profession have been determined to be ideal and internal motives. The facts, on the other hand, can be interpreted in a variety of ways. For example, whether altruistic motivation claimed by teachers has arisen as a coping strategy that motivates people to stay in the profession despite the fact that it is monetarily less lucrative.

REFERENCES

Beck, J. (2008). Governmental professionalism: Re-professionalising or de-professionalising teachers in England? British journal of educational studies, 56(2), 119–143.

Beijaard, D., Meijer, P.C., & Verloop, N. (2004). Reconsidering research on teachers' professional identity. Teaching and teacher education, 20(2), 107–128.

Bjork, C. (2004). Decentralisation in education, institutional culture and teacher autonomy in Indonesia. International review of education, 50(3–4), 245–262.

Bond, T. G., & Fox, C. M. (2015). Applying the Rasch Model: Fundamental Measurement in the Human Sciences (3rd ed.). New York: Routledge.

Boone, W. J., Yale, M. S., & Staver, J. R. (2014). Rasch analysis in the human sciences. In Rasch Analysis in the Human Sciences. https://doi.org/10.1007/978-94-007-6857-4

Chandratilake, M., McAleer, S., & Gibson, J. (2012). Cultural similarities and differences in medical professionalism: a multi-region study. Medical education, 46(3), 257–266.

Escarlos, G.S., & Tan, D. (2017). Motives, Attitudes and Performance of Teacher Education Students. International Journal of Scientific & Technology Research, 6(10), 20–25.

Evans, Linda. (2008). Professionalism, professionality and the development of education professionals. British journal of educational studies, 56(1), 20–38.

Evans, Linda. (2011). The 'shape' of teacher professionalism in England: Professional standards, performance management, professional development and the changes proposed in the 2010 White Paper. British educational research journal, 37(5), 851–870.

Hargreaves, A. (2000). Four ages of professionalism and professional learning. Teachers and teaching, 6(2), 151–182.

Hargreaves, A. (2001). The changing nature of teachers' professionalism in a changing world. New teacher education for the future: International perspectives, 89–107.

Hargreaves, L., Cunningham, M., Everton, T., Hansen, A., Hopper, B., McIntyre, D., . . . Wilson, L. (2006). The status of teachers and the teaching profession: Views from inside and outside the profession. Cambridge: University of Cambridge.

Hibbard, J.H., & Pope, C.R. (1987). Employment characteristics and health status among men and women. Women & Health, 12(2), 85–102.

Kasik, L., Guti, K., Gál, Z., Gáspár, C., Tóth, E., & Fejes, J. B. (2018). Development and psychometric properties of the Avoidance Questionnaire for Adolescents (AQA). International Journal of Emotional Education, 10(2), 59–76.

Ma, X., & MacMillan, R.B. (1999). Influences of workplace conditions on teachers' job satisfaction. The journal of educational research, 93(1), 39–47.

Menter, I., & Flores, M. A. (2021). Connecting research and professionalism in teacher education. European Journal of Teacher Education, 44(1), 115–127.

Mukminin, A., Kamil, D., Muazza, M., & Haryanto, E. (2017). Why teacher education? Documenting undocumented female student teachers' motives in Indonesia: A case study. The Qualitative Report, 22(1), 309–326.

Richardson, P.W., & Watt, H.M.G. (2016). Factors Influencing Teaching Choice: Why Do Future Teachers Choose the Career? International handbook of teacher education (pp. 275–304): Springer.

Sanaky, Hujair AH. (2005). Sertifikasi dan profesionalisme guru Di era reformasi pendidikan. Jurnal Pendidikan Islam, 1, 1–13.

Sauri, Sofyan. (2010). Membangun karakter bangsa melalui pembinaan profesionalisme guru berbasis pendidikan nilai. Jurnal Pendidikan Karakter, 2(2).

Suratman, B., Wulandari, S. S., Nugraha, J., & Narmaditya, B. S. (2020). Does teacher certification promote work motivation and teacher performance? A lesson from Indonesia. International Journal of Innovation, Creativity and Change, 11(10), 516–525.

Innovation on Education and Social Sciences – Maureen et al (Eds)

Teacher professional development in Indonesia: A comparative study with global practices

Silfia Asning Tias
Universitas Negeri Surabaya, Surabaya, Indonesia

Waraporn Tongjean
Rajamangala University of Technology, Phra Nakhon, Thailand

ABSTRACT: This review paper examines professional development (PD) for teachers in Indonesia in a comparative study with the global practices. The study explores teacher professional development from international literature in the teacher education and professional development field in light of PD practices in the Indonesian context. It explores the historical background of teacher professional development in Indonesia and how it shaped Indonesian education today. The underlying questions are on the professional development practice in Indonesia and its impact on Indonesia's education quality. Then, the paper undertakes a comparative study with global practices of teacher professional development activities. Moreover, the paper attempts to analyze the lessons learned from Singapore and the People's Republic of China on their efforts to improve teacher quality through professional development and how Indonesia can evade the isomorphic mimicry phenomenon of global education reform.

Keywords: Teacher Professional Development. Comparative study. Global practices. Indonesia

1 INTRODUCTION

Program for International student assessment (PISA), an international student assessment that measures numeracy and literacy skills of 15 years old students, which was released in December 2019, put Indonesian teachers and Indonesian education in general under tight scrutiny. The fact that the outcome of PISA 2019 was precisely the same as in the year 2000 indicates that schools in Indonesia have not improved for the last 18 years (Revina, Pramana, Fillaili, & Suryadarma, 2020). Students' performance remains poor, which has led to scrutiny of the Indonesian government's effort to improve the quality of teaching in Indonesia, which have taken place over the past ten years through the teacher certification policy and the government's commitment to allocate 20% of the national budget to the education sector. For this reason, the government was heavily criticized by education stakeholders such as education experts and researchers, political opponents, and the public in general. In their criticism, they harshly pointed out that the education system in Indonesia has "failed," and the government should be held accountable to perform its promise to improve the quality of education and the quality of teachers and teaching as well as student performance (Pritchett, 2020). Extensive research from international organizations indicates an urgent need to revisit the teacher education reform agenda and reassess the education policy that has shaped teacher education in Indonesia today (Tobias, Joseph Wales, & Syamsulhakim, 2014).

Professional Development (PD) for teachers has been widely researched as an essential mechanism to improve teaching quality. The most significant resources for this activity are undoubtedly teachers (Desimone, 2009). Soebari and Jill (2016) stated that an active and effective professional development program would improve the knowledge, belief, and skills of teachers, as well as influencing students' learning. Moreover, the Organization for Economic Co-operation and

DOI 10.1201/9781003265061-12

Development (OECD) (2005) refers to TPD as in-service education, teacher training practicums, in-service training, and other upgrading programs.

Tonga et al. (2019) further describe that PD could occur in formal and informal situations in which the formal will be more structured and systematically arranged to prepare teachers for their profession either provided by the government, district authorities, or the local school. In contrast the informal situations re personal activities that do not depend on the schools' program, including personal development involving lesson study training activities or teacher participation in classroom supervision of other teachers. In short, professional development is commonly agreed to be a tool to improve teachers' professional practice that ultimately aims to advance students' performance.

This paper aims to elaborate on TPD practice in Indonesia and how it has shaped Indonesian education. Also, by looking at the implementation of TPD in high achiever PISA countries in the region, it is expected to see some input on the type of TPD program and how Indonesia could learn from these countries. Two states are analyzed, namely, Singapore and the People's Republic of China, because the two countries were identified as the top two high achieving countries in the 2018 PISA. Also, the two countries are in the same geographical region as Indonesia and they share geographic and cultural similarities with Indonesia.

Although the phenomenon of isomorphic mimicry overshadows developing countries in building their reform policy (Pritchett, 2020), I argue that there is still a necessity to analyze the other country's success without imitating the practice of other countries. Isomorphic mimicry refers to the inclination of governments to copy the accomplishments of other governments, by imitating procedures, systems, and even goods from examples of best practices. This mimicking frequently conflates appearance and function, resulting in a scenario in which governments appear capable of imitating but are not able to apply those practices to their particular contexts (Andrews, Pritchett, & Woolcock 2017). Hence, the underpinning questions for this study are: (1) how does the TPD program in Indonesia impact the quality of education? (2) what can Indonesia learn from global practices to improve its teacher professional development program?

2 METHOD

This article is a review of international literature on professional development and how the Indonesian teacher professional development context compares to the global practices. It is a qualitative study approach by using a document review method of sources, such as reports by the government and other agencies, as well as government regulation on teachers and education in general (Tonga et al., 2019). This comparative study utilized Bereday's comparative study model (Brady, 1967). The model elaborates four steps in analysis: description, interpretation, juxtaposition, and comparison. This current study illustrates the descriptive data from three countries, including Indonesia, Singapore, and the People's Republic of China. The subthemes include teacher recruitment and its teacher education, the country's national budget in education, and the in-service teacher professional development model. These data were then analyzed according to Bereday's model of comparative study.

3 TEACHER PROFESSIONAL DEVELOPMENT IN INDONESIA

Professional Development (PD) activity has been part of education practices throughout the history of Indonesian education since the 1970s (Tanang & Abu, 2014; Thair & Treagust, 2003). The simple local training for equipping teachers with better teaching skills in the 1970s developed into a more organized and well-managed program such as *Pemantapan Kerja Guru* (PKG) or strengthened the work of teachers in the 1980s and *Kelompok Kerja Guru* (KKG) or teacher working group in 1990s. After the enactment of the teacher's law in 2005, teachers are also required to undergo Teacher Professional Education and Training for 12 days for certification purposes. The currently running PD model implemented by the government is called *Pengembangan Keprofesian Berkelanjutan*

(PKB) or continuing professional development as a follow-up of *Uji Kompetensi Guru* (UKG) or the teacher competency test in 2015 as part of the government's attempt to monitor whether teacher certification has a certain impact on teacher quality. After four decades of the government's attempt to improve teachers' pedagogic and professional skills, the current studies by Revina, Rezanti, Rizki and Daniel (2020) show that it has fallen short, including the recent TPD model. The fact that the students' literacy and numeracy skills in the 2018 PISA result were the same as the performance in 2000 and basic numeracy skills dropped between 2000 to 2014 (Beatty, Berkhout, & Suryadarma 2019) indicates that there is a strong relationship between teacher quality and student performance (Darling-hammond, Deborah, Su, & Julian, 2005), and Indonesia's teacher education reform, which included the PD program, was still inadequate to improve teacher's skills.

Based on longitudinal empirical studies on Indonesia's education reform (Chang et al., 2014; Jalal et al., 2009; Revina et al., 2020), the government's PD program's attempt to improve teacher quality has yet to attain its objectives. There are several reasons, nevertheless, the main ones are the problem with lack of coherence between policy and practices and its real implementation (Revina et al., 2020), in addition to the strong political agenda that is heavily involved in the education system (Rosser & Fahmi 2018). Although characteristics of PD effectiveness are feasible in Indonesia's PD program, there are substantial features that are missing. Studies have claimed that these weak points have been in existence since the beginning of PD implementation four decades back. The analysis unraveling the issues of PD ineffectiveness show that it goes beyond the technical aspects, rather it is more complex and systematic. Some suggestions from the study mentioned were pushing consistency in Indonesia's education system policy with regard to teacher quality and "to re-orient the system to produce high-quality teachers" (Revina et al., 2020, p. 28).

Looking at Indonesia's national budget on education, as much as 492.5 trillion IDR or 20% out of 2,461 trillion IDR is allocated for Education (World Bank, 2019). However, Indonesia's education expenditure is still the lowest in the Asia region. Since the national budget is 15% of GDP, the total education expenditure is only 3% (Yarrow, Rythia, Eema, & Bernard, 2020), compared to the countries of high-achiever PISA, which, according to OECD data, allocate around 10% of GDP (Tonga et al., 2019). All in all, Indonesia has shown a remarkable commitment to improving education quality through education expenditure by building physical infrastructure, although commitment to quality is still in question.

4 TPD COMPARATIVE STUDY WITH THE GLOBAL PRACTICES

A comprehensive study by Tonga et al. (2019) on PD of teachers in high achievers PISA countries is used mainly as comparative data. PD programs in Singapore and China are chosen due to the two countries being the top two in the 2018 PISA results and have been appraised for the best quality education in the world; likewise, these countries are in the same region as Indonesia and bear similarities in several aspects.

4.1 *Singapore*

Singapore has been acknowledged to have one of the best quality educations in the world. With a total population of 5.6 million people, Singapore has become one of the references in building a robust education system. With an allocation of 3% of GDP for education expenditure, Singapore has invested highly in teacher quality (OECD, 2018). For teacher education, Singapore's government mandates the National Institute of Education (NIE) as the only teacher training institute in Singapore. The reasons are to maintain a high-quality teacher education program and to prevent an oversupply of teachers in a country with a small population. Tonga et al. (2019) describe that Singapore's school graduates sit a Qualification Entrance Exam to be admitted in the Primary and Secondary School education training program. The duration of the study could be 2–4 years. To control the supply of graduates, a quota system is employed. Therefore, those who are accepted into the program will be automatically employed by the Ministry of Education.

Bautista, Wong, and Gopinathan (2015) argue that due to the high-quality PD implemented by the Ministry of Education of Singapore, Singapore is acknowledged as one of the best in terms of practices of education in the world. The features in the PD, among others, are: (a) subject matter specific that connect with classroom practice, (b) intensive and ongoing, (c) active learning, (d) collaborative learning, and e) coherent with the teacher's need and interest. Another model of PD to accelerate teachers' learning is the professional career track. Teachers in Singapore can choose to be in the teacher track, leadership track, or specialist track. These features undoubtedly have given the sense of development, both individual and professional. These PD programs are provided by institutions established by the Ministry. On top of the ones assigned by the government, in Singapore, schools play a crucial role by designing school-based PD as Professional Learning Communities (PLC) (Hairon & Dimmock 2012).

4.2 *People's Republic of China*

In PISA 2018, China has shown its domination on scoring number one in all three categories: reading, mathematics, and sciences. Although the sample of PISA was taken from schools in Shanghai only out of the population of China of more than 1.4 billion, the result shows that China's education is an example for other countries. In addition, China's government expenditure on Education is as much as 3.6% of the GDP total of (Tonga et al., 2019).

For the in-service teacher, especially for the novice, teachers are given the opportunity to undertake 120 hours of a continuing education program and 360 hours every five years. Teachers will do multiple evaluations, which then will decide on their promotion and increase in remuneration. The government also facilitates a web platform for teachers to expand their networks and improve their knowledge, as well as further elaborates that a career step is employed in China. Through this system, the Ministry ensures to maintain the quality of teaching.

In the past ten years, the Chinese government launched the National Teacher Training Program (NTTP), which primarily aims to bridge the gap between the quality of teaching in rural and urban China and improving students' learning (Lu et al., 2019). It further explained that in the heavily prescribed TPD program in NTTP, teachers focus on the ethics in teaching, subject-specific knowledge, and pedagogical practices. In addition, the NTTP systems employed blended learning, in which teachers will do online PD and continue with on-site training. From the study in Shaanxi province, there is evidence in increasing the teachers' content knowledge; however, there is no significant improvement in the students' learning, which means that content knowledge is not coherent with pedagogical content knowledge.

5 LESSON LEARNED FROM GLOBAL PRACTICES

Professional Development is undoubtedly one of the most critical aspects of the teaching quality improvement activity. Based on literature and empirical research evidence, the characteristics of countries with a high-quality education are: a large allocation of education expenditure in terms of GDP, a supporting government policy for quality education improvement starts from the admission to the professional development, and a strong government commitment to support these activities (Darling-hammond, 2017; Tonga et al., 2019). Although all elements are interrelated, studies show a correlation between government allocation for education and country quality. Most of the PISA high-achiever countries invest highly in improving the quality of education.

The Indonesian government has shown commitment towards quality education by allocating 20% of the national budget for the education sector. Through history, with 11 times changes of the curriculum, Indonesia has demonstrated that a professional education program for teachers is a crucial point (Rahman, Abdurrahman, Budi, & Nurlaksana, 2015). Although there have been doubts about whether action has been taken in response to low teacher and student's performance, experts believe that progress is taking place, as proven by the policies towards teacher regulation and management (*Ministry Decree* no. 18/2007, *Ministry Decree* no. 87/2013, *Ministry Decree* 37/2017).

With the enactment of Teacher Law 2005, the Indonesian government acknowledges teaching as a "profession" which makes it as highly regarded as other professions. Undeniably, Indonesia's commitment towards improving education is the same as its counterpart countries.

Another thing that Indonesia can learn from the global practice is in terms of teacher recruitment. In the whole process of teacher recruitment in the high-achiever countries, a candidate should hold a bachelor's degree regardless of whether they are from a teaching or non-teaching background. In Singapore, the candidate will then take graduate-level study; meanwhile, in China, candidate teachers will sit teacher training. To be admitted, the screening will involve academic achievement, characteristics, and skills (Tonga et al., 2019). In Indonesia, a similar procedure is taken to ensure the quality of the teacher candidates. The teaching program graduates will compete in an entrance exam and field study as the places are limited by quota. This way, the government can balance supply and demand. With the rigorous process, it is expected that the best candidate will be selected in the profession. While the policy on paper looks promising, teacher recruitment is considered one of the root problems in Indonesia's education system (Chang et al., 2014). According to findings from recent empirical studies by the RISE organization, there are three major factors behind the struggle for good teacher recruitment, namely, institutional design, political economy aspects, and social dynamic (Huang & Revina 2020).

Lastly, the PD programs in high-achiever countries are mostly put in the highest priority in the context of education reform. China, for example, set the National Teacher Training Project (NTTP) as a means for teachers' continuous development program. With a high investment, the Chinese government attempts to bridge the gap between urban and rural teaching and learning quality. Teachers are mobilized to attend on-site training as well as online. Similar activities are also conducted in Indonesia and Singapore. The difference is that in Singapore teachers' PD is more a bottom-up activity rather than top-down. The strong mentoring culture in Singapore's schools has proven to be effective for collaboration between the senior and junior teachers in the context of Professional Learning Communities (Hairon & Dimmock, 2012).

6 CONCLUSION

Teacher quality has been a concern for the Indonesian government. Many attempts have been made and reports have mentioned that Indonesia's education quality is improving, although some empirical evidence is showing the opposite in some aspects. This paper has presented the PD activity throughout the history of Indonesian education and in other countries in the region ranging from teacher education program to continuous professional development. There are similarities of practices, although, in Indonesia's cases, it is primarily top-down rather than bottom-up. Moreover, in the Indonesian context, many of the structured TPD programs in history were initiated and funded externally, and as the funding ran out the PD program also died out. Although several studies place in question the efficacy of the PD program that has been implemented for the past four decades, many believed that Indonesia is going in the right direction for improving the quality of education. Nevertheless, there are still many more things to work on improving the quality of teachers to improve students' learning.

ACKNOWLEDGMENT

We would like to express our gratitude to the editors and reviewers for their support to improve the manuscript.

REFERENCES

Andrews, Matt, Lant, Pritchett, and Michael, Woolcock. 2017. *Building State Capability: Evidence, Analysis, Action. Oxford University Press*. Oxford: Oxford University Press.

Bautista, Alfredo, Joanne, Wong, and Saravanan, Gopinathan. 2015. "Teacher Professional Development in Singapore: Depicting the Landscape." *Psychology, Society and Education* 7 (3): 311–26. https://doi.org/10.25115/psye.v7i3.523.

Beatty, Amanda, Emilie, Berkhout, and Daniel, Suryadarma. 2019. "Why Are Indonesian Students Getting Worse at Mathematics?" RISE. 2019.

Brady, George, Z. F. 1967. "Reflections on Comparative Methodology in Education, 1964-1966." *Comparative Education* 3 (3): 169–87. http://jstor.org/stable/3097986.

Chang, Mae Chu, Sheldon, Shaeffer, Samer, Al-Samarrai, Andrew, B. Ragatz, Joppe, de Ree, and Ritchie, Stevenson. 2014. "Teacher Reform in Indonesia: The Role of Politics and Evidence in Policy Making." *Teacher Reform in Indonesia: The Role of Politics and Evidence in Policy Making*. https://doi.org/10.1596/978-0-8213-9829-6.

Darling-Hammond, Linda. 2017. "Teacher Education Around the World?: What Can We Learn from International Practice??" *European Journal of Teacher Education*, 1–19. https://doi.org/10.1080/02619768.2017.1315399.

Darling-Hammond, Linda, Deborah, J. D. Holtzman, Su, Gatlin Jin, and Julian, Vasquez Heilig. 2005. "Does Teacher Preparation Matter? Evidence about Teacher Certification, Teach for America, and Teacher Effectiveness." *Education Policy Analysis Archives* 13 (42): 1–48.

Desimone, Laura M. 2009. "Improving Impact Studies of Teachers' Professional Development: Toward Better Conceptualizations and Measures." *Educational Researcher* 38 (3): 181–99. https://doi.org/10.3102/0013189X08331140.

Hairon, Salleh, and Clive, Dimmock. 2012. "Singapore Schools and Professional Learning Communities: Teacher Professional Development and School Leadership in an Asian Hierarchical System." *Educational Review* 64 (4): 405–24. https://doi.org/10.1080/00131911.2011.625111.

Huang, Aris R. and Shinta, Revina. 2020. "The Struggle to Recruit Good Teachers in Indonesia?: Institutional and Social Dysfunctions." RISE Working Paper Series. Jakarta.

Jalal, Fasli, Muchlas, Samani, Mae, Chu Chang, Ritchie, Stevenson, Andrew, B. Ragatz, and Sewage, D. Negara. 2009. "Teacher Certification in Indonesia?: A Strategy for Teacher Quality Improvement," no. September 2017: 1–219. http://documents.worldbank.org/curated/en/2009/04/10582109/teacher-certification-indonesia-strategy-teacher-quality-improvement.

Lu, Meichen, Prashant, Loyalka, Yaojiang, Shi, Fang, Chang, Chengfang, Liu, and Scott, Rozelle. 2019. "The Impact of Teacher Professional Development Programs on Student Achievement in Rural China: Evidence from Shaanxi Province." *Journal of Development Effectiveness* 11 (2): 105–31. https://doi.org/10.1080/19439342.2019.1624594.

OECD. 2005. "Teachers Matter: Attracting, Developing and Retaining Effective Teachers." https://doi.org/10.1787/9789264018044-sum-sv.

OECD. 2018. "Country Note: Singapore, PISA Result 2018."

Pritchett, Lant. 2020. "The Danger of Complacency from the Wrong Feedback: Indonesia, Schooling, and Learning." RISE. 2020.

Rahman, Bujang, Abdurrahman, Abdurrahman, Budi, Kadaryanto, and Nurlaksana, Eko Rusminto. 2015. "Teacher-Based Scaffolding as a Teacher Professional Development Program in Indonesia." *Australian Journal of Teacher Education* 40 (11): 66–78. https://doi.org/10.14221/ajte.2015v40n11.4.

Revina, Shintia, Rezanti, Putri Pramana, Rizki, Fillaili, and Daniel, Suryadarma. 2020. "Systemic Constraints Facing Teacher Professional Development in a Middle-Income Country?: Indonesia ' s Experience Over Four Decades." RISE Working Paper Series. Jakarta.

Rosser, Andrew, and Mohamad, Fahmi. 2018. "The Political Economy of Teacher Management Reform in Indonesia." *International Journal of Educational Development* 61 (December 2017): 72–81. https://doi.org/10.1016/j.ijedudev.2017.12.005.

Soebari, Titien, and Jill, M. Aldridge. 2016. "Investigating the Differential Effectiveness of a Teacher Professional Development Programme for Rural and Urban Classrooms in Indonesia." *Teacher Development* 20 (5): 701–22. https://doi.org/10.1080/13664530.2016.1185031.

Tanang, Hasan, and Baharin, Abu. 2014. "Teacher Professionalism and Professional Development Practices in South Sulawesi, Indonesia." *Journal of Curriculum and Teaching* 3 (2): 25–42. https://doi.org/10.5430/jct.v3n2p25.

Thair, M., and D. F. Treagust. 2003. "A Brief History of a Science Teacher Professional Development Initiative in Indonesia and the Implications for Centralised Teacher Development." *International Journal of Educational Development* 23 (2): 201–13. https://doi.org/10.1016/S0738-0593(02)00014-7.

Tobias, Julia, Joseph Wales, Ekki, and Syamsulhakim, Suharti. 2014. "Toward Better Education Quality: Indonesia's Promising Path." https://www.odi.org/sites/odi.org.uk/files/odi-assets/publications- opinion-files/9066.pdf.
Tonga, Funda Eda, Sümeyra, Eryiðit, Fatma, Ay Yalçın, and Feyza, Tantekin Erden. 2019. "Professional Development of Teachers in PISA Achiever Countries: Finland, Estonia, Japan, Singapore, and China." *Professional Development in Education* 00 (00): 1–17. https://doi.org/10.1080/19415257.2019.1689521.
World Bank. 2019. "The Promise of Education in Indonesia."
Yarrow, Noah, Rythia, Afkar, Eema, Masood, and Bernard, Gauthier. 2020. "Measuring the Quality of MoRA's Education Sevices." Jakarta. www.worldbank.org.

Innovation on Education and Social Sciences – Maureen et al (Eds)
© 2022 copyright the Author(s), ISBN 978-1-032-20752-0
Open Access: www.taylorfrancis.com, CC BY-NC-ND 4.0 license

Primary school principal perspective to strengthen Indonesian national identity

Sarmini & Warsono
Faculty of Social Sciences and Law, Universitas Negeri Surabaya, Surabaya, Indonesia

A.D.B.E. Rizaq
Faculty of Tarbiyah, IAIN Madura, Jawa Timur, Indonesia

ABSTRACT: National identity has been an important discussion topic worldwide since the loss of boundaries between countries in the current globalization era. Strengthening national identity can be done by utilizing the role of education. The focus of this study reveals the perspective of the Primary Schools Principal to strengthen Indonesian National Identity and the school programs related to the national identity. This study uses a used survey experiment based on a stratified three-stage probability proportion. Research subjects are 50 school principals from 18 sub-districts in Sidoarjo Regency. This research shows two findings; first, the source of the value of Indonesian national identity comes from Pancasila, the 1945 Constitution, the Unitary State of the Republic of Indonesia, Bhinneka Tunggal Ika, religious values?, and socio-cultural values. Second, several schools have implemented character education, but most do not yet have a model for strengthening the value of national identity for primary school students. So this study recommends developing a policy model to strengthen an Indonesian national identity as a guide in the era of globalization and modernization.

1 INTRODUCTION

It is vital for nations to maintain National identity in the face of the rapid flow of globalization and technological development that drive individuals to abandon their culture and replace it with global culture (Murphy 2018). Entering the era of globalization, racial, ethnic, and cultural diversity in nation-states in the world is getting higher. First, however, the concept of national identity must be maintained (Walton et al. 2018). National identity is a historical context that is "unique" to each country (Ha & Jang 2015) and which comes from religious identity (Bilali, Iqbal, and Elik 2018; Katarzi 2018), ideology, language (Boser & Brühwiler 2017; Lee & Ahn 2016; Octavian 2013), cultural identity (Katarzi 2018), history (Packer, Ballantyne, & Uzzell 2019), and self-esteem (Lucas et al. 2014).

National identity has several different perspectives based on its function and condition. National identity emerged as a form of modern nationalism (Liu & Turner 2018). The national identity is built on past collective memories (Choi et al. 2021; Sinkkonen 2013). In the micro condition, national identity is the recognition of being part of a group with similar emotions (David & Bar-Tal 2009) and the behavior of loving domestic products (Carvalho, Luna, & Goldsmith 2019).

National identities have different characteristics in every country. For example, in the UK, national identity is self-esteem (Georgiadis & Manning 2013). When people are treated with respect and tolerance, that's when they feel part of England, whereas in Scotland, the place of residence is the key to being "Scottish," and ancestry (ethnoreligious background) is also a marker of national identity (Bond 2017). In the Netherlands, national identity is related to limiting the consumption of foreign products (Carvalho, Luna, & Goldsmith 2019). China's national identity is marked by the belief in the historical triumph of the ancient Chinese civilization (Huang & Liu 2018).

 DOI 10.1201/9781003265061-13

Today, maintaining a national identity is a challenge in countries with diverse conditions (Harris et al. 2011), large immigrant populations (McAllister 2016), and multicultural countries (Sealy 2018). Internal problems in the form of weakening national spirit and waning regional cultural values have also triggered racist conflicts, injustice, and national identity crises (Wulandari 2012). The dissolution of the legacy of the past success stories between generations and the openness to change significantly shapes the youth, ignoring the urgency of national identity (Vlachova 2017).

Each nation-state has its efforts to maintain a national identity. For example, Russia conducts promotions through state-controlled media (Goble 2016). Research (Dai, Williams, & McGregor 2018) explores how Scottish Chinese children face challenges establishing their identity. The results found that Scottish Chinese children (children born and living in Scotland with Chinese parents) had judgments about their identity that changed with age. The strength of national identity was assessed in (Chinese) cultural identity, which was more robust in 8-year-olds than 14-year-olds. For countries with diversity, the integration of multicultural values in learning has strengthened national identity (Setyowati & Sarmini 2018). The Hong Kong Special Administrative Region (SAR) government introduces diverse cultural values into the school curriculum to maintain national education (Ho & Law 2004). In Indonesia, the implementation of multicultural values such as upholding justice, democracy, humanity, and pluralism has strengthened education character regulation (Masamah & Huda 2016).

The education system has been commonly used to strengthen national identity. For example, Europe strengthens national identity through national history propaganda and sometimes myths that glorify past national events (Berger 2009). In Australia, the education system develops a formal curriculum structure, which guides teachers to understand and maintain national identity (Walton et al. 2018). Israel resolves the conflicting narratives of Israeli and Palestinian citizens by developing a curriculum containing the national identity of students' countries of origin (Arar & Ibrahim 2016). The national identity in Japan is built through learning the Japanese language at school (Turnbull 2004). At the same time, the Chinese government has long made efforts to maintain national identity through traditional Chinese cultural education (TCC) (Xu 2017).

Instilling ideological doctrines content through the school curriculum will become a platform for maintaining the values of national identity (Dobrocká & Szórádová 2017). In Indonesia, Citizenship Education as early as possible is a means of strengthening national identity (Rizkiyani 2018). In addition, as the main facilitators in learning, teachers play an essential role in strengthening national identity by transferring knowledge of the history of a nation's struggle (Sarmini, Setyowati, & Rizaq 2020).

Then, the policy of strengthening character education has been carried out to maintain national identity (Arief, Prakoso, & Risman 2021). Furthermore, the values of Indonesian local wisdom that are commonly integrated into education have been proven to maintain national identity (Habibi, Pitana, & Susanto 2018). Therefore, this article explores the perspective of the primary school principals and the school programs policy to strengthen national identity.

2 METHOD

This research used a survey experiment in Sidoarjo conducted in August 2021. The survey was preceded by doing confirmation by phone to confirm theidentity of each primary school principal before giving 35-questions online questionnaire through a google form. The questions used in the "1–5 Likert scale" format. The population in this study was 467 primary school's principals in the Sidoarjo district. The sample was 50 principals from 18 sub-districts based on a stratified three-stage probability proportion to size random sample design (with additional onsite Global Positioning System/Geographical Information System remote sensing sampling strategy) to draw representative samples of the primary school principals to cover every sub-district. The focus of the research was to explore the efforts of strengthening the national identity in primary schools based on the principals' perspectives. The main focus includes two sub-questions: (1) the principals' perspectives (their understanding of the concept, urgency, and legal basis of National identity, their perceptive values of the National identity that has been implemented, and their genuine efforts regarding the national identity); and (2) the school programs that related to the national identity (implementation through activities that have been carried out through intracurricular, extracurricular, building school culture and community participation).

3 RESULTS AND DISCUSSION

3.1 *The primary schools principal perspective in maintaining national identity*

The perspective of the primary school principal in maintaining national identity consists of understanding the concept and importance of national identity in primary school students. Based on the results of the questionnaire, there were 50 respondents in this study. The details can be observed in Table 1.

Table 1. Recapitulation of principal perspective in instilling national identity for primary schools.

No.	Rated aspect	Scale/percentage				
		1	2	3	4	5
Understanding the concept of Indonesian national identity						
1	The identity of the Indonesian nation is a national identity that provides enthusiasm for the survival of the Indonesian nation.				10 (20%)	40 (80%)
2	The implementation of Pancasila values in the field of education can strengthen national identity and must start from elementary school students	4 (8%)	5 (10%)	2 (4%)	6 (12%)	33 (66%)
3	The implementation of the values of the 1945 Constitution in Elementary Schools can strengthen national identity.			3 (6%)	5 (10%)	42 (84%)
4	The Unitary State of the Republic of Indonesia (NKRI) can strengthen national identity and must be internalized since elementary school.	7 (14%)	3 (6%)	5 (10%)	7 (14%)	28 (56%)
5	The values of Bhinneka Tunggal Ika can strengthen national identity and must be internalized since elementary school students	4 (8%)	3 (6%)	3 (6%)	13 (26%)	27 (52%)
Urgency and legal basis						
6	Globalization and modernization supported by technological developments can shake the nation's identity.				8 (16%)	42 (84%)
7	Building a National identity is an absolute must so that the Indonesian nation will continue to stand firm amid the swift currents of globalization and modernization				5 (10%)	45 (90%)
8	The implementation of character education to build National identity in elementary schools already has a policy foundation.			10 (20%)	15 (30%)	25 (50%)
The Indonesian national identity values that have been implemented						
9	There is no related official policy for the implementation of certain values in building National identity for elementary schools	15 (30%)	5 (10%)	5 (10%)	10 (20%)	15 (30%)

(continued.)

Table 1. Continued.

No.	Rated aspect	Scale/percentage				
		1	2	3	4	5
10	Until now, the principal is allowed to choose the character values that will be implemented in building a National identity	11 (22%)	6 (12%)	4 (8%)	7 (14%)	22 (44%)
11	The Education Officers gives authority for school principals to choose character values in building National identity	22 (44%)	8 (16%)		10 (20%)	10 (20%)
12	The Education Officers should provide provisions for values that will be internalized in building a National identity for elementary school students.				5 (10%)	45 (90%)
	Efforts that have been made					
13	The school has a program for internalizing values to build a National identity for elementary school students	40 (80%)	3 (6%)	7 (14%)		
14	Although not clearly stated in the policy, schools have internalized values to build a national identity			1 (2%)	7 (14%)	42 (84%)
15	The school has conducted various competitions for students as a vehicle to build national identity		21 (42%)	3 (6%)	4 (8%)	22 (44%)
	Various existing policies					
16	The Education Officers has issued a Decree as a guideline for internalizing values in building National identity at the elementary school level	22 (44%)	8 (16%)		10 (20%)	10 (20%)
17	The school has issued a Decree that the internalization of values will guide to build national identity for students.	25 (50%)	7 (14%)		5 (10%)	13 (26%)
18	The Education Officers issues a Circular on the internalization of values in building National identity at the elementary school level	25 (50%)	5 (10%)		9 (18%)	11 (22%)
19	The school publishes a Circular as a guide in internalizing the values of National identity at the elementary school level	15 (30%)	13 (26%)		15 (30%)	7 (14%)
20	The Education Officers issues a Letter of Assignment to the principal to carry out the internalization of the value of National identity in his school	21 (42%)	4 (8%)		10 (20%)	15 (30%)
21	The principal has issued a Letter of Assignment to teachers to internalize values in intracurricular learning to build a national identity.	22 (44%)	8 (16%)		5 (10%)	15 (30%)

Information:
1. Strongly disagree
2. Disagree
3. Neither agree or disagree
4. agree
5. Totally agree

First, understanding the concept, urgency, and legal basis of the national identity. All school principals totally agree (80%) and agree (20%) with statements about the concept of national identity. Therefore, it can be said that the principal can understand the concept of national identity very well. However, it seems that the principals have various views regarding implementing the cultivation of a national identity for elementary school students. There are variants of arguments for planting national identity, namely the implementation of Pancasila values (66%), implementation of the values of the 1945 Constitution (84%), implementation of the values of the Unitary State of the Republic of Indonesia (NKRI) (56%), and through the implementation of the values of Bhinneka Tunggal Ika (52%). This percentage figure shows the need for an increase in the understanding of school principals about the source of the character values of national identity. Some of them even stated that in addition to the four things mentioned above, instilling a finger of the Indonesian self can be done by strengthening religious and socio-cultural values. In their view, a child who already has a religious character and behaves by the framework of values in society can be said to have a national identity.

Meanwhile, facing the current era of globalization, amid the open flow of information, the nation's children need a national identity that has character, upholds the values of unity, and is competitive. This condition is a valuable asset for the nation in fulfilling its independence and maintaining national security. The principal believes that globalization and modernization, supported by current technological developments, can shake the nation's identity (84%). Therefore, most of them (90%) believe that building a national identity is an absolute must so that the Indonesian nation remains strong amid the swift currents of globalization and modernization. Therefore, the implementation of character education to build a national identity that has been carried out in schools has a strong policy foundation (50%).

Second, the value of the national identity has been implemented. This substance discusses the policies of the Education Officers or school principals in building a national identity. Some of the principals (50%) admit that there is no policy so that they are allowed to choose the character values that will be implemented (58%) to build a national identity. This percentage figure shows differences in views between character education and the value of national identity. Most of the principals (90%) hoped that the Sidoarjo District Education Officers would provide values that will be internalized in building a national identity for elementary school students.

Third, efforts have been made in building the national identity. Most of the principals (80%) admitted that they did not have a clear and explicit program. However, they admit that the cultivation of national identity is carried out implicitly, without realizing it, and is integrated into character education (84%). If national identity is the same as character education, schools have conducted various competitions as a vehicle to build national identity for elementary school students (44%).

3.2 *The school programs that related to maintaining national identity*

This section explores the school programs related to the national identity through intracurricular activities, extracurricular activities, building school culture, and various public participation activities (Table 2).

Table 2. Recapitulation of the school programs that related to the national identity.

No.	Rated aspect	Scale/percentage				
		1	2	3	4	5
Implementation through intracurricular Activities						
1	In addition to social competence, schools ask teachers in lesson planning to include values in building a National identity	20 (40%)	10 (20%)		5 (10%)	15 (30%)
2	Schools have asked teachers to implement learning (initial, core, and final activities) to internalize values in building National identity.	20 (40%)	5 (10%)		5 (10%)	20 (40%)

(continued)

Table 2. Continued.

No.	Rated aspect	Scale/percentage				
		1	2	3	4	5
3	Schools set indicators of successful implementation of internalization of values in building National identity	26 (51%)	9 (18%)		5 (10%)	10 (20%)
4	Schools have an assessment model for internalizing values in building an National identity	23 (46%)	7 (14%)		11 (22%)	9 (18%)
Implementation through extracurricular activities						
5	The principal has a policy of internalizing values to build National identity in planning extracurricular activities	19 (38%)	11 (22%)		5 (10%)	15 (30%)
6	Extracurricular coaches have internalized the values of building National identity in their various activities	27 (52%)	3 (6%)		8 (16%)	12 (24%)
7	Extracurricular coaches have had an assessment of the implementation of values to build National identity	24 (48%)	6 (12%)		10 (20%)	10 (20%)
Implementation of activities in building school culture						
8	The principal has designed the internalization of values to build an National identity in every school cultural activity	21 (42%)	9 (18%)		3 (6%)	17 (34%)
9	The principal conducts monitoring and evaluation in the implementation of values that build National identity in school culture so that it runs well	23 (46%)	7 (14%)		5 (10%)	15 (30%)
10	Principals have indicators of successful implementation of the values of building an National identity that is guided by building school culture	24 (48%)	6 (12%)		3 (6%)	17 (34%)
Implementation in community participation activities						
11	The principal arranges the implementation of values to build National identity through community participation activities	29 (58%)	11 (22%)			10 (20%)
12	Principals use values to build National identity as a filter for various activities involving community participation	25 (50%)	5 (10%)		7 (14%)	13 (26%)
13	Principals already have indicators of successful community participation in the implementation of values to build National identity	28 (56%)	7 (14%)		5 (10%)	10 (20%)
14	The principal has placed community participation as a buffer for internalizing values to build an National identity for elementary school students.	37 (74%)	10 (20%)	3 (6%)		

Information:
1. Strongly disagree
2. Disagree
3. Neither agree or disagree
4. Agree
5. Totally agree

The implementation of the value of national identity has been carried out through intracurricular activities. Suppose character education is considered the same as the values of national identity. In that case, the principal acknowledges that the Education Officers of Sidoarjo Regency have issued a decree (44%) and circular letter (60%) as a guideline for internalizing values in building identity. Meanwhile, the majority of school principals (64%) admit that schools have not issued decrees or circulars (56%) of assignments to elementary school teachers (60%), in which the internalization of values will guide to build national identity for students in primary school.

Principals (60%) have not asked teachers in lesson planning (60%) to include, through initial, core, and final activities (50%), values in building national identity. In addition, schools have not set indicators of success (69%) and admit that they do not have an assessment model for internalizing values in building the national identity of elementary school students (60%).

Next is the implementation of national identity values through activities to build school culture and community participation. Related to this, principals (60%) have not designed, monitored, and have not had success indicators (60%) for implementing internalization of values that build national identity in every school cultural activity. Meanwhile, through community participation activities, the principal has not prepared an implementation plan (80%), a design as a filter for various activities (60%), and indicators in assessing the success (70%) of implementing values to build an national identity in various activities. They (94%) hope that they will place community participation as a buffer for internalizing values to build national identity.

4 CONCLUSION

Two things can be concluded. First, the source of the value of Indonesian national identity comes from Pancasila, the 1945 Constitution, the Unitary State of the Republic of Indonesia, Bhinneka Tunggal Ika, religious values, and socio-cultural values. Second, students who are obedient to worship and behave in accordance with the framework of values in society can be said to have a national identity.

In addition, if character education is the same as strengthening an Indonesian national identity, it is considered a policy basis. The school has strengthened the internalization of character values through various competitions. On the other hand, if it is considered different, the school does not yet have a clear program in strengthening a national identity, either through intracurricular, extracurricular activities, building school culture, or community participation. Therefore, it is important to have a policy model role to strengthen an Indonesian national identity in the era of globalization and modernization.

ACKNOWLEDGMENTS

The authors give thanks for the DRPM through the Decree of the Rector of the State University of Surabaya No. 409/UN38/HK/PP/2021 Dated March 18, 2021 concerning Determination of Research Recipients for Multi-Year Applied Research and Development/Capacity of DRPM Funds in 2021.

REFERENCES

Arar, Khalid, and Fadia Ibrahim. 2016. "Education for National Identity: Arab Schools Principals and Teachers Dilemmas and Coping Strategies." 31(6): 681–93. https://doi.org/10.1080/02680939.2016.1182647

Arief, Ruslan, Lukman Yudho Prakoso, and Helda Risman. 2021. "Understanding National Identity To Create Love And Proud Of Being A Part Of The Indonesian Nation." Journal of Research Innovation 1(11): 2549–56. https://stp-mataram.e-journal.id/JIP/article/view/518 (September 7, 2021).

Bilali, Rezarta, Yeshim Iqbal, and Ay°e Betül elik. 2018. "The Role of National Identity, Religious Identity, and Intergroup Contact on Social Distance across Multiple Social Divides in Turkey." International Journal of Intercultural Relations 65(April):73–85.
Bond, Ross. 2017. "Multicultural Nationalism? National Identity among Minority Groups in Scotland's Census." Journal of Ethnic and Migration Studies 43(7):1121–40.
Boser, Luke and Ingrid Brühwiler. 2017. "Languages, Script and National Identity: Struggles over Linguistic Heterogeneity in Switzerland in the Nineteenth and Twentieth Centuries." History of Education 46(3):306–23.
Cahyani, AD, and Sarmini. 2015. "National identity for Ethnic Chinese Descendants (In Kapasan Village In Kapasan Village, Surabaya City)." State University of Surabaya 01. https://core.ac.uk/download/pdf/230709283.pdf.
Carvalho, Sergio W., David Luna, and Emily Goldsmith. 2019. "The Role of National Identity in Consumption: An Integrative Framework." Journal of Business Research 103: 310–18.
Choi, Su Young, Magdalena Abel, Audrey Siqi-Liu, and Sharda Umanath. 2021. "National Identity Can Be Comprised of More Than Pride: Evidence From Collective Memories of Americans and Germans." Journal of Applied Research in Memory and Cognition 10(1): 117–30.
Dai, Qian, Joanne Williams, and Evelyn McGregor. 2018. "Am I 'Chinese' or 'Scottish'? Children's Perceptions of the Adaptive Nature of Chinese Scottish Children's Dual Identities." European Journal of Developmental Psychology 15(2):224–42.
David, Ohad, and Daniel Bar-Tal. 2009. "A Sociopsychological Conception of Collective Identity: The Case of National Identity as an Example:" http://dx.doi.org/10.1177/1088868309344412 13(4): 354–79. https://journals.sagepub.com/doi/abs/10.1177/1088868309344412 (September 8, 2021).
Dobrocká, Soòa and Eva Szórádová. 2017. "School Curriculum as a Means of Shaping National Identity: Music Education in the Slovak Region of Czechoslovakia in the Interwar Period (1918 – 1939)." Pedagogy, Culture & Society 1366:1–15.
Georgiadis, Andreas and Alan Manning. 2013. "One Nation Under a Groove? Understanding National Identity." Journal of Economic Behavior and Organization 93:166–85.
Goble, Paul. 2016. "Russian National Identity and the Ukrainian Crisis." Communist and Post-Communist Studies 49(1):37–43.
Ha, Shang E. and Seung Jin Jang. 2015b. "National Identity, National Pride, and Happiness: The Case of South Korea." Social Indicators Research 121(2):471–82.
Habibi, Titis Srimuda Pitana, and Susanto. 2018. "Protecting National Identity Based On The Value Of Nation Local Wisdom." International Journal of Malay-Nusantara Studies 1(2): 24–40. http://journal.unhas.ac.id/index.php/IJoM-NS/article/view/5516.
Harris, Melissa S., Knowlton Johnson, Linda Young, and Jessica Edwards. 2011. "Community Reinsertion Success of Street Children Programs in Brazil and Peru." Children and Youth Services Review 33(5): 723–31.
Ho, Wai-Chung, and Wing-Wah Law. 2004. "The Struggle between Globalisation, Nationalism and Music Education in Hong Kong." : 1–18.
Huang, Haifeng, and Xinsheng Liu. 2018. "Historical Knowledge and National Identity: Evidence from China:" https://doi.org/10.1177/2053168018794352 5(3). https://journals.sagepub.com/doi/full/10.1177/2053168018794352 (September 8, 2021).
Hung, Cheng Yu. 2017. "The Reformulation of National Identity in the New Taiwanese Citizenship Curriculum through the Lens of Curriculum Reformers." Asia Pacific Journal of Education 37(2):205–18.
Katarzi, Eugenia. 2018. "Unpacking Young People's National Identities: The Role of Ethno-Cultural and Religious Allegiances, History and 'Others."' Young 26(3):215–31.
Lee, Mun Woo and Sung Ho G. Ahn. 2016. "Relocation in Space, Language, and Identity: Dislocated North Korean Undergraduates in South Korean Universities." Language and Communication 47:43–52.
Liu, Qiang, and David Turner. 2018. "Identity and National Identity." https://doi.org/10.1080/00131857.2018.1434076 50(12): 1080–88. https://www.tandfonline.com/doi/abs/10.1080/00131857.2018.1434076 (September 7, 2021).
Lucas, Todd et al. 2014. "Political Affiliation, Collective Self-Esteem and Perceived Employability of Immigrants: Inducing National Identity Polarizes Host-Nation Employers." International Journal of Intercultural Relations 39(1):136–51.
Masamah, Ulfa, and Mualimul Huda. 2016. "Multicultural Education and the Nationalistic Reality (Photograph the Role of Teachers in Building a Multicultural Awareness in Indonesia)." QIJIS (Qudus International Journal of Islamic Studies) 4(1): 68–81. http://journal.stainkudus.ac.id/index.php/QIJIS/article/view/1578.

McAllister, Ian. 2016. "National Identity and Attitudes towards Immigration in Australia*." https://doi.org/10.1080/14608944.2016.1206069 20(2): 157–73. https://www.tandfonline.com/doi/abs/10.1080/14608944.2016.1206069 (September 9, 2021).
Murphy, Alison. 2018. "Charting the Emergence of National Identity in Children in Wales." Children & Society 32(4):301–13.
Octavian. 2013. "Language That Forms National Identity and Character." Arbitrary Journal 1(1):68–74.
Packer, Jan, Roy Ballantyne, and David Uzzell. 2019. "Interpreting War Heritage: Impacts of Anzac Museum and Battlefield Visits on Australians' Understanding of National Identity." Annals of Tourism Research 76: 105–16.
Rizkiyani, Fanny. 2018. "Strengthening National Identity through Civic Education for Young Children: A Case Study of Indonesia." International Journal of Engineering and Technology(UAE) 7(3): 291–94.
Sarmini and Warsono. 2012. "Nationalism Perspective in the Young People in Surabaya East Java," Adv. science. Lett., vol. 23, no. 12, pp. 11776–11780, 2017.
Sarmini and Warsono. 2018. "The role of education in the culture of four-pillar poverty to establish the nationalism of young generation," J. Phys. conf. Ser., vol. 953
Sarmini, Rr. Nanik Setyowati, and Agung Dwi Bahtiar El Rizaq. 2020. "Understanding of Multiculturalism Material: The CivicTeachers Way in Enhancing Nationalism Based on Multiculturalism Education for Young Generation in Surabaya." 473(Icss): 260–64.
Satori, Djam'an and Komariah, Aan (2009). Qualitative Research Methodology. Bandung: Alphabeta .
Sealy, Thomas. 2018. "Multiculturalism, Interculturalism, 'Multiculture' and Super-Diversity: Of Zombies, Shadows and Other Ways of Being." Ethnicities 146879681775157.
Setyowati, Rr. Nanik and Sarmini. 2018. "Analysis of Learning Model of Civic Education Based on Multicultural Education to Build National identity for Young Generation in Surabaya," Adv. soc. science. Educ. Humanity. Res., vol. 226, no. Icss, pp. 1567–1571, 2018.
Sinkkonen, Elina. 2013. "Nationalism, Patriotism and Foreign Policy Attitudes among Chinese University Students." China Quarterly (216): 1045–63.
Thiagarajan S, Semmel DS, Semmel MI. (1974). Instructional Development For Training Teachers of Exceptional Children. A Sourcebook. Indiana : Indiana University Bloomington
Trilaksana et al. 2018. "Integration of Moral Didactic Values and Patriotism of Hikayat Hang Tuah in Social Studies Learning to Build the Soul of Nationalism in Middle School Students in Surabaya," vol. 226, no. Icss, pp. 1542–1546, 2018.
Turnbull, Blake. 2004. "Learner Perspectives on National Identity and EFL Education in Japan: Report of a Questionnaire Study Cite This Paper Related Papers." http://dx.doi.org/10.18823/asiatefl.2017.14.2.1.211 (September 9, 2021).
Vlachova, Klara. 2017. "Significant Others and the Importance of Ancestry for Czech National Identity." https://doi.org/10.1080/14608944.2017.1362378 21(1): 57–72. https://www.tandfonline.com/doi/abs/10.1080/14608944.2017.1362378 (September 9, 2021).
Walton, Jessica et al. 2018. "Whiteness and National Identity: Teacher Discourses in Australian Primary Schools." Race Ethnicity and Education 21(1):132–47.
Wulandari, Sri. 2012. "Journal of Educational Development: Foundations and Applications." Journal of Educational Development: Foundations And Applications 1(1).
Xu, Shuqin. 2017. "Cultivating National Identity with Traditional Culture: China's Experiences and Paradoxes." Discourse 6306(April):1–14.

Innovation on Education and Social Sciences – Maureen et al (Eds)

Identification of character education values inherent to traditional games in Indonesia: A pilot study

N. Rusmana, L. Nur, P. Purwati & E.A. Mashudi
Universitas Pendidikan Indonesia, Bandung, Indonesia

A.A. Malik
Universitas Siliwangi, Tasikmalaya, Indonesia

ABSTRACT: This study aims at identifying character education values contained in Indonesian traditional games. In this study, researchers identified ten traditional games. The Delphi method was used by involving five experts. Researchers used a questionnaire with a Guttman scale given to experts to identify the character education values in each traditional game, including 13 essential character education values according to Davidson, Lickona, and Khmelkov's theory. The identification results were displayed in tabular form and then analyzed using a descriptive percentage technique to determine the most prominent character education values in each traditional game. The identification results showed that traditional games had all the mentioned character education values. According to the judge's view, the observed character education values were not limited to 13 essential character education values. Other character values, including thoroughness, responsibility, and confidence, can also be found in traditional games. The study concludes that a total of thirteen essential character education values appear in traditional games; however, in particular traditional games, especially individual games, the character value such as cooperation is not revealed.

1 INTRODUCTION

Character education is essential because morals and national identity are eroding due to modernization and globalization (Susanto, 2013). Sayektiningsih, Sumardjoko, and Muhibin (2017) revealed that one of the causes of moral and national identity degradation was the loss of character education in education practices. Furthermore, Purwanto, Susanto, and Pahalawidi (2014) revealed that education in Indonesia seems to prioritize intellectual development, while other aspects receive less attention. Therefore, the government is committed to building a character revolution and focusing on human resource development. In addition, government policy through Regulation of the Minister of Education and Culture of the Republic of Indonesia Number 20 of 2018 concerning strengthening character education in formal institutions has become one of the efforts to emphasize the importance of character education at schools.

Character education is a purposeful effort to make students recognize, care, and internalize values so that students behave as human beings (Harun & Sudaryant, 2020) who are perfect in form and knowledge (Mahmud, 2014). Character education aims to improve the quality of education implementation and outcomes at schools by shaping students' character in an intact, integrated, and balanced manner, based on the graduate competency standards (Julaeha, 2019). In addition, character education is positioned as a foundation to realize the vision of national development, attempting to create a society that is virtuous, dignified, ethical, cultured, and civilized based on *Pancasila* values composed of five principles, namely, (1) Belief in the One Supreme God; (2) A Just and Civilized Humanity; (3) The Unity of Indonesia; (4) Democracy led by the inherent wisdom of consensus arising from deliberation among popular representative; and (5) Social Justice for all the people of Indonesia (Ramadhani, 2018).

DOI 10.1201/9781003265061-14

The rapid development and use of technology is a double-edged sword. Besides its benefits in facilitating human activities, it might lead to negative impacts if it is not handled wisely. For example, the emergence of various modern games, such as video games and virtual games, can erode the existence of traditional games. At the same time, traditional games are a noble heritage and national identity containing the values ??of character education that are important to be transformed to the younger generations (Hapidin & Yenina, 2016; B. H. Susanto, 2017). Furthermore, Irman (2017) revealed that traditional games tend to promote social character education values, while modern games tend to boost individual character education values. In addition, some of the negative impacts of playing online games without self-control might appear, for instance, having a weak immune system due to the lack of physical activity, extended sitting, meal skipping, and radiation from a computer screen. Besides, it also can damage a person's mental development, inhibits the process of self-maturity, affects learning achievement, can be wasteful and dishonest, and lead to difficulties in socializing with others (Setianingsih, 2018). Another study revealed that playing MMOG (massively multiplayer online games) harms players' cognitive and psychological aspects (Liu & Peng, 2009).

The influence of modernization can cause less interest in traditional games even though traditional games positively affect children, namely escalating the intuition, training children's creativity, stimulating children's fine motor skills, and introducing and maintaining cultures amid the challenges of globalization. Therefore, the importance of preserving traditional games and utilizing their benefits to develop various aspects of life and children's character needs to be broadly investigated. It could be done through research activities, developing learning media, implementing appropriate learning methods, and other encouraging efforts, so that children get attracted and are eager to play traditional games (Ramadhani, 2018). Many previous studies have reviewed the character education values embedded in traditional games, such as *Mladok, Gom-pet, and Si-boi.* The research results revealed some character education values that were observed, namely honesty and discipline characters (B.H. Susanto, 2017). Besides, it was also found that some character education values in a traditional game called *Cim-ciman* were honesty, discipline, creativity, independence, responsibility, social care, hard work, and enthusiasm (Fauzi, 2016). However, many types of traditional games have not been identified regarding the character education values contained. Therefore, researchers are interested in examining ten traditional games commonly played in society: *Gobak Sodor, Engklek, Boy-boyan, Lari Balok, Bentengan, Tarik Tambang, Kasti, Balap Karung, Egrang, and Gatrik.*

The first traditional game is *Gobak Sodor*, a traditional game played in a rectangular field. Each line is guarded by one player (a guard). The player going to enter must cross the line, and if the guard hits him or her, he or she must take turns guarding the line. Secondly, it is *Engklek.* Engklek is another traditional game using a flat area or arena by making squares or rectangles and using tiles shards, so-called *Gaco*, as the playing tools. The next is *Boy-boyan*, a traditional game played in a team using a ball to hit the pile of stones or tiles. The fourth game is *Lari Balok,* a traditional game that measures the player's speed to reach a certain distance on four small blocks resembling bricks. The fifth game is *Bentengan*, a traditional game played by several people to catch the opposing players and defend the *Benteng* (fort) to win the game.

Another traditional game that might be popular in some countries is what is called *Tarik Tambang.* It is a traditional sports game using a specific size and length of rope to fight strength by pulling the rope between one team and another. The other game is *Kasti*, a ball game played in teams; one team is the batting team, and the other is the guard team. This game is almost similar to softball or baseball, the same as *Kasti, Benteng, Tarik Tambang, Balap Karung* is also a competitive game in which the participants must wear a sack reaching their waist or neck jump forward from the starting point to the finish line. Unlike other mentioned games, *Egrang* is mostly played individually by standing and walking using bamboo poles made in specific designs. The next game is *Gatrik,* a traditional game usually played in groups. One group is the thrower, and the other is the catcher. This game uses one long bamboo and one short bamboo. The group that gets the highest score in throwing and catching bamboo is the winner.

Apart from preserving the nation's culture, traditional games also have various embedded character education values. Therefore, the purpose of this pilot study is to find out the initial description of the character education values contained in the traditional game based on the results of the expert judgments.

2 METHODS

The study is a descriptive quantitative one measuring and identifying character values in various traditional games in elementary school. The measurement was conducted based on the assessment released by experts and practitioners using the Delphi technique, a technique used to collect ideas and perceptions of a group of experts to assess the extent of agreement and unravel the disagreement (Okoli & Pawlowski, 2004). This study employed a non-probability purposive sample to ensure whether or not the invited experts or practitioners meet the criteria, namely, being an expert in the field of physical education, being a traditional sports practitioner or physical education teacher with at least 2 years' teaching experience in the related field. This sampling technique concluded that five experts met the criteria and were selected to participate in this study. In reaching consensus, Delphi's sample size depends more on the dynamic of the groups rather than on statistical power (Vogel et al., 2019). Researchers used a questionnaire with a Guttman scale given to experts to find out the character education values that existed in each traditional game, including 13 essential character elements from Davidson, Lickona, and Khmelkov's theory (Davidson et al., 2008), namely, (1) diligence, (2) perseverance, (3) strong work ethic, (4) a positive attitude, (5) ingenuity, (6) self-discipline, (7) other character traits oriented to the ability in developing potency to be an outstanding individual, (8) Integrity, (9) justice, (10) caring, (11) respect, (12) cooperation, and (13) other character traits oriented to a good relationship with others. After the experts assessed what character elements existed in each traditional game, analysis was carried out by calculating the percentage based on the experts' answers and coding the characters as presented in Table 1.

Table 1. Code to identify character education values in traditional game.

Code	Character education values
K-1	Diligence
K-2	Perseverance
K-3	Strong work ethic
K-4	A positive attitude
K-5	Ingenuity
K-6	Self-discipline
K-7	Other character traits oriented to the ability in developing potency to be an outstanding individual
K-8	Integrity
K-9	Justice
K-10	Caring
K-11	Respect
K-12	Cooperation
K-13	Other character traits oriented to a good relationship with others

3 RESULTS AND DISCUSSION

Based on the questionnaire results given to the five experts, information on character education values promoted in each traditional game was obtained. A summary of the identification results of character education values in traditional games is presented in Table 2.

Table 2. Identification of character education values in the traditional game.

Traditional games	Character education values												
	K1	K2	K3	K4	K5	K6	K7	K8	K9	K10	K11	K12	K13
Gobak sodor	60%	60%	80%	100%	100%	60%	100%	60%	80%	100%	60%	100%	100%
Engklek	80%	80%	60%	100%	80%	80%	100%	60%	60%	20%	60%	20%	80%
Boy-boyan	60%	60%	100%	100%	100%	80%	100%	60%	80%	100%	60%	80%	100%
Lari balok	60%	60%	100%	80%	60%	60%	80%	80%	60%	20%	60%	20%	60%
Bentengan	60%	60%	100%	100%	100%	80%	100%	80%	80%	100%	80%	100%	100%
Tarik tambang	60%	60%	100%	80%	60%	60%	80%	60%	100%	100%	60%	100%	60%
Kasti	60%	60%	80%	100%	100%	80%	100%	80%	100%	100%	80%	100%	100%
Egrang	60%	60%	100%	100%	60%	80%	100%	60%	60%	20%	60%	20%	80%
Gatrik	60%	60%	80%	100%	100%	80%	100%	60%	80%	40%	80%	60%	100%

Table 1 summarizes the character education values inherent in the ten traditional games based on the results of experts. First, it can be seen that the most prominent character education values (100%) promoted by the *Gobak Sodor* game were K4 (a positive attitude), K5 (ingenuity), K7 (other character traits oriented to the ability in developing the potential to be an outstanding individual), K10 (caring), K12 (cooperation), and K13 (other character traits oriented to a good relationship with others). Second, it can be concluded that the most prominent character education values, which comprised 100% in *the Engklek* game, were K4 (a positive attitude) and K7 (other character traits oriented to the ability in developing the potential to be an outstanding individual). Third, it can be seen that K3 for solid work ethic trait, K4 for a positive attitude, K5 for ingenuity, K7 (other character traits oriented to the ability in developing the potential to be an outstanding individual), K10 for caring, and K13 (other character traits oriented to a good relationship with others) were the most prominent character education values (100%) appeared in *Boy-boyan* game. Fourth, it can be seen that the most prominent character trait with a percentage of 100% in *the Lari Balok* game was K3 which is responsible for a strong work ethic. Fifth, it can be seen that the most prominent character education values with a percentage of 100% in *the Bentengan* game were K3 (hard work ethic), K4 (a positive attitude), K5 (ingenuity), K7 (other character traits oriented to the ability in developing the potential to be an outstanding individual), K10 (caring), K12 (cooperation), and K13 (other character traits oriented to a good relationship with others). Sixth, it can be seen that the most prominent character education values with a percentage of 100% in *the Tarik Tambang* game were K3 which is responsible for solid work ethic, K9 for justice, K10 for caring, and K12 for cooperation. Seventh, it can be seen that the most prominent character education values with a percentage of 100% in the *Kasti* game were K4 (a positive attitude), K5 (ingenuity), K7 (other character traits oriented to the ability in developing the potential to be an outstanding individual), K9 (justice), K10 (caring), K12 (cooperation), and K13 (other character traits oriented to a good relationship with others). *Eighth*, it can be concluded that the most prominent character education values with a percentage of 100% in *the Balap Karung* game were K3 which was responsible for solid work ethic, and K4 for a positive attitude. Ninth, it can be concluded that the most prominent character education values with a percentage of 100% in *the Egrang* game were K3 strong work ethic, K4 a positive attitude, K7 other character traits oriented to the ability in developing the potential to be an outstanding individual. Tenth, it can be seen that the most prominent character education values with a percentage of 100% in *Gatrik* game were K4 a positive attitude, K5 ingenuity, K7 other character traits oriented to the ability in developing the potential to be an outstanding individual, and K13 other character traits oriented to a good relationship with others.

Generally, of the ten traditional games that the experts have identified, it can be seen that all of them had shown essential character education values. However, for an individual game, cooperation values would not be shown. One of the assessors commented that:

> *The character education values written by (Davidson, Lickona, & Khmelkov) were all presented by traditional games. More character education values that were promoted by these games include thoroughness, responsibility, and confidence. The character education values in traditional games are mostly complete. Besides affective, cognitive, and psychomotor aspects, they also involved a sense of belonging (love for the homeland).*

The results of this identification completed and supported the previous research findings in which the personal character development and social characters tend to be more dominant in traditional games (Ramadhani, 2018). In addition, traditional games teach about moral values and can also develop other aspects such as cognitive and psychomotor (Junaedah & Ahmad, 2020; Önder, 2018; Syamsurrijal, 2020). Specifically, from the result of identification of character education values in the traditional game, it was found that the *Boy-boyan* game can improve indicators of social skill, such as the initiative to engage in activities with peers, join games, maintain roles in a play, and resolve conflicts in the game being played (Saleh et al., 2017).

Based on the results of the identification of character education values in these ten traditional games, it showed that these traditional games contain various aspects of character education so that they have the potency to be developed and implemented optimally in educational institutions to prevent problems or fix the degradation of moral and national identity from an early age. In addition, with more comprehensive implementation and introduction of these traditional games, it is hoped to be an attempt to preserve the nation's culture.

4 CONCLUSIONS

It can be concluded from this study that of the ten traditional games that have been identified, they all have character education values. Therefore, the application of traditional games in developing student character at schools is highly recommended. Further research can be conducted to integrate a particular method or approach with traditional games to be effectively implemented to develop students' characters.

REFERENCES

Davidson, M., Lickona, T., & Khmelkov, V. (2008). Smart & Good Schools: A New Paradigm for High School Character Education. In L. P. Nucci & D. Narvaez (Eds.), *Handbook of Moral and Character Education* (pp. 370–390). Routledge Taylor & Franch Group.

Fauzi, F. (2016). Pembentukan Karakter Anak Melalui Permainan Tradisional Cim-ciman. *Jurnal Ilmiah VISI PPTK PAUDNI, 11*(2), 99–109.

Hapidin, H., & Yenina, Y. (2016). Pengembangan model permainan tradisional dalam membangun karakter anak usia dini. *Jurnal Pendidikan Usia Dini, 10*(2), 201–212.

Harun, A. J., & Sudaryanti, A. M. (2020). *Pengembangan Model Pendidikan Karakter Berbasis Multi Kultural dan Kearifan Lokal bagi Siswa PAUD*. UNY Press.

Irman, I. (2017). Nilai-Nilai Karakter pada Anak Dalam Permainan Tradisionan dan Moderen. *KONSELI: Jurnal Bimbingan Dan Konseling (E-Journal), 4*(2), 89–96.

Julaeha, S. (2019). Problematika Kurikulum Dan Pembelajaran Pendidikan Karakter. *Jurnal Penelitian Pendidikan Islam,[SL], 7*(2), 157–182.

Junaedah, S. B. T., & Ahmad, M. A. (2020). The outdoor learning modules are based on traditional games in improving the prosocial behavior of early childhood. *International Education Studies, 13*(10).

Liu, M., & Peng, W. (2009). Cognitive and psychological predictors of the adverse outcomes associated with playing MMOGs (massively multiplayer online games). *Computers in Human Behavior, 25*(6), 1306–1311.

Mahmud, A. (2014). Insan Kamil Perspektif Ibnu Arabi. *Sulesana, 9*(2), 33–45.

Okoli, C., & Pawlowski, S. D. (2004). The Delphi method as a research tool: an example, design considerations, and applications. *Information & Management, 42*(1), 15–29.

Önder, M. (2018). Contribution of Plays and Toys to Children's Value Education. *Asian Journal of Education and Training*, *4*(2), 146–149.
Purwanto, S., Susanto, E., & Pahalawidi, C. (2014). Pendidikan karakter dengan pendekatan sport education dalam perkuliahan di jurusan pendidikan olahraga UNY. *Jurnal Pendidikan Karakter*, *4*(2), 48–60.
Ramadhani, A. (2018). Identifikasi nilai-nilai pendidikan karakter dalam permainan anak tradisional. *Prosiding Seminar Nasional IPTEK Olahraga (SENALOG)*, *1*(1).
Saleh, Y. T., Nugraha, M. F., & Nurfitriani, M. (2017). Model permainan tradisional "boy-boyan" untuk meningkatkan perkembangan sosial anak SD. *ELSE (Elementary School Education Journal): Jurnal Pendidikan Dan Pembelajaran Sekolah Dasar*, *1*(2b), 127–138.
Sayektiningsih, S., Sumardjoko, B., & Muhibin, A. (2017). Penanaman Nilai-Nilai Karakter dalam Pembelajaran Pendidikan Pancasila dan Kewarganegaraan di Madrasah Aliyah Muhammadiyah Klaten. *Manajemen Pendidikan*, *12*(3), 228–238.
Setianingsih, A. (2018). *Game Online Dan Efek Problematikanya Terhadap Motivasi Belajar Siswa Pada Mata Pelajaran Pendidikan Agama Islam Di SMK Negeri 7 Bandar Lampung*. UIN Raden Intan Lampung.
Susanto, B. H. (2017). Model Pembelajaran Pendidikan Jasmani Melalui Permainan Tradisional Untuk Membentuk Karakter Pada Siswa Sekolah Dasar. *Jurnal Moral Kemasyarakatan*, *2*(2), 117–130.
Susanto, E. (2013). Pembelajaran pendidikan jasmani berbasis karakter untuk meningkatan nilai-nilai afektif di sekolah dasar. *Jurnal Pendidikan Karakter*, *3*(3), 288–301.
Syamsurrijal, A. (2020). Bermain Sambil Belajar: Permainan Tradisional Sebagai Media Penanaman Nilai Pendidikan Karakter. *ZAHRA: Research and Tought Elementary School of Islam Journal*, *1*(2), 1–14.
Vogel, C., Zwolinsky, S., Griffiths, C., Hobbs, M., Henderson, E., & Wilkins, E. (2019). A Delphi study to build consensus on the definition and use of big data in obesity research. *International Journal of Obesity*, *43*(12), 2573–2586.

Innovation on Education and Social Sciences – Maureen et al (Eds)

Global pandemic fear and international students: Negative thoughts, mental and physical well-being

S. Jamshaid, A. Olorundare, L. Wang, N. Lo-ngoen, M.I. Afzal & M. Bibi
School of Psychology, Northeast Normal University, Jilin, China

ABSTRACT: Being isolated for a long time in a dorm is causing pandemic fears in international students. Globally this pandemic fear is becoming burdensome and chronic for students in China. This study is aimed to investigate pandemic fear, negative repetitive thoughts, and the mental and physical well-being of international students in China. The research design used in this study was a mixed-method. Data were collected by conducting interviews with international students (N = 205) of different nations (i.e., Asia, Europe, America, and Africa) in China and with the Penn State Worry Questionnaire, Warwick Mental Wellbeing Scale, Perceived Wellbeing Scale. Participants were selected by using a convenience sampling technique. 56% of students were male (n = 115) and 44% (n = 90) were females. The participant's age (M = 1.35, SD = .48) ranged from 21 to 40 years, in which 94 were married, and 111 were single. The findings of the study revealed a negative association between worry and mental health, and worry was also a significant predictor of international students' mental health.

Furthermore, results indicate that students are afraid of contracting the coronavirus, isolation, and financial issues. This study also shows that people are not only afraid to go out in case they contract this contagious virus, but they are also afraid in their dormitories. This study shows that knowing one's own mental and physical well-being is beneficial in fighting the virus.

1 INTRODUCTION

Coronavirus cases were first discovered in the Chinese province of Wuhan (World Health Organization, 2020). Since December 2019, Wuhan has been in the news due to an outbreak of febrile respiratory syndrome caused by a novel coronavirus (Li et al., 2020), which has been linked to a wholesale seafood market in Huanan (Chen et al., 2020). On January 7, 2020, a new coronavirus (SARS CoV-2, initially named 2019-nCoV) was isolated from a patient in a short period of time, and its genome was sequenced (Lu et al., 2020). WHO recognized the SARS COV 2 genetic sequence in January 2020, resulting in the development of specific polymerase chain reaction PCR-based diagnostic tests to detect the new infection in various countries (Corman et al., 2020). In April 2020, a huge number of cases were increased in Europe; the World Health Organization declared the COVID-19 pandemic outbreak on March 11. This virus spread globally in all countries and affected around seven million people (WHO, 2020). The global COVID-19 pandemic and lockdown have taken the world by storm (Sundarasen et al., 2020). Overthinking is painful, but because no one has complete control over his or her thoughts, ruminative thoughts or fears negatively impact mental health. As a result, poor mental health impairs human immune systems, which affect the battle against the disease in the case of COVID-19 (Jamshaid et al., 2020).

The pandemic has disrupted the learning of more than one billion students in 129 countries worldwide, according to the United Nations Educational, Scientific, and Cultural Organization (UNESCO) (UNESCO, 2020). Many universities around the world have moved to emergency remote teaching (ERT) via an online platform, which has increased student anxiety. Literature

DOI 10.1201/9781003265061-15

on COVID-19 and lockdowns in Chinese colleges found significant negative effects on students' psychological well-being and high anxiety levels (Cao et al., 2020;; Bao et al., 2020; Wang et al., 2020). Worry refers to negative uncontrollable repetitive thoughts, emotions, images, and actions, outcomes from a practical consideration of intellectual risk embraced to stay away from or solve the expected threats and their results (Stöber & Borkovec, 2002). Mental health alludes to our emotional, social, and psychological well-being, which essentially impacts how we think, feel, and act.

Moreover, it is useful to decide, deal with our stress, interact with others, and make more decisions (Wortzel et al., 2020). Physical well-being refers to the ability to maintain a healthy lifestyle that allows us to get the most out of our daily activities while being energized and stress-free. It entails taking care of our bodies and comprehending how our daily habits and behaviors affect our overall health, happiness, and quality of life (Peterson & Bossio, 2001). In the face of danger, fear is an adaptive response. Fear can become chronic and stressful when the threat is unpredictable and ongoing, as it is in the present coronavirus illness (COVID-19) outbreak (Mertens et al., 2020).

The COVID-19 (Coronavirus, 2019) pandemic has caused unprecedented disruption to society, our healthcare system, and our economy. However, the pandemic's plethora of challenges may pose a significant threat to our psychological health (Holmes et al., 2020). This has resulted in unprecedented efforts to implement the practice of physical distancing (dubbed "social distancing" in most cases) in countries all over the world, resulting in changes in national behavioral patterns and shutdowns of normal day-to-day functioning. While these steps may be necessary to slow the spread of this disease, they will undoubtedly have long-term and short-term effects on mental health and well-being (Galea, 2020). The pandemic is increasing the risk for young adults (i.e., financial issues and isolation). During the COVID-19 pandemic young adults face psychological issues such as depression (25%) and report suicide ideation (26% versus 11%) and substance use (25% versus 13%), which is affecting their mental health. Before the pandemic, substance use and poor mental health were at risk, but the risk increased and further affected their mental health; many haven't received any treatment yet (Panchal et al., 2021). This demonstrates that, in the context of the COVID-19 lockout, substantial psychological health difficulties occur among university students (Rodríguez-Hidalgo et al., 2020). During the pandemics, students showed worry and anxiety due to the shift in educational approach (Browning et al., 2021; Villani et al., 2021). The countries implemented preventative measures such as lockdown and movement restrictions to stop the spread of coronavirus illness 2019 (COVID-19). This may have an impact on the population's mental health (Khan et al., 2020). Additional studies have demonstrated that globally COVID 19 is impacting mental health and causing different emotional reactions among people in many other countries (Villani et al., 2021).

Fears about contracting COVID-19 increased the signs of poor mental health by increasing their emotional reaction to the situation, either directly or indirectly (Okruszek et al., 2020). As a result, mental health is deteriorating in general, but it is deteriorating even more for people whose family members are infected with COVID-19, or whose loved ones feel unwell. According to a study, mental conditions worsened not just among COVID-19 patients but also among the general population worldwide (Bonsaksen et al., 2020). According to Kaiser Family Foundation research, nearly half of Americans say COVID-19-related events have damaged their mental health (Panchal et al., 2020). In April 2020, the mental health and drug addiction administration's emergency hotline reported that about 20,000 people had texted them (Achenbach, 2020). Similarly, research has shown that teenager and child fears work as a defensive thinking process, which is linked to a lower sense of well-being and anxiety. Furthermore, it was discovered that the amount of worry was higher throughout the adolescent age due to improved cognitive abilities that permitted defensive thoughts for potentially unfavorable consequences and a large number of social and personal issues (da Matos et al., 2013).

Worrying is common, but it has a detrimental influence on well-being, self-regulation, and resilience. A recent study found that when students' emotional and social skills were weak, their mental health was also affected (Reis et al., 2019). On the other hand, worrying about experiencing worse outcomes may be harmful to mental health (i.e., everyday associated worries experiences have been connected to depressive symptoms in long-term and short-term studies) (Charles et al., 2013). Conversely, according to a recent survey, negative reactions to any event and fears are more important in predicting well-being and mental health than the event itself (Furceri et al., 2020). The global spread of coronavirus disease 2019 (COVID-19) has impacted both physical and mental health. Stress-related reactions, if left unchecked, can lead to mental health issues (Chen et al., 2021). Although there appears to be a lower correlation between fear and depression, suicides have been documented in the general community as a result of COVID-19 fear (Mamun & Griffiths, 2020).

During lockdown and quarantine, people experienced a variety of stressors, including fear of infection and death, loss of social contacts, confinement, helplessness, stress, depression, anxiety, panic attacks, and even suicidality (Brooks et al., 2020; Jiménez-Pavón et al., 2020; Xiang et al., 2020). Pandemic-related anxieties and concerns (e.g., fear of infection and dread of being ill) can be particularly harmful to mental health and well-being during lockdowns and quarantines. According to researchers, over 20% of those who were quarantined during the severe acute respiratory syndrome (SARS) outbreak were afraid of falling ill (Reynolds et al., 2008). Furthermore, among the general population, fear of the current coronavirus was linked to higher levels of anxiety (Harper et al., 2020). The psychological concerns accompanying the global spread of the coronavirus (COVID-19) outbreak have quickly heightened the pandemic's burden on healthcare systems (Torales et al., 2020). According to new research examining the mental health risks of COVID-19, the general public has a higher incidence of moderate-to-severe self-reported depression and anxious symptomatology (Wang et al., 2020), indicating the general effects of uncertainty and health-related anxieties. This study intended to explore the pandemic fear among international students and investigate how negative repetitive thoughts impact their physical and mental health. The following were the research questions:

1: Will there be a significant association between study variables among international students?
2: Does worry have a significant impact on the physical and mental health of international students?
3: What are pandemic-related fears in international students?

2 METHODS AND MATERIALS

2.1 *Participants and research design*

In this study, a mixed method research approach was used to investigate study variables. Data were collected through questionnaires and interviews with young students of different nations (i.e., Asia, Europe, America, and Africa) in China. The sample of this study consisted of 205 international students from other universities from the province of China (Jilin). Participants were selected by using a convenience sampling technique. Data were collected from March 2020 to September 2020.

2.2 *Instruments*

Penn State worry questionnaire (PSWQ): The Penn state worry questionnaire was developed by Meyer et al. (1990) to measure the excessiveness, generality, and uncontrollable dimensions of worries. It comprises 16 items; however, we chose a shorter version form with 11 items in this study (Sandín et al., 2009). Each item is rated on a five-point Likert scale from always 1 = Not at all typical of me to 5 = Very typical of me. The total scores of this scale range from 16–80, with Cronbach's reliability of .92. A higher score on the scale indicates a high level of worry, and a lower score will indicate a low level of anxiety (Table 1).

Table 1. Percentage and frequency of study variables ($N = 205$).

Demographics		F	%
Gender	Male	115	56.1%
	Female	90	43.9%
Education	Undergraduate	45	22.0%
	Master	68	33.2%
	PhD	92	44.9%
Age	21-30 years	134	65.4%
	31-40 years	71	34.6%
Nationality	Asia	83	40.5%
	Europe	30	14.6%
	America	30	14.6%
	Africa	30	14.6%
	Others	32	15.6%
Marital status	Single	111	54.1%
	Married	94	45.9%

Note: f= frequency.

Warwick–Edinburg Mental Wellbeing: The Warwick–Edinburg Mental Wellbeing scale was developed by an expert panel drawing on current academic literature, qualitative research with focus groups, and psychometric testing of an existing scale (Tennant et al., 2007). This scale measures mental health. The scale was based on 14 items with a 5-point Likert scale ranging from 1 to 5 (none to all the time). This means that higher scores represent poor mental health in this subject. Cronbach's reliability of the questionnaire in this study was .66.

Perceived Well-being Scale (PWS): Physical well-being was measured by using a perceived well-being scale. This scale was developed by Reker and Wong (1984). It comprises 14 items to assess psychological, physical, and general well-being. This scale is based on three subscales (i.e., psychological, physical, and general well-being). Each item is scored on a 7-point strongly agree to disagree Likert scale. Higher scores demonstrate higher levels of well-being. In this study, we have adopted only the items of physical well-being. The score range is 6–42 for psychological well-being, 8–56 for physical well-being, and 14–98 for general well-being. The reliability of the scale in the study was .57.

2.3 *Interview outline*

The interview was based on open-ended questions designed by consulting old and the latest literature, expert seniors, and teachers' opinions. It was pre-interviewed first to see if there is a need for correction and changes. The main questions of the interview were the following: Are you scared of what another person might think of you when you cough in public? Are you afraid of being judged wrongly for underlying symptoms of other diseases? Are you afraid of being misdiagnosed? Are you self-medicating? Are you terrified of contracting the virus when you hear the sound of a cough or see people sneeze? Are you afraid of going to crowded places (e.g., markets, airports, train stations, and shopping malls, etc.)? Are you terrified of being isolated in an enclosed room? Do you feel fearful of setting off to the hospital because you may contract the coronavirus? Are you afraid of buying fresh fruits and meat physically or online? Are you afraid of eating out and in restaurants? Does this current situation affect your finance? Does this current situation affect your access to basic foodstuff or commodities?

2.4 *Procedure*

Data were collected in two different ways, by conducting interviews and by sending them an online survey. Interviews were conducted with international students by telephone because of the pandemic. Interviews were based on open-ended questions, which were designed to investigate pandemic-related fears in international students. All interviews were scheduled at the time of the participants' convenience. Before conducting the interview, they were debriefed for the purpose of the interview, and they were informed about the confidentiality of their responses. Each interview was recorded and kept confidential. Interviews took 15–20 minutes per participant. The online survey was conducted by using different instruments, i.e., demographics sheet and all questionnaires. Before data was collected, participants were given a brief overview of the study's goals. To keep the research confidential, participants gave their informed consent, and all ethical standards were followed. Data was collected from 238 participants, but after screening for missing values, outliers, and random responses, only 205 true responses were selected for data analyses. Data were collected from March 2020 to September 2020.

2.5 *Statistical analyses*

Descriptive statistics were used to show the frequency, percentage, mean, and standard deviation of study variables following collected data. The Cronbach's Alpha analysis was used to determine the internal consistency of the research instrument, while the Pearson correlation analysis was used to investigate the relationship between variables. Multiple linear regression analysis was used to test worry as a predictor for mental health and physical well-being.

3 RESULTS

Table 2. Mean, standard deviation, alpha reliabilities, and correlation among study variables ($N = 205$).

Variables	M	SD	α	1	2	3
Worry	29.64	9.47	.92	–	–.27***	.05
Mental health	36.57	3.17	.66		–	–.11
Physical well-being	22.54	3.17	.57			–

***p<.001.

Table 2 shows the study's means, alpha reliabilities, standard deviations, and correlations among variables. The alpha reliabilities coefficient of scales varies between 0.57 and 0.92. According to correlation analysis, worry has a significant negative association with mental health ($r = -.27$, p.000). Furthermore, the findings revealed that mental health does not affect physical well-being (Table 3).

Table 3. Multiple linear regression analysis for worry, mental health, and physical well-being ($N = 205$).

	Mental health			Physical well-being		
Variable	B	R2	F	β	R2	F
Worry	.27***	.07	15.87	.05	.00	.61

***p<.001

The findings show that worry has the power to predict mental and physical health. The R2 value of 0.07 shows a 7% variance in the dependent variable [$F(2, 203) = 15.87$], which might explain worry. Worry was found to be a significant predictor of mental health (R2 = .07, $p = .000$, and = .27) in this study. Worry was also found to be a non-significant predictor of physical well-being.

3.1 *Theme 1: afraid of health issues during the pandemic*

Participants ($n = 9$) in the interview reported that they are afraid to cough or sneeze in public. They are afraid to touch anything in public. International students have developed an obsession for cleanliness by washing all stuff multiple times. Every health-related problem makes them afraid of catching COVID-19.

3.2 *Theme 2: afraid of quarantine*

Some of the participants ($n = 15$) declare that they rely on self-medication rather than going to the hospital for treatment for any medical problem. They're afraid that any health issue can be symptoms of COVID-19, and for that, they'll have to quarantine themselves.

3.3 *Theme 3: terrified contagious corona-virus*

International students ($n = 7$) determined that this COVID-19 is contagious and fear it. Students are afraid to visit their friends because they don't want to contract this coronavirus. They are afraid to go to public places due to COVID-19.

3.4 *Theme 4: terrified of being isolated*

Some students ($n = 9$) are afraid to catch COVID-19 but also afraid to be isolated in the dorm. Fear of isolation makes them worried, and they're terrified of being isolated in an enclosed room. Isolation is a big source of mental and psychological health-related issues, i.e., depression, anxiety, stress, worry, etc.

3.5 *Theme 5: afraid of doing groceries*

Furthermore, findings declare that international students ($n = 18$) are afraid to go out for groceries and order online. Students are more concerned about food, so they like to buy by themselves by following complete sop's (standard operating procedure). It may be a rumor or real that this virus spread from seafood, so students are more afraid to buy seafood.

3.6 *Theme 6: financial problems*

Moreover, international students ($n = 21$) are facing financial crises during the pandemic. Prices are higher in the pandemic; even buying basic stuff is unaffordable for them. Safety is a priority, but sanitizing products are expensive to buy again.

4 DISCUSSION

The present study aims to explore the pandemic fears and to explore repetitive negative thoughts of international students during a pandemic and how it impacts their mental health and physical well-being. The study was based on different hypotheses. The first and second objectives were about the relationship between variables and how worry is impacting the mental health and physical well-being of international students in China. Findings suggested worry has a significant relationship with mental health but a non-significant relationship with physical well-being (Table 4).

Table 4. Coded answers regarding respondents' concerns about the COVID 19.

Themes	Subthemes	Example/Quotations
Afraid of health issues during the pandemic	Scared to cough in public	Since we got this COVID virus which has symptoms like coughing, sneezing so when I go out, I am scared to sneeze or cough in public.
	Fear of having symptoms	My headache makes me afraid that I'm having covid-19 symptoms.
Afraid of quarantine	Afraid to go to the hospital	For any health-related issues, I don't like to go to the hospital. I am afraid of being judged wrongly for underlying symptoms of other diseases.
	Self-medication	I always have medicine in my room for any health problem. I do feel fearful of setting off to the hospital because I may contract coronavirus from other patients.
Terrified contagious corona-virus	Terrified of contracting others	I'm terrified of contracting the virus when I hear the sound of a cough or see people sneeze.
	Afraid to touch anything in public	I wear gloves again and again so I won't catch coronavirus.
	Afraid of crowded places	I am afraid of going to crowded places (e.g., markets, airports, train stations, and shopping malls, etc.)
Terrified of being isolated	Terrified of being isolated	Quarantine and isolation make me worried. I'm terrified of being isolated in an enclosed room.
	Negative thoughts for going outside	I'm terrified to be isolated in the home because it's a tough part of this situation, leading to many negative gestures and thoughts.
Afraid of doing grocery	Afraid of online grocery	I am afraid of buying fresh fruits and meat physically or online.
	Afraid of physical grocery	I always like to do groceries by myself.
Financial problems	Expensive stuff during the pandemic	Due to pandemic stuff is way expensive for students to buy their daily basis need things. All sanitizing stuff is expensive for us to buy so we can stay safe.

Findings also indicated that worry is a significant predictor for mental health but not for physical well-being. Findings suggested that international students are worried about the pandemic COVID-19, which is affecting their mental health. The pandemic is critical for international students; this is why more repetitive negative thoughts increase their level of poor mental health. Physical well-being is not affected by worry; it may be because staying indoor students have more physical activities to make them active and boost their energy level. Previous studies support these findings. During the pandemics, students showed worry and anxiety due to the shift in the educational approach (Browning et al., 2021; Villani et al., 2021).

However, pandemics harm students' mental health. The previous study found that pandemics have negative influences that impact students' well-being, leading them to develop anxiety and depression (Aktekin et al., 2001). Another study by Cao et al. (2020) also investigated the psychological well-being of students studying in China during the third wave of COVID-19. They collected data from 7000 students and found that students had 2.7 percent moderate level of anxiety, 0.9 percent severe level of anxiety, and 21.3 percent mild level of anxiety. Meanwhile, females reported more psychological issues as compared to males (Wang et al., 2020).

Moreover, another purpose of this study was to explore the COVID-19 pandemic fears among international students. Findings suggested different themes related to COVID-19 fears. Pandemic-related anxieties and concerns (e.g., fear of infection and dread of being ill) can be particularly harmful to mental health and well-being during lockdowns and quarantines. According to researchers, over 20% of those who were quarantined during the severe acute respiratory syndrome (SARS) outbreak were afraid of falling ill (Reynolds et al., 2008). Furthermore, among the general population, fear of the current coronavirus was linked to higher anxiety levels (Harper et al., 2020). During lockdown and quarantine, people experienced a variety of stressors, including fear of infection and death, loss of social contacts, confinement, helplessness, stress, depression, anxiety, panic attacks, and even suicidality (Brooks et al., 2020; Jiménez-Pavón et al., 2020; Xiang et al., 2020).

Recent studies have demonstrated that the pandemic has had a huge impact on undergraduate students (Sanchi & Marasine, 2020). Students feel depression, anxiety, and stress because of technical issues in online study. Being alone, studying alone, and suffering financial issues are adversely affecting them. These factors also affect university education and their mental health (Aristovnik et al., 2020). Millions of people worldwide have been impacted by the novel coronavirus (COVID-19) (Malik et al., 2021). Doctors were particularly concerned about spreading the infection to their families and becoming infected with them. Doctors may develop worries, fear, and aversion to their tasks as a result of a lack of proven cure at the time the data was collected, a lack of a nurturing and accommodating workplace, and a lack of enthusiasm from concerned health departments (Jones, 2020;; Kumar et al., 2020; Liu et al., 2020; Urooj et al., 2020).

This study has its original contribution; despite that, it has some limitations. This study was based on a mixed-method research approach. The interview was based on an open-ended question that was not validated, so future researchers should use semi-structured interviews for more qualitative results. Another limitation was selecting data from one province of China, so results cannot be generalized to the wider population. Because the study sample was too small to predict more results, future researchers should gather data from more Chinese provinces to compare results. Cross-sectional research are not more validated, so researchers should select a longitudinal study design in the future to get more empirical results.

This study concludes that the global COVID-19 pandemic generates fears in international students, affecting their life activities. According to the findings, students' enhanced anxiety levels during the COVID-19 outbreak also raised their levels of poor mental health. Furthermore, the study concluded that worry has no relation to physical well-being. Moreover, the study concluded there are different pandemic COVID-19 fears in international students, i.e., fear of illness, fear of quarantine, fear of being isolated, fear of financial crises, fear of the crowd, and fear of contracting COVID-19, etc.

REFERENCES

Achenbach, J. (2020). Coronavirus is harming the mental health of tens of millions of people in the US, new poll finds. Washington Post, 2.

Aktekin, M., Karaman, T., Senol, Y. Y., Erdem, S., Erengin, H., & Akaydin, M. (2001). Anxiety, depression and stressful life events among medical students: a prospective study in Antalya, Turkey. Medical education, 35(1), 12–17.

Aristovnik, A., Keržié, D., Ravšelj, D., Tomaževié, N., & Umek, L. (2020). Impacts of the COVID-19 pandemic on life of higher education students: A global perspective. Sustainability, 12(20), 8438

Boshra, A. A., Al-Dabbagh, Z. S., Al Eid, N. A., Al Eid, M. A., Al-Musaibeh, S. S., Al-Miqtiq, M. N., & Al-Zeyad, G. M. (2020). The effects of Coronavirus (COVID-19) outbreak on the individuals' mental health and on the decision-makers: A comparative epidemiological study. International Journal of Medical Research & Health Sciences, 9(3), 26–47.

Bao, Y., Sun, Y., Meng, S., Shi, J., & Lu, L. (2020). 2019-nCoV epidemic: address mental health care to empower society. The Lancet, 395(10224), e37–e38.

Bonsaksen, T., Leung, J., Schoultz, M., Thygesen, H., Price, D., Ruffolo, M., & Geirdal, A. Ø. (2020). Cross-national study of worrying, loneliness, and mental health during the COVID-19 pandemic: a comparison between individuals with and without infection in the family.

Brooks, S. K., Webster, R. K., Smith, L. E., Woodland, L., Wessely, S., Greenberg, N., & Rubin, G. J. (2020). The psychological impact of quarantine and how to reduce it: rapid review of the evidence. The lancet, 395(10227), 912–920.

Browning, M. H., Larson, L. R., Sharaievska, I., Rigolon, A., McAnirlin, O., Mullenbach, L., & Alvarez, H. O. (2021). Psychological impacts from COVID-19 among university students: Risk factors across seven states in the United States. PloS one, 16(1), e0245327.
Cao, W., Fang, Z., Hou, G., Han, M., Xu, X., Dong, J., & Zheng, J. (2020).The psychological impact of the COVID-19 epidemic on college students in China. Psychiatry Research, 287, 112934.
Charles, S. T., Piazza, J. R., Mogle, J., Sliwinski, M. J., & Almeida, D. M. (2013).The wear and tear of daily stressors on mental health. Psychological science, 24(5), 733–741.
Chen, S. X., Ng, J. C., Hui, B. P., Au, A. K., Wu, W. C., Lam, B. C., & Liu, J. H. (2021). Dual impacts of coronavirus anxiety on mental health in 35 societies.Scientific reports, 11(1), 1–11.
Chen, Y., Liu, Q., & Guo, D. (2020). Emerging coronaviruses: Genome structure, replication, and pathogenesis. Journal of Medical Virology, 92, 418–423. https://doi.org/10.1002/ jmv.25681
Corman, V. M., Landt, O., Kaiser, M., Molenkamp, R., Meijer, A., Chu, D. K., & Drosten, C. (2020). Detection of 2019 novel coronavirus (2019-nCoV) by real-time RT-PCR. Eurosurveillance, 25(3), 2000045.
da Matos, M. G., Gaspar, T., Cruz, J., & Neves, A. M. (2013). New highlights about worries, coping, and well-being during childhood and adolescence. Psychology Research, 3(5), 252.
Furceri, D., Loungani, P., Ostry, J. D., & Pizzuto, P. (2020). Will Covid-19 affect inequality? Evidence from past pandemics. Covid Economics, 12(1), 138–157.
Galea, S., Merchant, R. M., & Lurie, N. (2020). The mental health consequences of COVID-19 and physical distancing: the need for prevention and early intervention. JAMA internal medicine, 180(6), 817–818.
Harper, C. A., Satchell, L. P., Fido, D., & Latzman, R. D. (2020). Functional fear predicts public health compliance in the COVID-19 pandemic. International journal of mental health and addiction, 1–14.
Holmes, E. A., O'Connor, R. C., Perry, V. H., Tracey, I., Wessely, S., Arseneault, L., & Bullmore, E. (2020). Multidisciplinary research priorities for the COVID-19 pandemic: a call for action for mental health science. The Lancet Psychiatry, 7(6), 547–560.
Jamshaid, S., Haider, N. I. M. A. A., Jamshed, K., & Jamshad, S. (2020, December). Overthinking Hurts: Rumination, Worry and Mental Health of International Students in China During Covid-19 Pandemic. In International Joint Conference on Arts and Humanities (IJCAH 2020) (pp. 17–24). Atlantis Press. https://doi.org/10.2991/assehr.k.201201.004.
Jiménez-Pavón, D., Carbonell-Baeza, A., & Lavie, C. J. (2020). Physical exercise as therapy to fight against the mental and physical consequences of COVID-19 quarantine: Special focus in older people. Progress in cardiovascular diseases, 63(3), 386.
Jones, D. S. (2020). History in a crisis—lessons for Covid-19. New England Journal of Medicine, 382(18), 1681–1683.
Khan, A. A., Lodhi, F. S., Rabbani, U., Ahmed, Z., Abrar, S., Arshad, S., & Khan, M. I. (2020). Impact of coronavirus disease (COVID-19) pandemic on psychological well-being of the Pakistani general population.Frontiers in psychiatry, 11.
Li, Q., Guan, X., Wu, P., Wang, X., Zhou, L., & Tong, Y. (2020). Early transmission dynamics in Wuhan, China, of novel Coronavirus – Infected pneumonia. New England journal of medicine. Advance online publication. HTTPS:// doi.org/10.1056/NEJMoa2001316.
Liu, Q., Luo, D., Haase, J. E., Guo, Q., Wang, X. Q., Liu, S., & Yang, B. X. (2020). The experiences of healthcare providers during the COVID-19 crisis in China: a qualitative study. The Lancet Global Health, 8(6), e790–e798.
Kumar, J., Katto, M. S., Siddiqui, A. A., Sahito, B., Ahmed, B., Jamil, M., & Ali, M. (2020). Predictive factors associated with fear faced by healthcare workers during COVID-19 pandemic: A questionnaire-based study. Cureus, 12(8).
Lu, H., Stratton, C. W., & Tang, Y. (2020). Outbreak of pneumonia of unknown etiology in Wuhan, China: The mystery and the miracle. Journal of Medical Virology, 92, 401–402. https://doi.org/10.1002/jmv.25678.
Malik, S., Ullah, I., Irfan, M., Ahorsu, D. K., Lin, C. Y., Pakpour, A. H., & Minhas, R. (2021). Fear of COVID-19 and workplace phobia among Pakistani doctors: A survey study. BMC Public Health, 21(1), 1–9.
Mamun, M. A., & Griffiths, M. D. (2020). First COVID-19 suicide case in Bangladesh due to fear of COVID-19 and xenophobia: Possible suicide prevention strategies. Asian journal of psychiatry, 51, 102073.
Mertens, G., Gerritsen, L., Duijndam, S., Salemink, E., & Engelhard, I. M. (2020). Fear of the Coronavirus (COVID-19): Predictors in an online study conducted in March 2020. Journal of anxiety disorders, 74, 102258.
Meyer, T. J., Miller, M. L., Metzger, R. L., & Borkovec, T. D. (1990). Development and validation of the Penn state worry questionnaire. Behaviour research and therapy, 28(6), 487–495.

Okruszek, L., Aniszewska-Stańczuk, A., Piejka, A., Wiśniewska, M., &Żurek, K. (2020).Safe but lonely? Loneliness, mental health symptoms and COVID-19.PsyArXiv. Preprint posted online on April, 4.
Panchal, N., Kamal, R., Orgera, K., Cox, C., Garfield, R., Hamel, L., & Chidambaram, P. (2020).The implications of COVID-19 for mental health and substance use. Kaiser family foundation, 21.
Peterson C., Bossio L. M., (2001). Optimism and physical well-being. Chang E. C., , ed.Optimism and pessimism: Implications for theory, research, and practice 127–145. American Psychological Association, Washington, DC.
Reis, M., Matos, M. G., & Ramiro, L. (2019).Worries, mental and emotional health difficulties of Portuguese University students. Advances in Social Sciences Research Journal, 6(7), 558–569.
Reker, G. T., & Wong, P. T. (1984). Psychological and physical well-being in the elderly: The Perceived Well-Being Scale (PWB). Canadian Journal on Aging/La Revue canadienne du vieillissement, 3(1), 23–32.
Reynolds, D. L., Garay, J. R., Deamond, S. L., Moran, M. K., Gold, W., & Styra, R. (2008). Understanding, compliance and psychological impact of the SARS quarantine experience. Epidemiology & Infection, 136(7), 997–1007.
Rodríguez-Hidalgo, A. J., Pantaleón, Y., Dios, I., & Falla, D. (2020). Fear of COVID-19, stress, and anxiety in university undergraduate students: a predictive model for depression. Frontiers in Psychology, 11, 3041.
Sandín, B., Chorot, P., Valiente, R. M., & Lostao, L. (2009). Validación española del cuestionario de preocupación PSWQ: Estructura factorial y propiedades psicométricas.
Sanchi S, & Marine NR. Impact of COVID-19 Impact of COVID-19 pandemic on mental health of the general population, students, and health care workers: A review. preprints.org. 2020. https://doi.org/10.20944/preprints202007.0616.v1
Stöber, J., & Borkovec, T. D. (2002). Reduced concreteness of worry in generalized anxiety disorder: Findings from a therapy study. Cognitive Therapy and Research, 26(1), 89–96.
Sundarasen, S., Chinna, K., Kamaludin, K., Nurunnabi, M., Baloch, G. M., Khoshaim, H. B., & Sukayt, A. (2020). Psychological impact of COVID-19 and lockdown among university students in Malaysia: implications and policy recommendations. International journal of environmental research and public health, 17(17), 6206.
Tennant, R., Hiller, L., Fishwick, R., Platt, S., Joseph, S., Weich, S., & Stewart-Brown, S. (2007). The Warwick-Edinburgh mental well-being scale (WEMWBS): development and UK validation. Health and Quality of life Outcomes, 5(1), 1–13.
Torales, J., O'Higgins, M., Castaldelli-Maia, J. M., & Ventriglio, A. (2020).The outbreak of COVID-19 Coronavirus and its impact on global mental health. International Journal of Social Psychiatry, 66(4), 317–320.
United Nations Educational, Scientific and Cultural Organization (UNESCO). Available online: HTTP: //en.unesco.org/covid19/educationresponse (accessed on 1 June 2020).
Urooj, U., Ansari, A., Siraj, A., Khan, S., & Tariq, H. (2020).Expectations, fears and perceptions of doctors during Covid-19 pandemic. Pakistan Journal of Medical Sciences, 36(COVID19-S4), S37.
Villani, L., Pastorino, R., Molinari, E., Anelli, F., Ricciardi, W., Graffigna, G., & Boccia, S. (2021). Impact of the COVID-19 pandemic on psychological well-being of students in an Italian university: a web-based cross-sectional survey. Globalization and health, 17(1), 1–14.
Wang, G., Zhang, Y., Zhao, J., Zhang, J., & Jiang, F. (2020). Mitigate the effects of home confinement on children during the COVID-19 outbreak. The Lancet, 395(10228), 945–947.
WHO, "Coronavirus disease (COVID-2019) situation reports," 2020, [Online]. Available: https://www.who.int/emergencies/diseases/novel- coronavirus-2019/situation-reports/).
Wortzel, J. R., Turner, B. E., Weeks, B. T., Fragassi, C., Ramos, V., Truong, T., & Lee, H. B. (2020). Trends in mental health clinical research: Characterizing the ClinicalTrials. gov registry from 2007–2018. PloS one, 15(6), e0233996.
Xiang, Y. T., Yang, Y., Li, W., Zhang, L., Zhang, Q., Cheung, T., & Ng, C. H. (2020). Timely mental health care for the 2019 novel coronavirus outbreak is urgently needed. The lancet Psychiatry, 7(3), 228–229.
Yang, J., Zheng, Y., Gou, X., Pu, K., Chen, Z., Guo, Q., & Zhou, Y. (2020). Prevalence of comorbidities and its effects in patients infected with SARS-CoV-2: a systematic review and meta-analysis. International Journal of Infectious Diseases, 94, 91–95.

Innovation on Education and Social Sciences – Maureen et al (Eds)

The relationship between psychological well-being and adversity quotient on fresh graduates during coronavirus pandemic

D.K. Dewi, P.N. Wijaya & A.P. Puteri
Universitas Negeri Surabaya, Surabaya, Indonesia

ABSTRACT: Psychological well-being can be recognized by the satisfaction level in the individual. The satisfaction of individuals who have become fresh graduates can be achieved by, one of many, having a job right away. However, the pandemic situation makes it difficult to get a job, so fresh graduates need the ability to face all obstacles to achieve happiness. This study aims to determine the relation between psychological well-being and the adversity quotient of 212 fresh graduates during Coronavirus. This study used the quantitative method and product-moment for the data analysis technique. The correlation coefficient obtained is 0.53, which means there is a significant relationship between psychological well-being and the adversity quotient. The higher the individual's psychological well-being, the higher the individual's adversity quotient. Based on the result, psychological well-being affects the adversity quotient level of fresh graduates during coronavirus pandemic and vice versa.

1 INTRODUCTION

The coronavirus pandemic has an effect on every part of life, including employment. Many businesses must lay off workers to keep afloat during the epidemic, which is also happening in Indonesia. According to figures from Indonesia's Ministry of National Development Planning, 3.5 million employees were laid off, and 3.7 million were jobless (Fauzia, 2020). This truth has a tremendous influence on recent grads.

Individuals who have completed their schooling, particularly at the higher education level, are known as fresh graduates, and they want to be financially independent. The job that each new graduate expects is unquestionably the one that can match the necessity and societal demand (Rachmady & Aprilia, 2018). The function of the adversity quotient is highly important for fresh graduates since the capacity to face and overcome obstacles faced will allow them to persist in their chosen choices. The high degree of adversity quotient in fresh graduates enables individuals to confront these challenges, and the aim of obtaining a job will be achieved (Aprilia & Khairiyah, 2018; Drahs, Haywood, & Schiprowski, 2018). Besides that, the adversity quotient effectively contributes to students' career maturity by almost 23%, which will help the fresh graduates as job seekers during the pandemic (Kurniawan, Daharnis, & Karneli, 2020).

Because many firms are laying off staff and regular hiring methods have been interrupted, finding a new job is especially challenging during the coronavirus pandemic. The existence of the adversity quotient in individuals who are looking for work will make them improve their ability to solve all obstacles in the job search process (Hardianto & Sucihayati, 2019). The strength of the ability to solve the difficulties makes individuals driven to achieve success in the business undertaken. Fresh graduates who want to have the adversity quotient ability must have several aspects, including the need for self-control, having a good interpretation of a problem, having a positive meaning to the problem, and the endurance of each individual.

A good adversity quotient in fresh graduates can be influenced by several factors, including competitiveness, optimism, talent, character, motivation, risk-taking, intelligence, and physical

DOI 10.1201/9781003265061-16

and psychological health (Stoltz, 1997). Adversity quotient itself includes 1) Self-control, namely the ability to manage and control the difficulties encountered so as not to have a bad impact on individuals; 2) Origin and ownership or origin-ownership, which is related to individual interpretation in interpreting the causes of obstacles and difficulties encountered; 3) Reach, namely the individual's meaning in considering certain difficulties and obstacles encountered in reaching out to other aspects of life; 4) Endurance, namely the individual's interpretation in assuming a difficulty encountered is only temporary or permanent (Stoltz, 1997).

The adversity quotient is extremely important for recent graduates seeking for job, particularly to boost their life happiness or satisfaction during the coronavirus epidemic. Individual life satisfaction is strongly connected to psychological well-being. Life satisfaction, as we all know, is a critical component of psychological well-being (Zhao, Sang, & Ding, 2021). Individuals require psychological well-being due to the existence of psychological health in order to boost their ability to overcome problems in life. Psychological well-being is commonly defined as a mix of pleasant affective emotions such as happiness (the hedonic perspective) and good functioning in individual and communal life (the eudaimonic perspective) (Decy & Ryan, 2008).

Psychological well-being, according to Ryff (1989a, 1989b), is a state in which individuals have fully achieved their psychological potential. Individuals with good psychological well-being will achieve full functioning according to Rogers' theory, self-actualization according to Maslow's theory, individuation according to Jung's theory, maturity according to Allport's theory, and integration rather than despair according to Erikson's theory. Psychological well-being has six dimensions that need to be fulfilled, namely self-acceptance, positive relationships with others, autonomy, environmental mastery, life goals, and personal growth (Ryff & Keyes, 1995). Based on this background, psychological well-being and adversity quotient influence each other. Furthermore, because of the coronavirus pandemic, recent graduates will be burdened with the dual task of finding work and coping with the epidemic. These facts make it important to do this research as soon as possible.

2 METHOD

The research method adopted is quantitative. Quantitative research is synonymous with numbers, beginning with data collection, analysis, and presentation (Mertens, 2015). In this study, the independent variable is psychological well-being, while the dependent variable is the adversity quotient. The purpose of this study is to examine the relationship between the adversity quotient and the participants' psychological well-being during the coronavirus pandemic.

2.1 *Participants*

In the academic year 2019–2020, 1419 students graduated from Universitas Negeri Surabaya. To be eligible for selection in the sample, we used the criteria of 1) the lack of work links with other organizations or agencies and 2) persons searching for employment. The purposive sampling approach was used to choose 212 graduates with the aforementioned criteria, 61 of whom were male and 151 of whom were female. They account for around 15% of the whole study population.

2.2 *Research instruments*

The scale used in this study is the adversity quotient scale and the optimism scale. The adversity quotient scale has been modified based on control, origin-ownership, reach, and endurance (Stoltz, 1997). The optimism scale in this study uses a modified Revised Life Orientation Test (LOT-R) scale developed by Scheier & Carver (Brissette, Scheier, & Carver, 2002). The research instruments of the study measure the psychological well-being condition of the participants. Psychological well-being contains six dimensions that need to be fulfilled, and it can be measured using the adversity quotient and the optimism scales.

Because the original scales were prepared in English, they were translated into Bahasa Indonesia by an English instructor at Universitas Negeri Surabaya's faculty of Education. Another English instructor in the faculty then reviewed the translation to verify that there were no issues with comprehending the questions. The researchers next examined the redesigned scale and delivered it to 20 graduates who were not chosen as a sample for this study. The goal of this pilot test was to see how long it took to fill out the scale and to ensure that no statements on the scale had a dual interpretation or were difficult to understand. The participants of the pilot study did not ask questions about the instrument and spent about 20-30 minutes filling in the two scales.

2.3 *Data Analysis*

The normality test was carried out to determine the distribution of the data obtained in the study using the Kolmogorov Smirnov test for normality. The result for the adversity quotient scale is 0.26, beside the optimism scale's result is 0.25. Both instruments showed that the result is higher than 0.05, indicating that the data is normally distributed. The linearity test in the study used the ANOVA table test to determine the direction of the two variables in the study. The result of the ANOVA table test is $0.19 > 0.05$, and it is shown that these instruments were linear. We used Pearson's product-moment to determine whether there is a relationship between optimism and the adversity quotient.

3 RESULT AND DISCUSSION

3.1 *Based on the results of the study are as follows*

According to Table 1, the psychological well-being instrument has a minimum score of 39.00 and a maximum value of 72.00, whereas the optimism instrument has a minimum score of 52.00 and a maximum value of 90.00. Based on the average value and the standard deviation value, the two variables indicate that the average value is bigger than the standard deviation value. It signifies that the data distribution in this study is homogenous.

Table 1. Psychological well-being and adversity quotient.

	Psychological well-being	Adversity quotient
Mean	54.10	71.59
Minimum	39.00	52.00
Maximum	72.00	90.00
Std. Deviation	5.49	7.75
Variance	30.11	60.07

Table 2. Correlation results between psychological well-being and adversity quotient.

		Psychological well-being	Adversity quotient
Psychological well-being	Pearson correlation	1	0.53
	Sig. (2-tailed)		0.00
	N	212	212
Adversity quotient	Pearson correlation	0.53	1
	Sig. (2-tailed)	0.00	
	N	212	212

According to Table 2, there is a positive association between psychological well-being and the adversity quotient. It indicates that the adversity quotient has a linear effect on an individual's psychological well-being. The correlation test result is 0.53, indicating a high association (Sugiyono, 2017) between adversity quotient and psychological well-being. It may also be argued that during the coronavirus pandemic, there is a linear link between psychological well-being and adversity quotient on young graduates. It suggests that if an individual's psychological well-being is high, it is safe to assume that the individual's adversity quotient is also high.

Based on the table above, it can be seen that the relationship between psychological well-being and the adversity quotient has a positive relationship. It means that the adversity quotient affects the individual's psychological well-being in a linear way. The correlation test's result is 0.53 which show the relationship between adversity quotient and psychological well-being is strong (Sugiyono, 2017). And it can be concluded there is a linear relationship between psychological well-being and adversity quotient on fresh graduates during the coronavirus pandemic. It means if the individual's psychological well-being is high, it can be concluded that the individual's adversity quotient is at the same level.

The purpose of this study was to determine the relationship between psychological well-being and the adversity quotient. The results above indicate that the relationship between the two variables is positive. So, individuals who have high psychological well-being will have a high adversity quotient also, and the relationship between these two variables is also at a strong level. The results of significant values ??in this study also indicate that the hypothesis in this study can be accepted so that the two variables have a significant relationship.

The strong relationship between these two variables shows that in this pandemic situation, finding a job is uniquely difficult. Which is as a job seeker, the fresh graduates will need the psychological well-being and the adversity quotient in ways to cope with the stress. Optimism is synonymous with positive expectations, where it helps individuals in dealing with problems. An optimistic individual, when facing problems, will see through a positive point of view so that they are not easily stressed (Brissette et al., 2002). Adversity quotient is an individual's ability to overcome difficult times. It is necessary for helping individuals overcome the problems that arise in their lives (Zainudin, 2011).

An unfavorable situation makes an individual's adversity quotient ability influenced by factors other than psychological well-being. There are several factors that can certainly affect the level of adversity quotient in individuals that are not described in this study. According to Stoltz (2007), several factors affect the adversity quotient, and individuals need to pay attention to what aspects are needed to improve the adversity quotient ability. These factors are 1) individual competitiveness; 2) Optimistic attitude; 3) Talent; 4) Character; 5) Motivation; 6) Risk-taking; 7) Intelligence; 8) Health; 10) Performance; 10) Education; 11) Perseverance; 12) Repair; 13) Study; 14) Environment. This is closely related to the dimensions of psychological well-being, where individuals are required to change situations that are less comfortable to be comfortable for themselves. In its application, individuals need to make environmental changes, positive relationships with others, self-acceptance, and self-development (Ryff & Keyes, 1995).

Aspects related to the adversity quotient certainly make individuals more resilient in facing the difficulties they will face. These aspects have a role that encourages individuals to remain strong and have endurance. Regarding psychological well-being, the aspect of the adversity quotient is directly related to the aspect of autonomy. It makes the individual able to struggle to achieve life goals even though the surrounding environment is less supportive (Hardianto & Sucihayati, 2019; Rahim, 2017). Increasing the adversity quotient can also be increased through the aspect of environmental control of the individual and the individual's way of achieving life goals.

From this aspect, psychological well-being certainly has a relationship with the adversity quotient so that the difficulties encountered can be resolved and the life goals of fresh graduates in looking for work can be achieved. Which it related to one dimension of psychological well-being is environmental mastery (Daretta, 2018). This dimension helps individuals to create an environmental condition that supports their psychological well-being. In comparison, the adversity quotient is an individual's ability to overcome a problem and survive in the situation (Parvathy & Praseeda, 2014). Both are related and support each other when an individual faces a problem.

According to Chadha (2021), the adversity quotient affects an individual's psychological well-being. The ability of an individual adversity quotient that is very helpful in dealing with problems will affect the steps taken by individuals for problem-solving and stress coping. It will affect the individual's level of psychological well-being. The results of this study are relevant to this study, where there is a significant positive relationship between psychological well-being and the adversity quotient.

According to the findings of this study, those who have a high psychological well-being also have a high adversity quotient. This is supported by data analysis results, which reveal a positive association between adversity quotient and psychological well-being. As a result, during the Coronavirus pandemic, there is a substantial association between psychological well-being and adversity quotient on young graduates. Despite the findings of this study, the researchers are also fully aware of the fact that this study has limitations. This study does not take into account the importance of external factors that could interfere psychological well-being but only focus to adversity quotient. Furthermore, the study is limited to the graduates of one public university in Indonesia, this might limit the generalizability of the findings of the present study to other conditions.

The adversity quotient is significant, especially for recent grads seeking for job in the middle of a coronavirus pandemic, because they are dealing with two difficulties at once: the coronavirus pandemic and how to find work. Individuals in the coronavirus situation require special elements and influencing variables, such as stress coping methods, to overcome the issues they face. As we know, this study found that the adversity quotient has an impact on people's psychological well-being. The more their adversity quotient, the greater their psychological well-being. The higher the adversity quotient they have, the higher their psychological well-being. All in all, it would be a proposal to emphasize the need of implementing problem-solving approach in order to increase students' resilience.

REFERENCES

Aprilia, E. D., & Khairiyah, Y. (2018). Optimisme menghadapi persaingan dunia kerja dan adversity quotient pada mahasiswa. SEURUNE: Jurnal Psikologi UNSYIAH, 1(1), 18–33. https://doi.org/https://doi.org/10.24815/s-jpu.v1i1.9922

Brissette, I., Scheier, M. F., & Carver, C. S. (2002). The role of optimism in social network development, coping, and psychological adjustment during a life transition. Journal of Personality and Social Psychology, 82(1), 102–111. https://doi.org/https://doi.org/10.1037/0022-3514.82.1.102

Chadha, N. (2021). Adversity quotient: Surviving rather than giving up. Psychology and Education Journal, 58(2), 5942–5947. https://doi.org/10.17762/pae.v58i2.3068

Daretta, S. (2018). Psychological well being pada korban kekerasan dalam rumah tangga [Universitas Medan Area]. http://repository.uma.ac.id/bitstream/123456789/10668/1/148600073 - sabrina daretta – Fulltext.pdf

Deci E. L., & Ryan R. M: Hedonia, eudaimonia, and well-being: An introduction. Journal of Happiness Studies 2008, 9: 1–11. 10.1007/s10902-006-9018-1

Drahs, S., Haywood, L., & Schiprowski, A. (2018). Job search with subjective wage expectations. SSRN Electronic Journal, 1–35. https://doi.org/10.2139/ssrn.3146764

Fauzia, M. (2020). Akibat covid-19, jumlah pengangguran RI bertambah 3,7 juta. KOMPAS.Com. https://money.kompas.com/read/2020/07/28/144900726/akibat-covid-19-jumlah-pengangguran-ri-bertambah-3-7-juta

Hardianto, Y., & Sucihayati, R. B. (2019). Hubungan adversity quotient dengan career adaptability pada koas angkatan 2015 FKG "X" di RSGM. Psibernetika, 11(2), 79–90. https://doi.org/10.30813/psibernetika.v11i2.1433

Kurniawan, W., Daharnis, D., & Karneli, Y. (2020). Contribution of adversity quotient, self awareness and demographic factors to student career maturity. International Journal of Research in Counseling and Education, 4(1), 70. https://doi.org/10.24036/00261za0002

Kusbandini, W., & Suprapti, V. (2014). Psychological Well Being Perempuan Dewasa Awal yang Pernah Mengalami Kekerasan Dalam Pacaran. Jurnal Psikologi Kepribadian Dan Sosial, 3(2), 80–92. http://www.journal.unair.ac.id/download-fullpapers-jpksb96837bbd0full.pdf
Mertens, D. M. (2015). Research and evaluation in education and psychology (H. Salmon, E. Oettinger, K. Guarino, & J. Haenel (eds.); 4th ed.). SAGE Publications.
Parvathy, U., & Praseeda, M. (2014). Relationship between adversity quotient and academic problems among student teachers. *IOSR Journal of Humanities and Social Science*, 19(11), 23–26. https://doi.org/10.9790/0837-191172326
Rachmady, T. M. N., & Aprilia, E. D. (2018). Hubungan adversity quotient dengan kecemasan menghadapi dunia kerja pada freshgraduate Universitas Syiah Kuala. Psikogenesis, 6(1), 54–60. https://www.academicjournal.yarsi.ac.id/index.php/Jurnal-Online- Psikogenesis/article/view/632
Rahim, A. (2017). Pengaruh konsep diri dan adversity quotient terhadap kemandirian santri. Fenomena, 16(1), 61–78. http://ejournal.iain-jember.ac.id/index. php/fenomena/article/view/669/547
Ryff, C. D. (1989a). Choosing Wisely campaign builds momentum. Journal of Personality and Social Psychology, 57(6), 1069–1081.
Ryff, C. D. (1989b). Happiness is everything, or is it? Explorations on the meaning of psychological well-being. Journal of Personality and Social Psychology, 57(6), 1069–1081. https://doi.org/ 10.1037/0022-3514.57.6.1069
Ryff, C. D, & Keyes, L. M. (1995). The structure of psychological well-being revisited. Journal of Personality and Social Psychology, 69(4), 719–727. https://doi.org/10.1037/0022-3514.69.4.719
Stoltz, P. G. (1997). Adversity quotient: Turning obstacles into opportunities. Jon Wiley & Sons, Inc.
Sugiyono. (2017). Metode Penelitian Pendidikan: Pendekatan Kuantitatif, Kualitatif, dan R&D. Alfa Beta.
Zainudin. (2011). Pentingnya adversity quotient dalam meraih prestasi belajar. Guru Membangun, 26(2), 1–10. https://media.neliti.com/media/publications/218112-pentingnya-adversity- quotient-dalam-mera.pdf
Zhao, Y., Sang, B., & Ding, C. (2021). The roles of emotional intelligence and adversity quotient in life satisfaction. Current Psychology. https://doi.org/10.1007/s12144-021-01398-z

Innovation on Education and Social Sciences – Maureen et al (Eds)

Neuroscience can contribute to change management: STREAP-Be model

M. Nguyen-Phuong-Mai
Amsterdam University of Applied Sciences, Amsterdam, The Netherlands

ABSTRACT: In our world of volatile, uncertain, complex, and ambiguous (VUCA), a capacity for change and adaptation is vital. However, changing successfully has been a challenging task for both individuals and organizations. Taking into account the insights of neuroscience, this chapter introduces a framework of change management called STREAP-Be. The acronym represents seven factors that could significantly influence the effect of change: safety, trigger, reward, emotion, alignment, people, and behavior.

1 CHANGE MODELS AND OBSTACLES IN THE CHANGE PROCESS

1.1 *Change Management Model*

Change management refers to approaches that respond to external and internal forces (Moran & Brightman, 2001). Many change models have been developed along with two broad categories. Firstly, there are models that aim at institutional change (e.g., Kotter, 1996; Lewin, 1947; Waterman, Peters, & Phillips, 1980). Secondly, there are models that support both individuals and organizational change processes (e.g., Duhigg, 2012; Kübler-Ross, 1969; Nguyen-Phuong-Mai, 2019; Thaler & Sunstein, 2009).

This chapter introduces the STREAP-Be model – a framework that incorporates insights from studies in social neuroscience, psychology, health behavioral change, and cognitive behavioral therapy. It's argued that such an interdisciplinary approach can provide new insight, hybrid solutions, and highlight areas where future research may focus on.

1.2 *Obstacles to change process*

Change is challenging. Only 25% of change initiatives are successful in the long term (Emerman, 2013). From the evolutionary neuroscience's point of view, the obstacles to change may come from three characteristics: the human brain tends to (1) save energy, (2) avoid uncertainty, and (3) imitate others.

Firstly, converting activities into habits saves energy. Doing something effortlessly can (1) free up the processing power for the brain's prefrontal cortex for complex tasks and (2) turn vital reactions into automatic routines that help us to stay alive. However, strong neural patterns also mean that once the brain sees a familiar signpost, it may revert to the old way because it's easier to follow an old route. This happens even when the habit is harmful (Seligman, 1972). Because change costs energy, it means past investment is lost (Kahneman & Tversky, 1992) and a switch cost has to occur to suppress the old and make way for the new neural pathway (Loose, Wisniewski, Rusconi, Goschke, & Haynes, 2017). This all costs energy. And thus, no pain no change, no pain no gain.

Secondly, avoiding uncertainty helps humans to survive. Safety means status quo and ambiguity implies potential dangers. Humans experience more stress when we don't know than when we know for sure that bad things will happen (de Berker, 2016). When we speculate with fear, we tend to imagine the worst, making ambiguity even more stressful.

DOI 10.1201/9781003265061-17

Finally, imitation is an essential part of social learning, which is vital for survival. In the brain, mirror neurons fire in both situations: when we act and when we see someone else acting (Ramachandran, 2012). These neurons help us to pick up behaviors and emotions, to communicate, and learn from each other. This also means change has to win over the tendency of mirroring old habits.

To conclude, change can be challenging because the brain tends to save energy by routines, prioritizes certainty, and is wired for imitation. The next section introduces a change model that takes these roadblocks for change into account. STREAP-Be stands for: Safety, Trigger, Reward, Emotion, Alignment, People, and Behavior.

2 STREAP-BE CHANGE MODEL

2.1 *The "S" of STREAP-Be: Safety*

Both interest and fear trigger change. However, due to survival purposes, the brain prioritizes fear to the extent that it may register it before consciousness (Burra et al., 2013). Fear is a powerful driver of change as it increases vigilance to avoid danger (Baumeister, Bratslavsky, Finkenauer, & Vohs, 2001). But it also impairs proactive and enhances reactive control (Yang, Miskovich, & Larson, 2018). In the long run, threatening messages may not lead to sustainable change (Ruiter et al., 2014), and change is more likely when people have the ability to respond (Peters et al., 2013).

This insight suggests that it is vital to identify fear issues and create a safe environment. A potential candidate to address this issue comes from the field of neuro-leadership: the SCARF model by Rock (2008). In general, there are five primary concerns that tap into the brain's emotional system: (1) Status: position in a hierarchy; (2) certainty: ability to predict future; (3) autonomy: sense of control; (4) relatedness: sense of attachment with other; and (5) fairness: perception of fair exchange.

An indication from this framework is that safety strategies may include clarity, which promotes one's "certainty" about the future. In fact, workforce engagement improves when supervisors have frequent communication with direct reports (Gallup, 2015). Similarly, the ability to make a choice is associated with "autonomy", a sense of being in charge, and hence, safety. Nearly half of employees would give up a 20% raise for greater control over how they work (Feintzeig, 2014).

Creating a safe environment for change also involves building trust. The brain subconsciously picks up mistrust signals without us knowing it (Freeman, Stolier, Ingbretsen, & Hehman, 2014). "Trust" and "change success" are closely associated (Vosse & Aliyu, 2018). A critical prerequisite for trust is fairness. The famous ultimatum game suggests that when a change is perceived as unfair, people can sabotage to punish, even if that means their own loss. However, by administering intranasal oxytocin – a hormone associated with both trust and distrust – the right social context could trigger people to become more generous and offered 80% more compared to placebo (;ikolajczak et al., 2010; Zak, Stanton, & Ahmadi., 2007).

2.2 *The "T" of STREAP-Be: Trigger*

Triggers are cues that activate habits (Duhigg, 2012). Even when the habit is gone, a trigger (e.g., cigarette) can bring back the old routine (smoking). Triggers are also explicit drivers of change such as competitors, technology, efficiency, or policies (Dawson, 2002).

With regard to triggers as habit activators, Duhigg (2014) referred to the "habit loop" which consists of a trigger, a routine, and a reward. For example, a mobile phone (trigger) activates the urge to check social media (routine) because it feels great (reward). To change this habit, one needs to identify the trigger, then experiment with different rewards. In short, to get rid of a habit, we need to acquire new ones: "Keep the same triggers and rewards as before, and feed the craving by inserting a new routine."

With regard to triggers as change drivers, new triggers can be created to promote change. For example, in bias management, people change their behaviors when their brain is primed with counter-stereotypes. Priming Asian women to think about gender decreased their math performance, but priming them to think about race increased it (Gibson, Losee, & Vitiello, 2014). This malleability of identities is supported by both behavioral (see Cameron, Brown-Iannuzzi, & Payne, 2012 for a review) and neural studies (e.g., Chiao et al., 2010).

However, a trigger is just one element of the habit loop. A trigger is significant because it's cued for a reward – which is the focus of the next session.

2.3 *The "R" of STREAP-Be: Reward*

Seeking pleasure and avoiding pain are important for survival. The brain constantly maximizes reward and minimizes punishment (Leknes & Tracey, 2008). Rewards provide positive reinforcement and shape desired behaviors (Milkovich & Newman, 2002).

In the brain, the anterior cingulate cortex (ACC) is associated with detecting errors and conflict. But when we are in pain and do not focus on it, the ACC detects a conflict, gets activated, and due to this distraction, we feel less pain (Sprenger et al., 2012). On the other hand, focusing on reward through positive activity interventions such as writing gratitude journals could rewire the brain and even treat depression (Layous, Chancellor, Lyubomirsky, Wang, & Doraiswamy, 2011). Furthermore, brain studies suggested that humans seem to learn better from success than from failure (Histed et al., 2009). Making a mistake is only rewarding when people have a chance to review the mistakes and learn from them in a safe environment because the brain sees this learning opportunity as a reward (Palminteri, Khamassi, Joffily, & Coricelli, 2015).

This insight suggests that focusing on the positive side and reward can help the brain recode the meaning of failure. Furthermore, change should be recoded as an opportunity to master new skills rather than a threat of exposing a lack of skill. In the same vein, people should be encouraged to imagine, visualize, talk about, and discuss the desirable result of the change. It's good to imagine possible roadblocks too, and how to overcome them. Mental rehearsal is a technique that taps into the power of self-fulfilling prophecy. Imagination acts as a precursor to action. Both stimulate the same brain region, thus enhancing motivation to achieve it (Clark et al., 2014; Vasquez & Buehler, 2007).

Finally, some studies in neuroscience suggest that time of reward is essential. Dopamine is a neurotransmitter associated with attention and motivation. It spikes because we anticipate rewards, but not necessarily when we receive it (cited in Kumar & Meenakshi, 2009). This indicates that fixed ratio and fixed variable rewards such as salary is less interpreted as "reward" in the brain. On the contrary, a surprised prize, bonus, or compliment can be effective because people don't know what reward and when they will receive it. So they continue to show effort, just in case.

2.4 *The "E" of STREAP-Be: Emotion*

The Latin root of emotion is movere (to move). In essence, emotion is the basis of motivation and the very reason why we want to change (Cameron & Green, 2019). Rationality does not separate from emotion and needs input from emotion to reach the final destination (Damasio, 2005). Evolutionary speaking, the limbic system was developed millions of years before the cerebral cortex, so the latter relies heavily on the former to act effectively.

This insight supports strategies associated with storytelling. Data can persuade people, but it may not inspire them to act (Monarth, 2014). The brain activates the same region when people hear the story and experience the event themselves (Gonzales et al., 2006). For survival purposes, humans have evolved to remember experiences better than unrelated facts. Storytelling is powerful because it influences us the way our actual life experience does. Hence, change should be in the form of stories.

However, emotions in the wrong context or with the wrong intensity can backfire. This is why dealing with undesired emotion is as important as strategies of seeing emotion as a resource. For example, while empathy creates trust and mutual understanding (Head, 2012), it can also lead to empathy fatigue (Stebnicki, 2007), confuse others' feelings with own feelings, and distort clear thinking (Galinsky, Maddux, Gilin, & White, 2008). For this reason, mindfulness is a strategic tool in emotion management, both at individual and organizational level.

2.5 *The "A" of STREAP-Be: Alignment*

In change management, goal alignment is a critical approach because conscious goals affect actions (Locke & Latham, 2002). Humans are motivated to achieve three main goals: "communion", "meaning" and "agency" (Talevich, Read, Walsh, Iyer, & Chopra, 2017). Alignment refers to the first goal, indicating the way people instinctively want to align with an in-group (Brewer, 1979). The brain is highly sensitive to ingroup-outgroup boundaries, to the extent that it could react automatically and subconsciously (van Bavel, Packer, & Cunningham, 2008). However, in-group boundary is "soft-wired". It takes a flip of a coin to prime the human's mind to feel belong to a new in-group (Tajfel, Billig, Bundy, & Flament, 1971), and it takes only a few minutes to deflate the tendency to categorize others by race built in a life's time (Kurzban, Tooby, & Cosmides, 2001).

Aligning with an in-group can be (re)created through a process called "recategorization" (see a review by Gaertner & Dovidio, 2005). Some neural studies reported that sharing a common ground help people to compete with other outgroups, influence their empathy (Han, 2015) and interactions (White, Abu-Rayya, & Weitzel, 2014). Creating alignment with an in-group can be based on many kinds of social connections: backgrounds, work benefits, hobbies, interests, emotions, values, and so forth. Habermacher (2011) even suggested an external competitor or threat as a common enemy in order to unite people.

2.6 *The "P" of STREAP-Be: People*

A critical aspect of change management focuses on people as "change agents" (Ulrich & Brockbank, 2005). These people are pro-active, goal-aligning, and action-oriented. They have skills, tools, and purpose to promote the transformation and rally forces for change (;artunek, 2014; Lunenburg, 2010).

The insight from studies on mirror neurons suggests a neural basis for the adage "walk the talk". People naturally look up to role models and modify their thoughts and behaviors as change agents demonstrate desired behaviors and lead the change by examples, they prime and trigger the imitation from others in the organization. This indicates that choosing the right change leaders and change agents is vital. Many organizations tend to appoint HR manager for the task. But by regarding the change as an HR project, this approach may involve the wrong change leaders and risk the lack change agents coming from both the top positions and the rank and file (Want, 2006). Other candidates for change agents are up-and-coming managers who see changes as opportunities to reach their ambition and career development.

2.7 *The "Be" of STREAP-Be: Behavior*

Studies in cognitive psychology suggested that once making a behavioral choice, people are more likely to associate their attitudes and behaviors with that choice (Nakamura & Kawabata, 2013; Harmon-Jones et al., 2015). The underlying concept for this is the brain's tendency to avoid cognitive dissonance, i.e., a conflict between thoughts and behaviors. A tactic in marketing called "foot-in-the-door" tells us that once people agree to go with the first step, they would feel an inner need to go all the way through, making their attitude consistent with their behavior (Harmon-Jones & Mills, 1999).

In mental health research, the power of strategic actions has a rich history under the term "behavioral activation". The theory underlying this approach is that rewarding activities is powerful for change (see Forbes, 2020 for a review in a clinical context). Change management can benefit from a vast resource of strategies associated with this approach, with concrete structures such as activity monitoring, assessment of goals and values, activity planning, skills training contingency management, and many other processes aiming at (non)verbal behaviors (see a review by Kanter et al., 2010).

Some specific strategies that involve the power of behaviors include: (1) short-term wins in which a goal is split up into smaller tasks, micro-deadlines, and mini rewards (Amabile & Kramer, 2017); (2) Intentionally break the pattern and act "as-if" one has achieved the desired aspect of oneself, therefore, opening up possibilities that one didn't know existed (Puett & Gross-Loh, 2016). In other words, fake it until you become it; (3) Identify desired behaviors and reward them consistently until they become a habit (Power, 2014); (4) Make "good enough" and not necessarily "perfect" decision can help to bring about control, calm the brain's region associated with emotion and engage the region associated with rationality (Korb, 2015), thus, focus on progress, not perfection; and (5) Use action oriented approach with a clear plan of what to do can help us see the choice we made more favourably, which in turn, will help us to follow through the decision easier (Harmon-Jones, Harmon-Jones, Fearn, Sigelman, & Johnson, 2008). In short, the mind leads the action, but action can change the mind as well. We are what we repeatedly do - Aristotle.

To conclude, this chapter has incorporated insights from neuroscience and briefly discussed the challenges in change management. It introduced the STREAP-Be model, which consists of 7 fundamental aspects of change and strategies that could be considered in a change process. This model can be applied at both individual and organizational level. Collective culture is similar to a person's habit in the sense that culture and habits are both persistent and evolving. Humans may find it difficult to change, but we are built to adapt, and we are the only the species that can do so deliberately.

REFERENCES

Amabile, T., & Kramer, S. (2017). "The Power of Small Wins." Accessed October 4. https://hbr.org/2011/05/the-power-of-small-wins.

Arun, K., & Meenakshi, N. (2009). *Organizational behaviour: A modern approach*. Vikas Publishing House. 186

Bartunek, J. M. (2014). *Organizational and educational change: The life and role of a change agent group*. Psychology Press.

Baumeister, R. F., Bratslavsky, E., Finkenauer, C., & Vohs, K. D. (2001). Bad is stronger than good. *Rev. Gen. Psychol.* 5, 323–370. doi: 10.1037/1089-2680.5.4.323

Brewer, M. B. (1979). In-group bias in the minimal intergroup situation: A cognitive-motivational analysis. Psychological Bulletin, 86, 307–324. doi:10.1037/0033-2909.86.2.307

Burra, N., Hervais-Adelman, A., Kerzel, D., Tamietto, M., De Gelder, B., & Pegna, A. J. (2013). Amygdala activation for eye contact despite complete cortical blindness. *J. Neurosci.* 33, 10483–10489. doi: 10.1523/JNEUROSCI.3994-12.2013

Cameron, C. D., Brown-Iannuzzi, J. L., & Payne, B. K. (2012). Sequential priming measures of implicit social cognition: A meta-analysis of associations with behavior and explicit attitudes. *Personality and Social Psychology Review, 16*, 330–350.

Cameron, E., & Green, M. (2019). *Making sense of change management: A complete guide to the models, tools and techniques of organizational change*. Kogan Page Publishers.

Chiao, J. Y., Harada, T., Komeda, H., Li, Z., Mano, Y., Saito, D., …& Iidaka, T. (2010). Dynamic cultural influences on neural representations of the self. *Journal of cognitive neuroscience, 22*(1), 1-11.

Clark, B. C., Mahato, N. K., Nakazawa, M., Law, T. D., & Thomas, J. S. (2014). The power of the mind: the cortex as a critical determinant of muscle strength/weakness. *Journal of neurophysiology, 112*(12), 3219–3226.

De Berker, A. O., Rutledge, R. B., Mathys, C., Marshall, L., Cross, G. F., Dolan, R. J., & Bestmann, S. (2016). Computations of uncertainty mediate acute stress responses in humans. *Nature communications, 7*(1), 1–11.

Duhigg, C. (2014). *The Power of Habit*. Random House Trade.

Emerman, (Ed.). (2013). Quarter of Employees Gain from Change Management Initiatives. *Willis Towers Watson*.

Feintzeig, R. (2014). Flexibility at Work: Worth Skipping a Raise? *The Wall Street Journal*. October 31.

Freeman, J. B., Stolier, R. M., Ingbretsen, Z. A., & Hehman, E. A. (2014). Amygdala responsivity to high-level social information from unseen faces. *Journal of Neuroscience, 34*(32), 10573–10581.

Gaertner, S. L., & Dovidio, J. F. (2005). Categorization, recategorization, and intergroup bias. *On the nature of prejudice: Fifty years after Allport*, 71–88.
Galinsky, A. D., Maddux, W. W., Gilin, D., & White, J. B. (2008). Why it pays to get inside the head of your opponent: The differential effects of perspective taking and empathy in negotiations. *Psychological science*, *19*(4), 378–384.
Gallup. 2015. "State of the American Manager: Analytics and Advice for Leaders report." *Gallup*. http://www.gallup.com/services/182138/state-american-manager.aspx.
Gibson, C. E., Losee, J., & Vitiello, C. (2014). A replication attempt of stereotype susceptibility (Shih, Pittinsky, & Ambady, 1999): Identity salience and shifts in quantitative performance. *Social Psychology*, *45*(3), 194.
González, J., Barros-Loscertales, A., Pulvermüller, F., Meseguer, V., Sanjuán, A., & Belloch, V. ávila, C.(2006). Reading cinnamon activates olfactory brain regions. *Neuroimage*, *32*, 906–912.
Habermacher, A. (2011). *Leading 100 Billion Neurons*. Smashwords.
Han, S. (2015). Intergroup relationship and empathy for others' pain: a social neuroscience approach. In *Neuroscience in intercultural contexts* (pp. 31–47). New York, NY: Springer.
Harmon-Jones, E., & Harmon-Jones, C. (2002). Testing the action-based model of cognitive dissonance: The effect of action orientation on postdecisional attitudes. *Personality and Social Psychology Bulletin*, *28*(6), 711–723.
Harmon-Jones, E., & J. Mills. (1999). *Cognitive dissonance: Progress on a pivotal theory in social psychology*. Washington, D. C.: American Psychological Association.
Harmon-Jones, E., Harmon-Jones, C., Fearn, M., Sigelman, J. D., & Johnson, P. (2008). Left frontal cortical activation and spreading of alternatives: tests of the action-based model of dissonance. *Journal of personality and social psychology*, *94*(1), 1.
Head, N. (2012). Transforming conflict: Trust, empathy, and dialogue. *International Journal of Peace Studies*, 33–55.
Kahneman, D., & Tversky, A. (1992). *Advances in prospect theory, cumulative representation of uncertainty*, 297–323.
Korb, A. (2015). *The upward spiral: Using neuroscience to reverse the course of depression, one small change at a time*. New Harbinger Publications.
Kotter, J. P. (1996). Leading change. Boston, MA: Harvard Business School Press.
Kübler-Ross, E. (1969). On death and dying: What the dying have to teach doctors, nurses, clergy and their own families. New York, NY: Macmillan.
Kurzban, R., Tooby, J., & Cosmides, L. (2001). Can race be erased? Coalitional computation and social categorization. Proc. Natl. Acad. Sci. U.S.A. 98, 15387–15392. doi: 10.1073/pnas.251541498
Layous, K., Chancellor, J., Lyubomirsky, S., Wang, L., & Doraiswamy, P. M. (2011). Delivering happiness: Translating positive psychology intervention research for treating major and minor depressive disorders. *The Journal of Alternative and Complementary Medicine*, *17*(8), 675–683.
Leknes, S., & Tracey, I. (2008). A common neurobiology for pain and pleasure. *Nature Reviews Neuroscience*, *9*(4), 314–320.
Lewin, K. (1947). Change management model. New York, NY: MacGraw Hill.
Locke, E. A., & Latham, G. P. (2002). Building a practically useful theory of goal setting and task motivation: A 35-year odyssey. *American psychologist*, *57*(9), 705.
Loose, L. S., Wisniewski, D., Rusconi, M., Goschke, T., & Haynes, J. D. (2017). Switch-independent task representations in frontal and parietal cortex. *Journal of Neuroscience*, *37*(33), 8033–8042.
Lunenburg, F. C. (2010). Managing change: The role of the change agent. *International journal of management, business, and administration*, *13*(1), 1–6.
Mikolajczak, M., Gross, J. J., Lane, A., Corneille, O., de Timary, P., & Luminet, O. (2010). Oxytocin makes people trusting, not gullible. *Psychological science*, *21*(8), 1072–1074.
Milkovich, G. T., & Newman, J. M., (2002). *Compensation*, 7th ed., McGraw Hill Irwin.
Monarth, H. (2014). The Irresistible Power of Storytelling as a Strategic Business Tool. *Harvard Business Review*. March 11. https://hbr.org/2014/03/the-irresistible-power-of-storytelling-as-a-strategic-business-tool.
Nguyen-Phuong-Mai, M. (2019). Change management: Creating culture and evolving yourself. *Cross-cultural Management: With Insights from Brain Science* (pp. 126–148). Routledge.
Palminteri, S., Khamassi, M., Joffily, M., & Coricelli, G. (2015). Contextual modulation of value signals in reward and punishment learning. *Nature communications*, *6*(1), 1-14.
Puett, M., & Gross-Loh, C. (2016). *The path: What Chinese philosophers can teach us about the good life*. Simon and Schuster.

Ramachandran, V. S. (2012). *The Tell-Tale Brain: A Neuroscientist's Quest for What Makes Us Human*. W. W. Norton & Company.

Rock, D. (2008). SCARF: A brain-based model for collaborating with and influencing others. *Neuro Leadership Journal*, no. 1.

Seligman, M.E. P. (1972). Learned helplessness. *Annual Review of Medicine* 23 (1): 407–12. doi:10.1146/annurev.me.23.020172.002203.

Sprenger, C., Eippert, F., Finsterbusch, J., Bingel, U., Rose, M., & Büchel, C. (2012). Attention modulates spinal cord responses to pain. *Current Biology*, *22*(11), 1019–1022.

Stebnicki, M. A. (2007). Empathy fatigue: Healing the mind, body, and spirit of professional counselors. *American Journal of Psychiatric Rehabilitation*, *10*(4), 317–338.

Tajfel, H., Billig, M. G., Bundy, R. P., & Flament, C. (1971). Social categorization and intergroup behaviour. *European Journal of Social Psychology*, 1, 149–178. doi:10.1002/ejsp.2420010202

Talevich, J. R., Read, S. J., Walsh, D. A., Iyer, R., & Chopra, G. (2017). Toward a comprehensive taxonomy of human motives. *PLoS One*, *12*(2), e0172279.

Thaler, R. H., & Sunstein, C. R. (2009). *Nudge: Improving decisions about health, wealth, and happiness*. Penguin.

Ulrich, D., & Brockbank, W. (2005). *The HR value proposition*. Harvard Business Press.

Van Bavel, J. J., Packer, D. J., & Cunningham, W. A. (2008). The neural substrates of in-group bias: A functional magnetic resonance imaging investigation. *Psychological Science*, *19*(11), 1131–1139.

Vasquez, N. A., & Buehler, R. (2007). Seeing future success: Does imagery perspective influence achievement motivation?. *Personality and social psychology bulletin*, *33*(10), 1392–1405.

Vosse, B. J. F., & Aliyu, O. A. (2018). Determinants of employee trust during organisational change in higher institutions. *Journal of Organizational Change Management*.

Want, J. H. 2006. *Corporate culture: Illuminating the black hole*. New York: St. Martins Press.

Waterman Jr, R. H., Peters, T. J., & Phillips, J. R. (1980). Structure is not organization. *Business horizons*, *23*(3), 14–26.

White, F. A., Abu-Rayya, H. M., & Weitzel, C. (2014). Achieving twelve-months of intergroup bias reduction: The dual identity-electronic contact (DIEC) experiment. *International Journal of Intercultural Relations*, *38*, 158–163.

Yang, Y., Miskovich, T. A., & Larson, C. L. (2018). State anxiety impairs proactive but enhances reactive control. *Frontiers in Psychology*, *9*, 2570.

Zak, P. J., Stanton, A. A., & Ahmadi, S. (2007). Oxytocin increases generosity in humans. *PLoS One*, *2*(11), e1128.

Innovation on Education and Social Sciences – Maureen et al (Eds)

Brief report: Visualizing scientific landscape on learning disabilities in Scopus between 2012 and 2021

C. Boonroungrut
Faculty of Education, Department of Psychology and Guidance, Silpakorn University, Bangkok, Thailand

W.P. Saroinsong
Faculty of Education, Department of Childhood Education, Universitas Negri Surabaya, Surabaya, Indonesia

N. Eiamnate
Administrative Management, Faculty of Public Administration, Huazhong University of Science and Technology, Wuhan, P.R. China

Sujarwanto
Special Education, Faculty of Education, Universitas Negeri Surabaya, Surabaya, Indonesia

ABSTRACT: The report aimed to investigate the Scopus database's scientific landscape on learning disabilities (LD). Two thousand seven hundred forty-eight research articles released between 2012 and 2021 were retrieved and reviewed. Bibliometric network analysis with VOSviewer was performed to create co-occurrence and co-authorship mappings. There are eight research clusters related to autism, mathematic disabilities, online, and COVID-19, which indicate a new trend after 2017. This report provides a complementary research review approach to receive a comprehensive view of what researchers have already done. A future review of the impact of COVID-19 and online learning on LD. students is suggested.

1 INTRODUCTION

People with learning disabilities (LD) have been defined by reading, writing, and mathematics difficulties. These difficulties could not be explained by physical conditions (APA, 2013). Therefore, the learning environment at all educational levels is the most important for creating inclusive and supportive learning. Several countries set the legal requirement to support students to study at a higher level. Thus, research papers examine learning difficulties during their secondary school, high school, and higher education levels, especially in the US. Therefore, Perelmutter, McGregor, and Gordon (2017) found that those countries published papers on how assistive technology can serve the adult learner's needs.

The number of research in the previous decade focuses on reading and writing difficulties more significant than mathematic difficulties (Doabler & Fien, 2013; Pirttimaa, Takala, & Ladonlahti, 2015; Quinn, 2018). Kang, McKenna, Arden, and Ciullo (2016) explain that literacy difficulties occur in early grades and exacerbate higher levels. The research found the correlation among those difficulties that most students have comorbid difficulties, not only a single one (Dechprom & Jermsittiparsert, 2018; Flanagan, Alfonso, Costa, Palma, & Leahy, 2018; Quinn, 2018). These difficulties can be continued in all grades and receive lower scores in various subjects.

To explore our literacy in LD within the current decade, LD experiences new issues at the forefront of changes in all education levels. This study presents a bibliometric network analysis for an initial step for reviewing all databases. This technique helps us determine which knowledge

 DOI 10.1201/9781003265061-18

clusters are effective for LD knowledge and the present research trend. Furthermore, reviewing research articles with this method provides an overview of extensive load documents leading readers to understand what the researchers have done or what is missing in our literacy. This technique is accepted in various fields, including sciences and art (Boonroungrut & Toe-Oo, 2017; Knutas, Hajikhani, Salminen, Ikonen, & Porras, 2015). Here this study will describe LD knowledge clusters and discussing the essential LD trend in each category.

2 METHODOLOGY

2.1 *Data sample*

The samples were obtained from the Scopus database from January 1, 2012 to June 30, 2021. The search term was "Learning Disabilities" OR "LD" which is explicit in article titles. The sample search was limited to English and journal or conference proceeding publication. They were published most in the social science and psychology area (approximately 69.5% of all records), followed by the medical and nursing area (23%). The samples were most published in 2019, as shown in Figure 1. Those records were downloaded as a CVS file. Silpakorn University digital resources were used to connect and access records from the Scopus database. Notably, even if this study included records from 2012, the first record was 1935 with terms indicating reading disabilities and word-blindness.

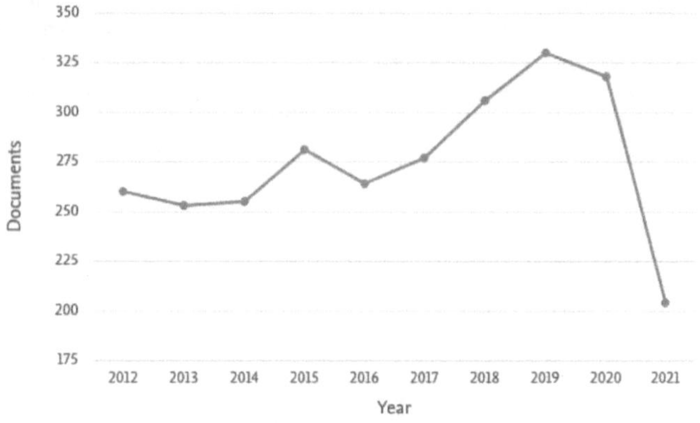

Figure 1. Number of documents by publication year.Source: Scopus.

2.2 *Analytical method*

The data was visualized by bibliometric analysis using various techniques, including co-occurrence analysis and co-authorship. VOSviewer software (version 1.6.13) was applied for the analysis with all default normalization settings. (to see how VOSviewer term calculation, please study Van Eck and Waltman's (2010) work and visit the website www.vosviewer.com for detail). No ethical approval is required for bibliometric studies.

3 FINDINGS

Two thousand seven hundred forty-eight records were focusing on the social sciences and psychology area. They were cited from 6,474 documents. The average citation was 2.35 citations per article. The publication growth was increasing overall, and the rate was polynomial cubic increasing ($r^2 = .64, p < .05$). Notably, the research publication number in 2021 reached the lowest point during this decade.

3.1 *Knowledge clustering*

In the co-occurrence mapping, there were 6,182 different terms used from researcher worldwide. This study set the minimum number of occurrences of a term at 10, then the result of 152 threshold was met. The term 'learning Disabilities' was removed for better clustering. The top 5 terms were student (occurrence = 604, relevance = .84), child (occurrence = 437, relevance = .69), person (occurrence = 393, relevance = 1.08), adult (occurrence = 149, relevance = .96), intellectual Disabilities (occurrence = 124, relevance = .59). The other terms indicate occurrence lower than 100.

The authors' co-occurrence term mapping has visualized eight recognized clusters, as shown in Figure 2. The red cluster ($n = 21$) at the left includes terms on care, service, treatment, view, woman, adult, access, change, community, and COVID-19. It was the most significant cluster compared to other clusters. The light green cluster ($n = 17$) at the right were terms related to school systems, including elementary school, secondary school, and high school, and terms related to specific learning Disabilities and mathematic learning and reading. At the top center, the blue cluster ($n = 12$) seemed related to psychological approaches, including attention deficit hyperactivity disorder, executive function, and loneliness. At the top left, the yellow cluster ($n = 11$) is related to several disorders such as autism and language disorder, including the nonverbal response. The violet cluster ($n = 11$) under the blue related to the terms in commentary, depression, severe disabilities, and trouble. Besides, other small 3 clusters, orange, light blue, and brown, included voice, reflection, machine, and online learning.

As presented in Figure 2, the latest terms were principally presented in the green and yellow clusters and a portion in those small clusters compared to Figure 3. Autism spectrum disorder was the newest term from the yellow and red clusters. The favored term of the orange cluster is multiple learning disabilities. Another cluster consists of the latest terms, such as mathematics Disabilities disorder. Online issues received significant interest from researchers worldwide. Lastly, research interest in learning disabilities during COVID-19 has been concluded as a current trend.

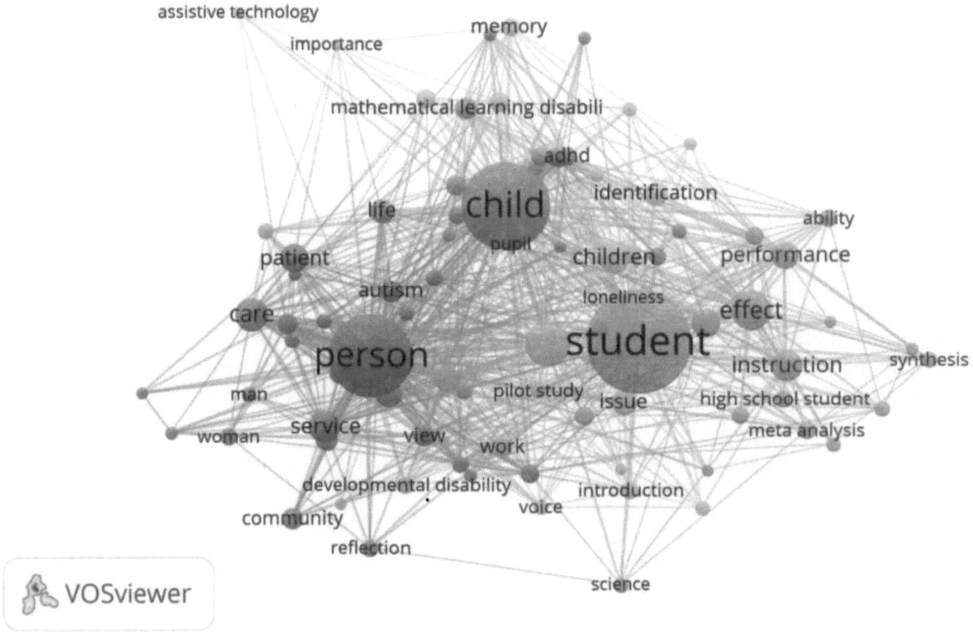

Figure 2. Co-occurrence mapping.

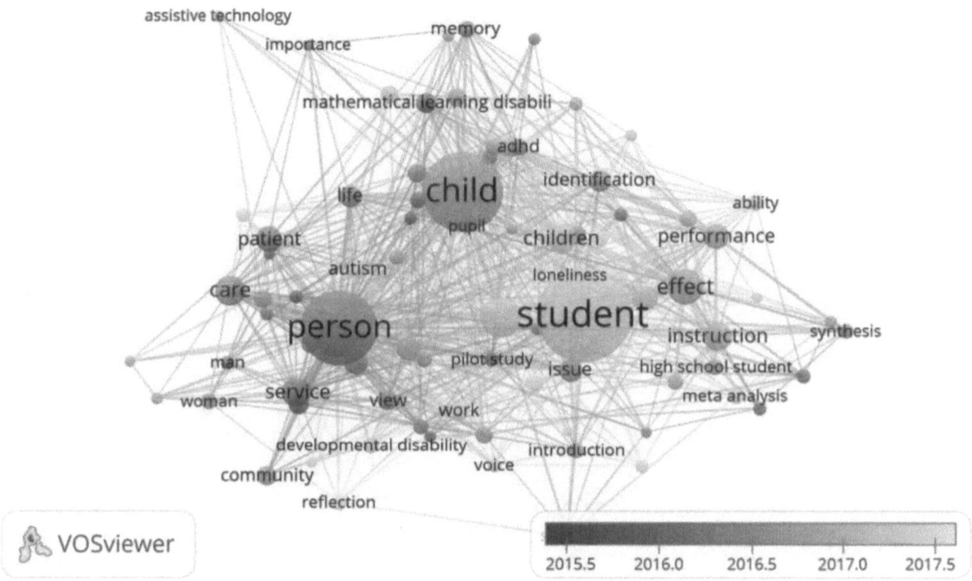

Figure 3. Publication year overlay mapping.

Table 1. List of top active authors.

No	Keywords	Occurrences	Citation	Link Strength
1	Bryant, D. P.	20	152	27
2	Hotton, C.	20	83	10
3	Gates, B.	20	40	0
4	Ciullo, S.	16	217	12
5	Mammarella, I. C.	16	293	8
6	Vaughn, S.	14	327	8
7	Bryant, B. R.	11	136	20
8	Shin, M.	11	96	14
9	Cornoldi, C.	11	252	10
10	Ok, M. W.	10	119	16

3.2 *Active co-authorship*

Total 6,588 authors and co-authors have published articles related to learning disabilities, as shown in Table 1. When setting the minimum number of documents of an author at 5, there were 71 that meet the thresholds. Among them, Bryant, D. P. from the University of Texas was the most productive author who focuses on mathematics difficulties and assistive technology for people with disabilities. The main collaborator with him is Bryant, B. R. which ranked in the 7th. The total like strength was 27. The main research clusters were the group of Bryant, D. P., Bryant, B. R., and Ok, M. W. and the group of Ciullo, S., and Shin, M., which Shin, M. was the linker of those two clusters consisted only 14 items. The co-authorship mapping declared a non-smell world effect in which almost all terms in the network were not connected as shown in Figure 4.

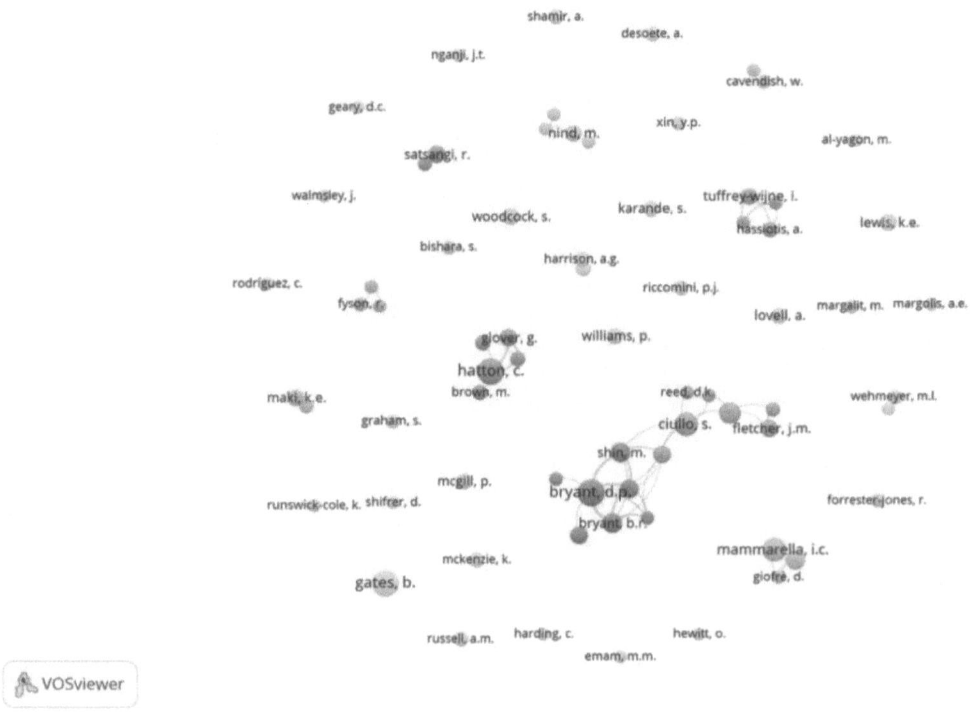

Figure 4. Co-authorship.

4 DISCUSSION

This study provides a bibliometric network overview of the literature on learning disabilities (LD) between 2012 and 2021 published in the Scopus database, which is the most extensive social science abstract index database. The number of publications related to LD has been increasing since 2016; however, the number of publications has been decreased since 2020. The authors from the US are the most active in publication. The note from co-authorship mapping showed the absence of small world collaboration among researchers worldwide.

Among the latest terms as the research trend, several topics could develop from review to meta-analysis in the future. Autism spectrum disorder (ASD) with LD seems to be growing. There are suggested that ASD and LD, predominantly non-verbal learning disabilities, could be the same spectrum disorders. However, the evidence is not sufficient to prove possible neuropsychological differences between these disorders. Several studies indicated that commonalities in deficits visual-spatial skill and nonverbal social cue (Kiln, Sparrow, Marans, Carter, & Volkmar, 2000; Semrud-Clikeman, Walkowiak, Wilkinson, & Christopher, 2010). However, neurocognitive research indicated different drawings fundamentally (Booth, Charlton, Hughes, & Happé, 2003; Booth & Happé, 2010). An autism-specific impairment was discovered. The findings show that children with ASD do not have a general lack of imagination; rather, their impairment is linked to planning demands (Allen & Craig, 2016). These dilemmas could be systematically explained.

During the COVID-19 pandemic, as shown in Figure 3, people with learning disabilities also found difficulties. The LD group may face more high risk during the pandemic, leading to restrictions on lives and mental health, which causes intense distress (Frankova, 2020; Picciani et al., 2020; Saroinsong, Reza, Khotimah, Sidiq, & Boonroungut, 2020). Speech perception is considered a process intergraded between auditory from sound generation and visual from lip and face movements, creating the McGurk effect. Children with normal hearing have been investigated, but less often among children with LD during online learning. In contrast, some researchers hypothesized that online teaching with computer-based material could provide a more suitable face and voice for the lecturer. It might be a solution to speech difficulties due to noise background and distance from the speaker (Kkese, 2020; Summerfield, 1992). In addition, the pandemic might enhance opportunities for social inclusion with increased power for families and key stakeholders for children with learning disabilities (Beaton, Codina, & Wharton, 2021). Finding from this report is crucial that further study of online benefit enhanced learning opportunities for them.

Notably, this brief report declared several limitations. First, this report included records only in the Scopus database. Publication in English is included in this analysis, so the other language articles are excluded. In conclusion, publications on learning disabilities have been increased in the present decades between 2012 and 2021. Most of the publications on terms of care, service, treatment, woman, adult, access, change, community, and COVID-19 with the latest terms on autism spectrum disorder, viewed online during COVID-19 as the current trend. Moreover, the small-world effect of research collaboration in this field is still absent.

REFERENCES

Allen, M. L., & Craig, E. (2016). Brief report: Imaginative drawing in children with autism spectrum disorder and learning disabilities. *Journal of Autism and Developmental Disorders*, *46*(2), 704–712. https://doi.org/10.1007/s10803-015-2599-y

American Psychiaristic Association. (2013). *Diagnostic and statistical manual of mental disorders* (5th ed). https://doi.org/10.1176/appi.books.9780890425596

Beaton, M. C., Codina, G. N., & Wharton, J. C. (2021). Decommissioning normal: Covid-19 as a disruptor of school norms for young people with learning disabilities. *British Journal of Learning Disabilities*, 1–10. https://doi.org/10.1111/bld.12399

Boonroungrut, C., & Toe-Oo, T. (2017). Dark triad trends in personality studies: Systematic review with bibliometric network analysis. *Journal of Humanities and Social Sciences Mahasarakarm University*, *36*(6), 63.

Booth, R., Charlton, R., Hughes, C., & Happé, F. (2003). Disentangling weak coherence and executive dysfunction: Planning drawing in autism and attention-deficit/hyperactivity disorder. Philosophical Transactions of the Royal Society of London. *Series B: Biological Sciences*, *35*(1), 387–392. https://doi.org/10.1098/rstb.2002.1204

Booth, R., & Happé, F. (2010). "Hunting with a knife and…fork": Examining central coherence in autism, attention-deficit/hyperactivity disorder, and typical development with a linguistic task. *Journal of Experimental Child Psychology*, *107*(4), 37–393. https://doi.org/10.1016/j.jecp.2010.06.003

Dechprom, S., & Jermsittiparsert, K. (2018). Foreign aid, foreign direct investment and social progress: A cross-countries analysis. *Opcion*, *34*(86), 208–2097.

Doabler, C. T., & Fien, H. (2013). Explicit mathematics instruction: What teachers can do for teaching students with mathematics difficulties. *Intervention in School and Clinic*, *48*(5), 276–285.

Flanagan, D. P., Alfonso, V. C., Costa, M., Palma, K., & Leahy, M. A. (2018). Use of ability tests in the identification of specific learning disabilities within the context of an operational definition. In D. P. Flanagan & E. M. McDonough (Eds.), *Contemporary intellectual assessment: Theories, tests, and issues* (pp. 608–642). The Guilford Press.

Frankova, H. (2020). The impact of covid-19 on people with autism, learning disabilities and mental health conditions. *Nursing and Residential Care*, *22*(6), 1–3. https://doi.org/10.12968/nrec.2020.22.6.10

Kang, E. Y., McKenna, J. W., Arden, S., & Ciullo, S. (2016). Integrated reading and writing interventions for students with learning disabilities: A review of the literature. *Learning Disabilities Research & Practice*, *31*(1), 2–33. https://doi.org/10.1111/ldrp.12091

Kiln, A., Sparrow, S., Marans, W., Carter, A., & Volkmar, F. (2000). Assessment issues in children and adolescents with Asperger syndrome. *Asperger Syndrome*, 30–339.

Kkese, E. (2020). McGurk effect and audiovisual speech perception in students with learning disabilities exposed to online teaching during the covid-19 pandemic. *Medical Hypotheses*, *144*, 110233. https://doi.org/10.1016/j.mehy.2020.110233

Knutas, A., Hajikhani, A., Salminen, J., Ikonen, J., & Porras, J. (2015). Cloud-based bibliometric analysis service for systematic mapping studies. *Proceedings of the 16th International Conference on Computer Systems and Technologies*, 184–191. https://doi.org/10.1145/2812428.2812442

Perelmutter, B., McGregor, K. K., & Gordon, K. R. (2017). Assistive technology interventions for adolescents and adults with learning disabilities: An evidence-based systematic review and meta-analysis. *Computers & Education*, *114*, 139–163. https://doi.org/10.1016/j.compedu.2017.06.005

Picciani, B. L. S., Bausen, A. G., Dos Santos, B. M., Marinho, M. A., Faria, M. B., Bastos, L. F., & Dziedzic, A. (2020). The challenges of dental care provision in patients with learning disabilities and special requirements during covid-19 pandemic. *Special Care in Dentistry*, *40*(5), 52–527. https://doi.org/10.1111/scd.12494

Pirttimaa, R., Takala, M., & Ladonlahti, T. (2015). Students in higher education with reading and writing difficulties. *Education Inquiry*, *6*(1), 5–23. https://doi.org/10.3402/edui.v6.2427

Quinn, J. M. (2018). Differential identification of females and males with reading difficulties: A meta-analysis. *Reading and Writing*, *31*(5), 1039–1061. https://doi.org/10.1007/s11145-018-9827-8

Saroinsong, W. P., Reza, M., Khotimah, N., Sidiq, B. A., & Boonroungut, C. (2020). A stress immunity system of covid-19 through academic stress. *Proceedings of the International Joint Conference on Arts and Humanities* (IJCAH 2020), 45–462. https://doi.org/10.2991/assehr.k.201201.082

Semrud-Clikeman, M., Walkowiak, J., Wilkinson, A., & Christopher, G. (2010). Neuropsychological differences among children with Asperger syndrome, nonverbal learning disabilities, attention deficit disorder, and controls. *Developmental Neuropsychology*, *35*(5), 58–600. https://doi.org/10.1080/87565641.2010.494747

Summerfield, Q. (1992). Lipreading and audio-visual speech perception. Philosophical Transactions of the Royal Society of London. *Series B: Biological Sciences*, *335*(1), 7–78. https://doi.org/10.1098/rstb.1992.0009

Van Eck, N. J., & Waltman, L. (2010). Software survey: VOSviewer, a computer program for bibliometric mapping. *Scientometrics*, *84*(2), 52–538. https://doi.org/10.1007/s11192-009-0146-3

Innovation on Education and Social Sciences – Maureen et al (Eds)

Power cards digital: An advanced strategy in enhancing communication skills of students with autism

M.N. Ashar, Wagino & A.O.N. Beny
Universitas Negeri Surabaya, Surabaya, Indonesia

ABSTRACT: This study was intended to examine the use of digital Power Cards for children with ASD to enhance their communication skills. A single-subject research design was applied. Data from one participant from one special school were collected through the frequency recording sheet. Data were then analyzed through the comparison between desirable behaviors and undesirable behaviors. The results reported showed that the undesirable behaviors were more frequent than desirable behaviors on the first week before the Digital Power Cards were used. In the next week, the frequency was different and the desirable behaviors were more frequent than the undesirable behaviors. In the last week, the same pattern as week 2 was evident. Therefore, it can be concluded that the recent innovation of a Power Cards strategy involving the use of technology could help a student with ASD in developing his communication skills.

1 INTRODUCTION

The term autism term was initially mentioned by Kanner (1943). His idea was developed from the Greek word "auto" which has meaning self. He reported that some individuals were socially indifferent and/or aloof (Kanner, 1943). Moreover, they were hesitant to change and had struggles in communication (Kanner, 1943). The following year, in 1944, Asperger (as cited in Wing, 1997) described a group consisting of children and teenagers who had similar traits as Kanner's group. Moreover, it was intriguing that many of them had superior intelligence, too (Wing, 1997).

The formal diagnosis for autism was officially launched 40 years after Kanner's first publication in which the criteria for diagnosing autism were explained in the Diagnosis and Statistical Manual of Mental Disorders – Third Edition (DSM-III; American Psychiatric Association, [APA], 1987). Those criteria were social disorder, speech difficulties, and stereotypical behavior (APA, 1987). The newest edition of the DSM-V criteria explains the criteria for diagnosing autism spectrum disorder (ASD) under problems associated with restrictive and repetitive patterns of behaviors, activities, interests, and social communication (APA, 2013).

ASD is prevalent in many countries (Centers for Disease Control & Prevention [CDC], 2014; MacDermot, Williams, Ridley, Glasson, & Wray, 2008; Saracino, Noseworthy, Steiman, Reisinger, & Fombonne, 2010; Sun & Allison, 2010). Specifically, in Asian territories, Sun and Allison (2010) explained that the number of children with ASD had risen remarkably, especially in Japan, Taiwan, and Israel. Moreover, in Indonesia, which has 260 million people, the Ministry of Women Empowerment and Child Protection (2018) explained that approximately there are 2.4 million people with ASD, with a growth rate of at least 500 individuals in a year.

One of the biggest challenges to be solved for children with ASD is improving their social interaction. Social interaction is hugely regarded as the vital aspect of the life quality of people with ASD (Verdugo, Navas, Go'mez, & Schalock, 2012). Likewise, the World Health Organization [WHO] (2001) said that social participation is a preferrable goal for individuals with health and behavior conditions, including children with ASD. However, some recent studies found that people with ASD rarely socialize with others (Chen, Bundy, Cordier, Chien, & Einfield, 2017; Hintzen,

DOI 10.1201/9781003265061-19

Delespaul, van Os, & Myin-Germeys, 2010). They tend to use their time in either parallel or solitary activities (Bauminger et al., 2008). For example, they prefer to play in a separate area from a friend during similar activities (Bauminger et al., 2008). Given this condition, it does not mean that people with ASD cannot interact with each other. Instead, they still prefer to socialize with others (Philip et al., 2012). Although people with ASD struggle to engage in social activities, some of them are eager to socialize and develop relationships with others (Howard, Cohn, & Orsmond, 2006; Müller, Schuler, & Yates, 2008). Indeed, one of the vital aspects supporting social interaction abilities is communication skill.

Communication skills among children with ASD are ranging. For example, some may have good receptive skills but lack vocabulary or speak many words and struggle to comprehend sentences (Fletcher-Watson & Happe, 2019). Therefore, children with ASD need an intervention to develop their communication skills. There are some interventions available to support communication development. One of those is Power Cards. A Power Card is a research-based strategy developed by Elisa Gagnon (2001) that applies a child's special interest, favorite celebrity, or character to help them understand goals, confirm choices, teach consequences for a specific behavior, or use relevant social responses.

A huge number of studies shows that the Power Cards strategy is relevant to enhance an ASD student's abilities in social communication. A study by Spencer, Simpson, Day, and Buster (2008), which applied the single-subject research design, showed that it is applicable to use Power Cards in improving the social abilities of ASD children. Moreover, a study on enhancing conversational skills among adolescents with ASD who were high-functioning reported that Power Cards significantly enhanced their skills in making conversation. Other studies have also investigated the use of Power Cards. For example, a study by Kryzak and Jones (2015) showed that by using both visual materials and prompting based on the students' interests, three participants with ASD could do a joint attention.

Furthermore, another study shows that the ability of two students with ASD in turn-taking improves because they were given the intervention which involved the Power Cards strategy combined with prompting (most-to-least prompting, gestural to verbal) (Daubert, Hornstein, and Tincani (2015). However, no study has examined the use of digital Power Cards in developing the communication skills of children with ASD. Therefore, this study intended to examine the use of digital Power Cards for children with ASD to enhance communication abilities, particularly in terms of expressive communication skills.

2 METHOD

2.1 *Research design*

This study applied a single-subject research design, which focused on examining one specific behavior of a small number of participants (Kazdin, 1982). In general, there were three stages to this study. The study started with preparation, including the design of the instrument to measure the target behavior, deciding the target behavior, i.e., saying "could you be silent please" while hit by friends, and designing the digital Power Cards. The next stage was evaluating the target participant behavior before, during, and after using digital Power Cards. The last stage was analyzing the data.

2.2 *Ethical consideration*

The Human Research Ethics Committee (HREC) of The Faculty of Education Universitas Negeri Surabaya granted the ethical approval. Materials were validated before they were given to participants. Firstly, the school's principal was given the Participant Information Statement (PIS) to gain permission. After the principal gave the permission, the principal asked students who met the inclusion criteria of this study to be the participants. Then, teachers helped the researcher to look for participants based on the previously designed criteria. Afterward, students and parents were given the PIS. Lastly, parents and students were provided the space to think about their involvement in this study. Once they agreed, the study was conducted.

2.3 *Participants*

This study applied the purposive sampling technique, in which researchers chose participants purposively based on designed criteria (Guest, Namey, & Mitchell, 2013). In addition, this study was applied the single-subject research design. Therefore there was one student selected as the participant in this study. He is a student with ASD who was studying in a special elementary school. He was in grade five and was considered a high-functioning student by his teacher. In the classroom, he had five classmates. All of his friends also had the ASD diagnosis. He usually learned in school from 7.30 AM. to 11.30 AM from Monday to Friday.

2.4 *Settings*

This study took place in one public special school in one large city in Indonesia. Data were collected during the morning circle session every day from Monday to Friday for three weeks.

2.5 *Data collection*

Data were collected through the frequency sheet (Table 1). In addition, data were collected on the researcher's second stage of the study during the Morning Circle activity for about three weeks.

Table 1. Frequency sheet.

FREQUENCY SHEET				
Participant Name	AA			
Date, Day	Minutes of the Morning Circle Activity	Undesirable Behaviour		Desirable Behaviour
		Number of hitting	Number of saying, "could you be silent please." and hitting	Number of saying: "could you be silent please"
Weekly-recap				

2.6 *Data analysis*

Data from the frequency recording sheet were counted and compared between the undesirable and desirable behaviors in each week.

3 FINDINGS AND DISCUSSION

3.1 *Preparation stage of the study*

During the preparation stage, the target behavior was decided, and the Power Cards Digital was created. In deciding the target behavior, the previous assessment result from the student (AA) was used to get the information. Based on the assessment, AA usually hit classmates in the Morning Circle session. Moreover, referring to the assessment, the cause of his behavior was while their classmates were being noisy. The behavior aimed to escape from sensory load (loud noises) and attract his classmates. Therefore, AA needs to be taught how to escape from that load and show attention by saying: "Could you be silent, please".

Front of Card	Back of Card
 **Mater loves his friends**	Mater loves to interact with his friends. Sometimes, his friends are noisy. When Mater feels annoyed, he likes to remind his friends to be silent. He loves his friends. Mater wants you to love your friends too. He wants you to say to your friends "Could you be silent please" If you are being annoyed

Figure 1. Printed power cards.

Figure 2. Digital power cards.

In creating the Power Cards Digital, teachers said that AA's favorite character was Matter from Cars animated movie. Then, by using the Pictello, the Power Cards Digital was designed (see Figures 1 and 2). The steps to be taken on using the Power Cards digital were as follows:

a. First Meeting
   1) Researchers read stories from the Power Cards to all students during the Morning Circle activity.
   2) Researchers taught AA to use Pictello, so that, when AA was being annoyed by his classmates, he could open up the Pictello app on his gadget and then do what his classmate said to AA.
b. Followed Meetings
   1) The researcher told AA right before AA entered the class on using the Power Cards on Pictello.
   2) The researcher helped AA when AA was annoyed by his classmates by saying, "What did he say to you that you feel annoyed?" followed by inviting him to open up the Pictello if he did not react.

3.2 *Evaluation stage of the study*

During this stage, researchers took data from the three different weeks. In the first week, AA was assessed with the frequency recording sheet without getting any Digital Power Cards. Then, the same instrument was used to evaluate AA's behavior while using the Digital Power Cards. On the last week. On each day, during the Morning Circle session, AA was observed for about 30 minutes.

3.3 *Analysis stage of the study*

Based on data from the frequency recording sheet, the undesirable behavior was more frequent than the desirable behavior in the first week, before the Digital Power Cards was used (figure 3). While, in the next week, the frequency was different; the desirable behavior was more frequent than the undesirable behavior (Figure 4). In the last week, the same pattern as week 2 was evident (Figure 5).

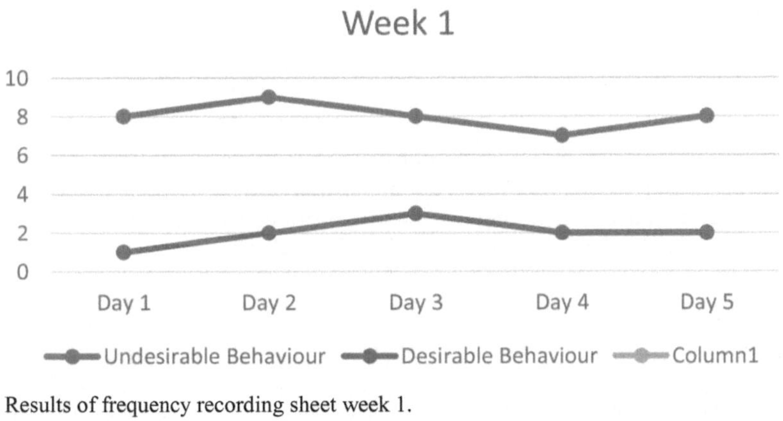

Figure 3. Results of frequency recording sheet week 1.

Figure 4. Results of frequency recording sheet week 2.

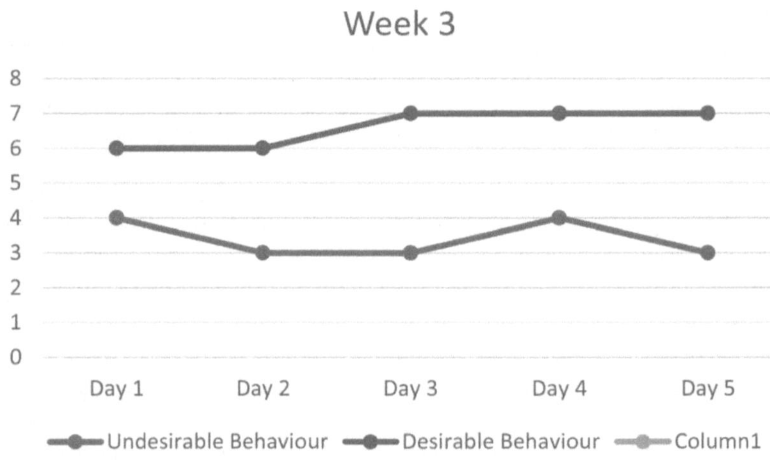

Figure 5. Results of frequency recording sheet week 3.

Therefore, it can be seen that the Digital Power Cards could help AA reach the target behavior, which is the role of the pragmatic communication. The statement is in line with the study from Kryzak and Jones (2015), which reported using both visual materials and prompting strategies with reference to the students' interest. All participants, the three children with special needs, were able to initiate the joint attention. Furthermore, research by Daubert, Hornstein, and Tincani (2015) showed that the skills of two students with ASD in turn-taking improved after they were given the intervention with the Power Cards strategy combined with a prompting strategy.

4 CONCLUSION

This study showed that the recent innovation of the Power Cards strategy involving technology could help a student with ASD develop his communication skills. It is evident that the frequency of desirable behaviors was higher during the use of the strategy and in the week after. However, some developments need to be made to improve the Digital Power Cards.

REFERENCES

American Psychiatric Association. (1987). Diagnostic and statistical manual of mental health disorders: DSM-III-R (3rd ed. rev.). Washington, DC: American Psychiatric Association.

American Psychiatric Association. (2013). Diagnostic and statistical manual of mental health disorders: DSM-V (5th ed.). Washington, DC: American Psychiatric Association.

Bauminger, N., Solomon, M., Aviezer, A., Heung, K., Gazit, L., Brown, J., & Rogers, S. (2008a). Children with autism and their friends: A multidimensional study of friendship in high-functioning autism spectrum disorder. Journal of Abnormal Child Psychology, 36(2), 135–150. DOI: 10.1007/s10802-007-9156-x

Centers for Disease Control and Prevention. (2014). Summary of autism spectrum disorder (ASD) prevalence studies. Morbidity and Mortality Weekly Report, 61, 1–6.

Chen, Y., Bundy, A., Cordier, R., Chien, Y., & Einfeld, S. (2017). A cross-cultural exploration of the everyday social participation of individuals with autism spectrum disorders in Australia and Taiwan: An experience sampling study. Autism, 21(2), 231–241. DOI: 10.1177/1362361316636756

Daubert, A., Hornstein, S., & Tincani, M. (2015). Effects of a modified power card strategy on turn-taking and social commenting of children with autism spectrum disorder playing board games. Journal of Developmental and Physical Disabilities, 27(1), 93–110. DOI: 10.1007/s10882-014-9403-3

Davis, K., Boon, R., Cihak, D., & Fore, C. (2010). Power cards to improve conversational skills in adolescents with Asperger syndrome. Focus on Autism and Other Developmental Disabilities, 25(1), 12–22. DOI: 10.1177/1088357609354299

Fletcher-Watson, S., & Happé, F. (2019). Autism: A new introduction to psychological theory and current debate. Routledge.

Gagnon, E. (2001). Power Cards. Using special interests to motívate children and youth with Asperger Syndrome and Autism.(Third Printing 2006). Shawnee Mission: Autism Asperger Publishing, Co.

Howard, B., Cohn, E., & Orsmond, G. (2006). Understanding and negotiating friendships: Perspectives from an adolescent with Asperger syndrome. Autism, 10(6), 619–627. DOI: 10.1177/1362361306068508

Hintzen, A., Delespaul, P., van Os, J., & Myin-Germeys, I. (2010). Social needs in daily life in adults with Pervasive Developmental Disorders. Psychiatry Research, 179(1), 75–80. DOI: 10.1016/j.psychres.2010.06.014

Kanner, L. (1943). Autistic disturbances of affective contact. Nervous Child, 2(3), 217–250.

Kazdin, A. E. (1982). Single-case research designs: Methods for clinical and applied settings. New York, NY: Oxford University Press

Kryzak, L., & Jones, E. (2015). The effect of prompts within embedded circumscribed interests to teach initiating joint attention in children with autism spectrum disorders. Journal of Developmental and Physical Disabilities, 27(3), 265–284. DOI: 10.1007/s10882-014-9414-0

MacDermott, S., Williams, K., Ridley, G., Glasson, E., & Wray, J. (2008). The prevalence of autism in Australia. Can it be established from existing data? Journal of Paediatrics and Child Health, 44, 504–510. DOI: 10.1111/j.1440-1754.2008.01331.x

Ministry of Women Empowerment and Child Protection Republic of Indonesia. (2018). Kenali dan deteksi dini individu dengan spektrum autisme melalui pendidikan keluarga untuk tingkatkan kualitas hidupnya [Knowing and early detecting individuals with autism spectrum through family-based education to improve their life]. Retrieved from https://www.kemenpppa.go.id/index.php/page/read/31/1682/hari-peduli-autisme- sedunia-kenali-gejalanya-pahami-keadaannya

Müller, E., Schuler, A., & Yates, G. (2008). Social challenges and supports from the perspective of individuals with Asperger syndrome and other autism spectrum disabilities. Autism, 12(2), 173–190. DOI: 10.1177/1362361307086664

Philip, R.C., Dauvermann, M.R., Whalley, H.C., Baynham, K., Lawrie, S. M., & Stanfield, A. C. (2012). A systematic review and meta-analysis of the fMRI investigation of autism spectrum disorders. Neuroscience and Biobehavioral Reviews, 36(2), 901–942.

Saracino, J., Noseworthy, J., Steiman, M., Reisinger, l., & Fombonne, E. (2010). Diagnostic and assessment issues in autism surveillance and prevalence. Journal of Developmental and Physical Disabilities, 22, 317–330. DOI: 10.1007/s10882-010-9205-1

Spencer, V., Simpson, C., Day, M., & Buster, E. (2008). using the power card strategy to teach social skills to a child with autism. TEACHING Exceptional Children Plus, 5(1), 1-10.

Sun, X., & Allison, C. (2010). A review of the prevalence of autism spectrum disorder in Asia. Research in Autism Spectrum Disorders, 4, 156–167. DOI: 10.1016/j.rasd.2009.10.003

Verdugo, M. A., Navas, P., Go ìmez, L. E., & Schalock, R. L. (2012). The concept of quality of life and its role in enhancing human rights in the field of intellectual disability. Journal of Intellectual Disability Research, 56(11), 1036–1045. doi:10.1111/j.1365-27 88.2012.01585.x.

World Health Organization. (2001). International classification of functioning, disability, and health. Geneva: World Health Organization.

Wing, L. (1997). The history of ideas on autism: Legends, myths and reality. Autism, 1(1), 13–23. DOI: 10.1177/1362361397011004

Linguistics and discourse analysis

Innovation on Education and Social Sciences – Maureen et al (Eds)
© 2022 copyright the Author(s), ISBN 978-1-032-20752-0
Open Access: www.taylorfrancis.com, CC BY-NC-ND 4.0 license

From crisis to opportunity: Promoting culture and language learning online

J.L. Hill
University of Szeged, Szeged, Hungary

ABSTRACT: In January 2020, no one could have predicted the rapid changes the world would undergo or the far-reaching effects it would have on education, predicating the adoption of innovative delivery methods. This paper investigates the published responses of more than 148 educators from 33 countries, analyzing their perspectives concerning the crises precipitated by the COVID-19 pandemic, documenting the opportunities the changes afforded them, and speculating on the sustainability of the new techniques and tools that came of age during the worldwide lockdown in promoting culture and equality in English language teaching.

1 INTRODUCTION

March 30, 2020, my journal might have read something like this:

> A month ago, I had not heard of coronavirus. I knew about SARS because I had just moved to Western China at its first outbreak. My school had not closed down, but, in an effort to restrict the movement of its foreign students, canceled our semester break. I am once again out of harm's way during a deadly virus, preparing to teach from home. A month ago, I had not heard of Zoom. Over the past two weeks, I have learned how to use it and organized a forum on Zoom in which my colleagues and I can swap the tips and tricks we have learned and share ideas for lesson plans. Today, I will teach my first class on Zoom, for which I will dedicate a significant portion of the lesson time to bringing the students up to speed on how to use technology. Nearly a year ago, when I was promoted to distance learning coordinator, my duties required no use of my training in curriculum development for virtual learning. I signed students up for distance learning programs that had been developed commercially and approved by the state. I had face-to-face workshops in which I prepared them to use the curriculum and encouraged them to be persistent. At the end of the month, I monitored their use and wrote a report. A month ago, if we had received a call from someone who wanted to enroll in fully online courses, we would have encouraged them to join face to face. Today, we are trying to move enrollment and all face to face classes online. So much has changed in just thirty days.

Ours was not the only school that underwent swift change during the global pandemic. All over the world, university students were sent home (Crawford et al., 2020) and schools for primary and secondary students were shut down (UNESCO, 2021). At first, administrators thought that it would be for only a few weeks. A few weeks stretched out to a few months. We had to learn to do testing remotely. We devised ways to do registration remotely. A year later, in the adult education center in Texas where I worked, most enrollments had gone back to face to face, but half of testing was still online and more than seventy-five percent of the students participated in classes remotely. What had begun as an emergency response on March 30, 2020 had become the school's new routine. Would this new routine be more commendatory towards increasing the students' cultural understanding and intercultural competence and promoting equality among them? If so, would it be sustained by the teachers when they safely returned to the physical classrooms?

DOI 10.1201/9781003265061-20

2 PURPOSE OF STUDY

The purpose of this study is to explore the impact of the pandemic on education around the globe from the perspective of teachers and researchers who lived and worked through it. This study seeks to synthesize reports of the challenges teachers faced and the tools and techniques that they implemented in their classrooms. Hopefully, the findings will inspire English teachers to expand their teacher toolbox and equip them to help their students acquire the digital skills needed to open the door to other opportunities as a means to increase their English proficiency.

3 METHODOLOGY

To define the global experience of English teachers during the pandemic and to explore their responses to it, this study took a thematic approach in analyzing published research.

3.1 *Data collection*

The published work of 148 authors, that is sixty-nine articles, were examined in this study. Thirty-five articles were authored by language teachers; twenty of those, English language teachers. The authors come from thirty-three nations.

The investigation began with the Educational Resource Information Center (ERIC.ed.gov) database. The keywords COVID and ESL/EFL registered four hits. The articles were downloaded and read, with special attention given to the references, which often provided other articles referencing the pandemic, though not always referencing English language teaching. While reading each downloaded article, the researcher made note of the different themes authors mentioned as concerns brought about by the pandemic, recording these on a spreadsheet. Each article was coded by the nationality of its author(s); whether they taught English, another language, or another subject; and the topics that most concerned them.

3.2 *Data analysis and guiding questions*

After the first reading of the sixty-nine articles, the researcher identified the most common themes: connectivity, professional development, student engagement, and anxiety. She reread the twenty articles related to ESL a second time, making note of when, how frequently, and how they dealt with the four common topics. Specific techniques for online language teaching and learning were also noted and organized into an outline.

1. Which challenges brought about by the lockdown most affected English language teachers? Did they respond in similar ways as non-language teachers?
2. What innovative techniques did language teachers develop during the pandemic? Did these address issues of culture and equality?
3. Can English teachers continue to incorporate the use of these techniques and resources when they return to the physical classroom?

4 RESULTS AND DISCUSSION

Most of the English teachers surveyed in these studies are supportive of the use of technology. An English teacher in Thailand, Kawinkoonlasate (2020) insists that technology is incredibly important: "EdTech is not some small oft overlooked business sector, it is a thriving sector of the economy and it should surprise nobody that EdTech provides a multitude of online and offline resources capable of improving the educational paradigm" (p. 24) Shahzad and his team of university professors from linguistics, business, and engineering professors seeking to promote e-learning in Pakistan (2020) assert, "E-teaching and learning are trending in the world" (p. 8). It is the perceived importance of distance learning that allows some to endure the lack of funding or computers (Wen & Tan, 2020) or find work-arounds for pre-service teachers whose access to the classroom has been cut off by the pandemic (Cho & Clark-Gareca, 2020).

4.1 *Challenges*

The pandemic has brought about a new term in virtual education: emergency response teaching (ERT). This word was coined to differentiate the type of online teaching that resulted from pandemic protocols and carefully planned distance education curricula (Hodges, Moore, Lockey, Trust, & Bond, 2020). Although the concept is present throughout the twenty articles by English language teachers, it is only mentioned twice by its new name. Talidong (2020) emphasizes the need for more speed-planning that is, having a Plan B and Plan C ready just in case the first plan falls through. Korean researchers refer to it as a sudden transition:

> The sudden transition to online learning has posed various challenges and difficulties for language instructors and administrators at GILC who needed to not only quickly switch to online course delivery but also create interactive and sustainable online language learning environments, with little to no time to prepare and with limited professional preparation for an online approach to teaching (Chung & Choi, 2021, p. 7).

It should be noted that these researchers were not concerned with the inconvenience of swift planning but in building a sustainable system of assessment to complement online learning.

Like Chung and Choi, compatriot researchers (Yi & Jiang, 2020) also responded in a way that illustrates their flexibility and shows that they regard the change as opportunity rather than a threat: "The shift from face-to-face to remote teaching and learning has challenged all of us—teachers, students, parents, and administrators. Such challenges reassert the opportunities to leverage our experience, interest, and knowledge as resources" (p. 4).

The response of the Korean researchers is characteristic of the attitude shown in the twenty articles by researchers that will be examined in this investigation. It contrasts sharply with a Filipino educator (Alvarez, Jr., 2020) who used words like disruption and suicide to describe the "sudden shift" from traditional classrooms to online learning. English teachers simply referred to connectivity and professional development as challenges. Problems that other teachers seemed to face in student engagement and anxiety were hardly mentioned in the research about ERT for English classes.

4.1.1 *Connectivity and access*

Six of the twenty articles mentioned difficulties in getting on the internet because it was down, the teacher or student didn't know how to log on, or the teacher or student did not have a device to connect to the internet (Talidong, 2020) For example, Mahyoob's (2021) Saudi students struggled with using the LMS. Malaysian teachers lacked resources (Cho & Clark-Gareca, 2020) as did American immigrants (Sayer & Braun, 2020).

While connectivity was a problem both for English teachers and their students, researchers mentioned it in passing. For example, Talidong, a Filipino teacher in China, was much more concerned about how well the teachers were able to implement remote teaching (Talidong, 2020). Students in other studies noticed issues of poor internet connections, but these issues did not detract from online learning as being a "personal and meaningful experience" (Pasaribu & Dewi, 2021, p. 420) for some Indonesian students or a "new but happy experience" (Shahzad et al., 2020, p. 8) for a group of Pakistani students.

Concerning connectivity or access to the internet, the disparity was not between one culture and another or between English teachers and other educators. Not surprisingly, the marginalization occurred between rural and urban schools. Wen and Tan (2020) observed, "Teachers who are teaching in rural schools might suffer from weak internet connections" (p. 392). This situation in a rural county in Mexico was such that teachers working in these underserved communities called their students on the phone and emailed those worksheets (Padilla Rodriguez et al., 2021).

4.1.2 *Professional development*

Four of the twenty articles alluded to professional development. Non-English teachers and English teachers alike encountered a need for teachers to be trained. Indian students recognized that their teachers needed training. English teachers recognized the need for themselves. Korean teachers trying out a new method for assessment, surmised that with a little more training they could more closely align learning and assessment (Chung & Choi, 2021). Malaysian teachers recognized that with training in online learning tools they would gain more confidence in conducting online teaching (Wen & Tan, 2020).

Two other researchers took another approach. Ghouname (2020) suggested a need to persuade the Algerian government and administrators to accept social networking. Kawinkoonlasate (2020) argued that daily technology should be integrated into English language programs in Thailand.

4.1.3 *Student engagement*

Five of the articles considered student engagement. Unlike the results of studies in general education which cite barriers to engagement, for online language classes, engagement does not seem to be a challenge. The issues discussed are completely different. Findings regarding student engagement are quite different.

Teachers of other subjects in Turkey had students who had difficulty paying attention. One researcher found that teachers of the Turkish language considered online language classes as being inefficient (Aiden & Erol, 2021). In another study, Turkish teachers reported "a lack of distance education due to students' internet connection problems, the inability of students to maintain their learning motivation, the inability of parents to create a learning environment at home, and their inadequate support for their children" (Aytac, 2020, p. 416).

Students of the English language reported enjoying online learning. Omani students "are interested in online learning and hold a positive attitude towards it" (Salih, Ferdinansyah, Sembilan, Sembilan, & Zahro, 2020, p. 72). Eighty-five percent of the Filipino teachers surveyed in China had no trouble engaging their students (Talidong, 2020). Indonesian researchers found that in spite of technical problems, Indonesian students had a positive attitude toward language learning (Ferdiansyah, Ridno, Sembilan, Sembilan, & Zahro, 2020). Pakistani students also enjoyed online learning (Shahzad et al., 2020).

Kawinkoonlasate (2020) surmised that technology increases engagement in language learning:

> When technology integration in the classroom is seamless, thoughtful, and well prepared, students become more engaged, begin to take more control over their own learning, and have a tendency to enjoy the learning process as a whole more (p. 23).

4.1.4 *Anxiety*

Another theme that was uncovered in the texts of general education but not for English teaching is that of anxiety. A Filipino researcher found that students were anxious about their family's financial situation and the lack of emotional support (Alvarez, Jr., 2020). Indian students felt uneasy about lack of privacy on the internet (Bhaumik & Priyadarshini, 2020)

Only two studies of the twenty articles from English teachers mentioned anxiety. English students seemed to report less anxiety. Pasaribu and Dewi (2021) found that Indonesian students' anxiety lessened and asserted, "The dominant use of Affect markers shows that online learning has given the students a deeply personal and meaningful experience during this exceptional upheaval" (p. 420). ESL students in the UK also coped well (Thomas, Lucksi, & McCulloch, 2020).

4.2 *Techniques and tools*

"As we find ourselves being gently nudged and pushed to explore new practices and tools, we might be surprised by the innovation and creative solutions that can result from this time of uncertainty and change," wrote Lomika (2020, p. 311) in a special issue of the *Foreign Language Annals*. Indeed, from the thirty-five articles by language teachers, five innovative techniques appear to have come of age: telecollaboration, tandem language learning, virtual literature circles, video conferencing, and games.

4.2.1 *Telecollaboration*

Telecollaboration is not a particularly novel development in language education. During the pandemic, however, it was one technique that could be easily incorporated without having to make major adjustments. Freiermuth and Huang (2021), English teachers in Japan and Taiwan, respectively, assigned their students the task of introducing their country to the other. They found that telecollaboration aided in the development of intercultural competence.

Telecollaboration has also been employed in teaching writing. In Turkey, Balaman (2021) examined the collaboration of students completing a writing assignment using Google Docs. The researcher suggested that collaborative writing helps students put their knowledge of grammar and vocabulary into practice.

4.2.2 *Tandem language learning*

Tandem learning also employs partnering students of different cultures together. However, whereas in telecollaboration one language predominates, in tandem language learning partners take turns communicating in their native tongues. At the Harvard Language Center before the pandemic, the technique was used pairing international students and visiting professors and their families with students who wanted to learn their language. During the lockdown, some of these partnerships moved online (Ross & DiSalvo, 2020).

Thor Sawin of Middlebury Institute in California (cited in Guillen, Sawin, & Aveniri, 2020) demonstrates how tandem partners have gone one step further into the digital age. Apps are now available for tandem language learning. This has the added benefit of moving language learning into the learner's court. They can take the initiative to find their own tandem partner.

4.2.3 *Literature circles*

In Indonesia, the lockdown forced both teachers and students to get creative (Ferdiansyah et al., 2020). To build students' reading comprehension, teachers have long used literature circles to help students process the English texts. The Literature Circles moved online and were arranged through WhatsApp because of the pandemic. To further ensure that the students understood the texts, Ferdiansyah et al. (2020) also incorporated an assignment in which students made digital posters about the texts, which they shared with their classmates.

4.2.4 *Video-conferencing*

The pandemic thrust video-conferencing to the forefront for language education. Through Skype, it was possible to meet in small groups before the pandemic, but Zoom made it possible to meet with larger groups. Digital whiteboards, screen-sharing capabilities, breakout rooms, and accessibility made Zoom a natural choice for language teachers. In Hong Kong, English teachers created asynchronous activities for students to do before and after the video-conference (Moorhouse & Beaumont, 2020).

4.2.5 *Games*

When a French class in Carnegie Mellon had to move online, Dubreil (2020) had to make some adjustments to the curriculum. The theme of the semester was games and the students were all assigned the project of creating party games and board games in French. Through a combination of video-conferencing tools and apps for creating games, Dubreil's class went digital.

5 IMPLICATIONS

As we English educators enter physical classrooms in the aftermath of the pandemic, which of the practices that we learned in our journey of online teaching will we continue? Technology offers us many opportunities to flexibly address the needs of the individual student, equalizing their access to knowledge about culture and language in ways that are not possible in a traditional teacher-fronted classroom.

5.1 *Culture*

Fifty years ago, access to other cultures was limited to reading about the culture, traveling if one had the resources, and perhaps pen pals. Nowadays, through technology, language teachers can connect students to real people in real time. Could students in Taiwan connect with students in Japan while sitting in a traditional brick and mortar classroom? The push precipitated by the pandemic could potentially open a whole new world of learning about culture in which students do not have to depend on the interpretation of teachers or textbooks, but can draw their own conclusions about culture.

5.2 *Equality*

Through technology, students have equal access to knowledge about language and culture. Unfortunately, people do not yet have equal access to reliable internet services or devices to connect them to that knowledge. There are many apps and open educational resources available to users of the internet. It may be that we need to find ways toward equitable access and develop means of professional development to give teachers the competence and confidence they need to connect language learners to the resources that are out there.

5.3 *Sustainability*

John Hattie, professor emeritus at the University of Melbourne wrote, “Perhaps the greatest tragedy to come from COVID-related distance learning would be *not* learning from this experience to improve our teaching when we physically return to the classrooms” (Hattie, 2021, p. 14).

What happens next? Having experienced the possibilities of telecollaboration, videoconferencing, and language apps will language teachers go back to the normal of two years ago, or will they incorporate what they have learned from the experience, inviting students to bring their devices into the classroom to explore how to make them into educational tools?

REFERENCES

Aiden, E., & Erol, S. (2021). The views of Turkish language teachers on distance education and digital literacy during Covid-19 Pandemic. *International Journal of Education and Literacy Studies*, *9*(1), 60–71. DOI: http://dx.doi.org/10.7575/aiac.ijels.v.9n.1p.60

Alvarez, Jr., A. (2020). The phenomenon of learning at a distance through emergency remote teaching amidst the pandemic crisis. *Asian Journal of Distance Education*, *15*(1), 144–153. DOI: https://doi.org/10.5281/zenodo.3881529

Aytac, T. (2020). The problems faced by teachers in Turkey during the COVID-19 Pandemic and their opinions. *International Journal of Progressive Education*, *17*(1), 404–420. DOI: http://dx.doi.org/10.7575/aiac.ijels.v.9n.1p.60

Balaman, U. (2021). The interactional organization of video-mediated collaborative writing: Focus on repair practices. *TESOL Quarterly*, *55*(1). DOI: ttps://doi.org/10.1002/tesq.3034

Bhaumik, R., & Priyadarshini, A. (2020). E-readiness of senior secondary school learners to online learning transition amid COVID-19 lockdown. *Asian Journal of Distance Education*, *15*(1), 244–256. http://www.asianjde.com/ojs/index.php/AsianJDE/article/view/456

Cho, S., & Clark-Gareca, B. (2020). Approximating and innovating field experiences of ESOL preservice teachers: The effects of COVID-19 and school closures. *TESOL Journal*, *11*, e548. DOI: https://doi.org/10.1002/tesj.548

Chung, S., & Choi, L. (2021). The development of sustainable assessment during the COVID-19 Pandemic: The Case of the English Language Program in South Korea. *Sustainability*, *13*(8), 4499. DOI: https://doi.org/10.3390/su13084499

Crawford, J., Butler-Henderson, K., Rudolph, J., Malkawi, B., Glowatz, M., Burton, R., Magni, P., & Lam, S. (2020). COVID-19: 20 countries' higher education intra-period digital technology responses. *Journal of Applied Learning & Teaching*, *3*(1), 9–28. DOI: https://doi.org/10.37074/jalt.2020.3.1.7

Dubreil, S. (2020). Using games for language learning in the age of social distancing. *Foreign Language Annals*, *53*(2), 250–259. DOI: https://doi.org/10.1111/flan12465

Ferdiansyah, S., Ridno, M. A., Sembilan, F. D., Sembilan, F. D., & Zahro, S. F. (2020). Online literature circles during the COVID-19 pandemic: engaging undergraduate students in Indonesia. *TESOL Journal*, *11*, e544. DOI: https://doi.org/10.1002/tesj.544

Freiermuth, M., & Huang, H. (2021). Zooming across cultures: Can a telecollaborative video exchange between language learning partners further the development of intercultural competences? *Foreign Language Annals*, *54*(1), 185–206. DOI: https://doi.org/10.1111/flan.12504

Ghouname, N. (2020). Moodle or Social Networks: What Alternative Refuge is Appropriate to Algerian EFL Students to Learn during Covid-19 Pandemic. *Arab World English Journal*, *11*(3), 21–41. DOI: https://dx.doi.org/10.24093/awej/vol11no3.2

Guillen, G., Sawin, T., & Aveniri, N. (2020). Zooming out of the crisis: Language and human collaboration. *Foreign Language Annals*, *53*, 320–328. DOI: http://doi.org/10.1111/flan.12459

Hattie, J. (2021). What can we learn from COVID-Era instruction? *Educational Leadership*, *78*(8), 14–17. http://www.ascd.org/publications/educational_leadership/may21/vol78/num08/What_Can_We_Learn_from_COVID-Era_Instruction%C2%A2.aspx

Hodges, C., Moore, S., Lockey, B., Trust, T., & Bond, A. (2020, March 27). The difference between emergency remote teaching and online learning. Educause. *Educause*. DOI: https://er.educause.edu/articles/2020/3/the-difference-between-emergency-remote-teaching-and-online-learning

Kawinkoonlasate, P. (2020). Online language learning for Thai EFL learners: an analysis of effective alternative learning methods in response to the COVID-19 Outbreak. *English Language Teaching*, *13*(12), 15–26. https://files.eric.ed.gov/fulltext/EJ1279886.pdf. DOI: https://doi.org/10.5539/elt.v13n12p15

Lomika, L. (2020). Creating and sustaining virtual language communities. *Foreign Language Annals*, *53*, 306–313. DOI: https://doi.org/10.1111/flan.12456

Mahyoob, M. (2021). Challenges of e-Learning during the COVID-19 Pandemic experienced by EFL learners. *Arab World English Journal*, *11*(4), 351–362. DOI: https://dx.doi.org/10.24093/awejvol1no4.23

Moorhouse, B., & Beaumont, A. (2020). Utilizing video conferencing software to teach young learners in Hong Kong during the COVID-19 class suspensions. *TESOL Journal*, *11*, e545. TESOL Journal 11 (2020) e545. DOI: https://doi.org/10.1002/tesj.545

Padilla Rodriguez, B. C., Armellini, A., & Traxler, J. (2021). The forgotten ones: How rural teachers in Mexico are facing the COVID-19 pandemic. *Online Learning*, *25*(1), 253–268. DOI: https://doi.org/10.24059/olj.v25i1.2453

Pasaribu, T. A., & Dewi, N. (2021). Indonesian EFL students' voices on online learning during COVID-19 through appraisal analysis. *LEARN Journal: Language Education and Acquisition Research Network*, *14*(1), 399–426. https://files.eric.ed.gov/fulltext/EJ1284575.pdf. https://so04.tci-thaijo.org/index.php/LEARN/index

Ross, A., & DiSalvo, M. (2020). Negotiating displacement, regaining community: The Harvard Language Center's response to the COVID-19 crisis. *Foreign Language Annals*, *53*, 371–379. DOI: https://doi.org/10.1111/flan.12463

Salih, A. A., Ferdinansyah, S., Sembilan, F. D., Sembilan, F. D., & Zahro, S. F. (2020). Online literature circles during the COVID-19 pandemic: engaging undergraduate students in Indonesia. *TESOL Journal*, *11*, e544. DOI: https://doi.org/10.1002/tesj.544

Sayer, P., & Braun, D. (2020). The disparate impact of COVID-19 remote learning on English learners in the United States. *TESOL Journal*, *11*, e546. DOI: https://doi.org/10.1002/tesj.546

Shahzad, S. K., Hussain, J., Sadaf, N., Sarwat, S., Ghani, U., & Saleem, R. (2020). Impact of virtual teaching on ESL learners' attitudes under COVID-19 circumstances at postgraduate level in Pakistan. *English Language Teaching*, *13*(9), 1–9. DOI: https;//doi.org/10.55539/elt.v13n9p1

Talidong, K. J. (2020). Implementation of emergency remote teaching (ERT) among Philippine teachers in Xi'an, China. *Asian Journal of Distance Education*, *15*(1), 196–203. DOI: https://doi.org/10.5281/zenodo.3881825

Thomas, N., Lucksi, G., & McCulloch, E. (2020). Resilience in the face of emergency remote teaching: EAL pupils' experiences during the COVID-19 pandemic. *TESOL Journal, 11*, e591. DOI: https://doi.org/10.1002/tesj.591

UNESCO. (2021, September 30). *COVID-19 Impact on Education*. Education: From disruption to recovery. . https://en.unesco.org/covid19/educationresponse

Wen, K., & Tan, K. (2020). ESL Teachers' intention in adopting online educational technologies during COVID-19 Pandemic. *Journal of Education and E-learning Research*, *7*(4), 387–394. DOI: http://doi.org/10.20448/journal.509.2020.74.387.394

Yi, Y., & Jiang, J. (2020). Envisioning possibilities amid the COVID-19 pandemic: Implications for English language teaching in South Korea. *TESOL Journal*, *11*, 543e. DOI: https://doi.org/10.1002/tesj.543

Innovation on Education and Social Sciences – Maureen et al (Eds)

When legacy meets modernity, is there any pride? A linguistic perspective

S. Setiawan
Universitas Negeri Surabaya, Indonesia

ABSTRACT: The phenomena of language use are mushrooming among younger generations, who tend to use mixed and made-up language when communicating on social media. Thus, this study aims to explore the norm of language use, the primary reasons for using the language, and the implications. Through the qualitative method, this study employs documentation and virtual interviews as research instruments. Fourteen members of younger generations were selected as informants. The findings reveal that the youths frequently use integrated bi- or multilingualism norms in which English is employed in their conversation. Three reasons for using the selected language are social and traditional factors, the prestige and power of the language, and the language-internal system. Hence, it sheds light on their attitude towards language use in social media. Surprisingly, a negative attitude towards local languages is detected, and it threatens the local language's existence. Consequently, language loss might bring about the loss of culture, identity, and value.

1 INTRODUCTION

Language represents the personality of a nation (Darwati & Santoso, 2017). Specifically, language represents the personal identity of the person. Nowadays, the mushrooming of language use on social media can define identity among the younger generation (Greenfield & Subrahmanyam, 2003). Despite its popularity, it is discovered that the younger generation tends to use a variety of language on social media. They make up their language creatively based on the users in the community (Crystal, 2006). In this case, the younger generation frequently uses more than one language in communication. In the Indonesian context, people are considered multilingual due to the varieties of norms and general areas in which there is only one dominated by the language (Musgrave, 2014).

In this modern era, social, linguistic, and cultural activities can be denoted from the younger generation's involvement with contemporary mass media of communication. Their involvement is no longer focused on local or national recognition but has expanded globally (Leppanen, Pitkanen-Huhta, Piirainen-Marsh, Nikula, & Peuronen, 2009). As a lingua franca, English is widely used by people worldwide, so that the local and national languages are less desirable to use. This phenomenon leads to the issue of regional or national identity and language shift. For instance, Botswana's younger generation is more likely to use the Setswana national language than Sekgalagadi, Ndebele, and Otjiherero, which are local languages (Arua & Magocha, 2002). Another case also indicates that the younger generation's preference is English rather than the local or national languages (Maghfur, Ali, & Indah, 2018). As a result, the potential of losing their cultures and languages, both local and national, gradually emerges.

There have been a considerable number of previous studies that concentrate on the language shift among the younger generation (David, Tien, Meng, & Hui, 2009; Leppanen et al., 2009; Sagimin, 2020; Setiawan, 2020; Sevinc, 2016). As shown in earlier studies conducted by Leppanen et al. (2009) and David et al. (2009), the results found that the younger generation is likely to employ English as a language of communication over their first language and switch between the use of

DOI 10.1201/9781003265061-21

several languages to interact. The second previous study done by Sevinc (2016) revealed that the third-generation Turkish immigrants put much effort into maintaining the Turkish language since they use Dutch as their primary means of communication in the Netherlands. The previous studies conducted by Setiawan (2020) and Sagimin (2020) indicated that the younger generation primarily uses Indonesian as a communication tool. Yet, they constantly utilized their local language even in the low variety, and the positive attitudes were shown by their parents (Setiawan, 2020). In addition, it is undeniable that they were also influenced by modernity and the rapid development of technology (Sagimin, 2020).

Different from the aforementioned studies, this present study is triggered by the appealing phenomenon which concerns the variant of language used by the younger generation in various social media, whether they still preserve their local and national language or not. It should be noted that the meeting between the heritage language and the cultural values are still in proper condition with the modern era. On the other hand, nowadays, the younger generation tends to use made-up language. With this phenomenon, the pride of the local and national languages is still being investigated. Therefore, this current study seeks to answer the three research questions: (1) What is the evidence of legacy and modernity in language use? (2) Why do these phenomena happen? and (3) What is the implication of these phenomena?

1.1 *The norm of language use*

In the communication process, there are norms of language use ??to be applied. People tend to employ these norms of language use based on the interlocutor and context (Holmes & Wilson, 2017, p. 21). The emergence of English is one reason people use more than one language when interacting. In such a case, the people who initially use one language must be separated since the norms for using the language used are dissimilar.

Jorgensen (2012) proposed four kinds of language use norms: monolingualism norm, double or multiple monolingualism norm, integrated bi- or multilingualism norm, and poly-lingualism norm. The monolingualism norm denotes the people who master one language before using another language. The double or multiple monolingualism norm refers to the people who require two languages but only use one for communication. The integrated bi- or multilingualism norm indicates that the people proficient in two languages use code-switching in the communication. The poly-lingualism norm points to the people who employ any linguistic features to attain their communicative goals.

1.2 *Language attitudes*

Language attitudes are closely connected to the individual or community's subjective (Setiawan, 2013). It has an impact on the existence of the language in the future. The language's viability could be seen from the positive attitude revealed by the users, while the shifting of the language is recognized through the negative attitude possessed by the users (Setiawan, 2013). The appearance of attitude is associated with the values inside the language, such as 'good,' 'beautiful,' 'simple,' and 'efficient.' Concerning the Javanese and Indonesian languages, the values such as 'modern' and 'simple' (Sneddon, 2003, as cited in Setiawan, 2013), 'cool' (Smith-Hefner, 2007) can be deliberated.

According to Jendra (2010), language attitude is influenced by four factors. The first factor is the prestige and power of the language, which denotes the speakers' prestige. In this case, a language can quickly obtain its prestige if the speakers have prestige towards the language, and vice versa. The second factor is the historical background of nations which is closely related to colonization and independence. The third factor of the language attitude is the social and traditional factor. In connection with that, language is the means of communication within society. The relationship between society and tradition is quite tight since tradition is carried out within the community. The last factor is the language-internal system related to the grammar, pronunciation, and vocabulary of the language. Therefore, this internal system could be the consideration to reveal the positive and negative attitude towards the language.

1.3 *Language, youth, and lifestyle*

According to Blommaert (2010, p. 10), language is closely related to broader cultural and social processes. Currently, the younger generation is confronted with the profound changes that have occurred several years ago. They are encountered with diversity and mixed ethnicities, languages, and cultures spreading faster than ever before (Nortier, 2018). As a result of globalization, nowadays, the younger generation dominantly utilizes English as the means of communication in social media. For example, a Malaysian child used code-switching while communicating with her grandparents due to the lack of proficiency (mother tongue) (David et al., 2009). This phenomenon indicates the shifting of the language from the local or national to the foreign language. It is in accordance with Pauwels (2016), who defines language shift as a situation where the users ignore the local or national language and use other languages.

One of the factors that influence the occurrence of language shift is because young generations do not use the heritage language when they notice that they are not proficient in that language (McCarty et al., 2009). When younger people are told that they have made a mistake in uttering the language, they directly utilize another language as another choice (Zentz, 2015). This phenomenon also leads to the lack of motivation to learn that language. In other words, these statements can facilitate a shift from the traditional language of society to the national language of prestige and education, which gradually switches *krama* to formal contexts (Errington, 1998, as cited in Zentz, 2015). In such a case, it can contribute to the loss of the people's identity. Language loss is quite influential in changing the identity of people since there is a belief that language, identity, and culture are closely interrelated (Guardado, 2002). It is also in harmony with Nortier (2018), which stated that language creates and expresses identity and vice versa. Therefore, the loss of the language indicates the loss of identity, culture, and value.

2 METHODS

A qualitative method was employed in this study. The procedure of the qualitative approach has unique stages in analyzing the data, relies on the image and text data, and draws a variety of investigative strategies (Creswell & Creswell, 2018). As instruments, documentation and virtual interview were utilized in this study. A total of fourteen members of younger generations from various regions in Indonesia were selected as informants for this study since they have dissimilar cultural backgrounds. Nine of them were Javanese, three Madurese, and the rest were Sundanese and Buginese. In the case of gender, twelve of them were females, and the rest were males aged around 20–25 years old.

The data were gained from various social media, such as Instagram, Twitter, and WhatsApp communications in July 2021. The data analysis is presented as follows: (1) classifying the conversation based on the norms of the language use, (2) categorizing the younger generations' responses according to their attitude of the languages, and (3) connecting the findings of the first and the second research questions as well as presenting the selected responses of younger generations to lead to the implication of these phenomena.

3 RESULTS AND DISCUSSION

3.1 *The evidence of legacy and modernity in language use*

As evidence of legacy and modernity in language use, several conversations from social media are displayed to answer the first research question. It is discovered that there are five monolingualism norms, one double or multiple monolingualism norm, seven integrated bi- or multilingualism norms, and one poly-lingualism norm. The details of the data are presented as follows.

(01) *Gusti nu agung semester ieu meuni hareudang pisan.*
'God, this semester is so hot (complicated).'

(02) *Lho saestu merguru teng gus.*
'Well, I really learned from you, brother.'

The data above shows that each conversation only uses one language, which is considered a monolingualism norm. (01) is the evidence of the use of Sundanese language, while (02) is the evidence of the use of Javanese language. It is good that the local language is still employed by young generations on social media, especially Instagram. This phenomenon is also in line with Setiawan and Nita (2018), who stated that young netizens still often use Javanese in their Instagram accounts. Another evidence of legacy and modernity in language use is presented in (03) (Figure 1).

Figure 1. Double or multiple monolingualism norm.

The following data denotes that the conversation consists of two languages, English and Indonesian. Thus, double or multiple monolingualism norm can be seen in (03). It is included as a double or multiple monolingualism norm since the youth uses two languages, but she only uses one for communication. English is used to make a caption, and Indonesian is used to respond to the comment from another youth. In this case, the Indonesian refers to the excerpt *sini dong main-main biar ketemu,* which means 'come here and play so that we can meet.' Another evidence of legacy and modernity in language use is presented in (04)–(06).

(03) *Iyup... Yak risik. Mek lok toman juduh kita mon mau lanjalan yeeee.*
'Yes, it's drizzle. Why are there obstacles when we want to go for a walk?'
(04) *Janganlah kamu katakan baraccung nanti orang tertakutkan.*
'Don't make a fuss; people will be afraid.'
(05) *Makin di swipe, makin ngga proper.*
'The more you swipe, the less proper it is.'

The three conversations in (04)–(06) above denote the integrated bi- or multilingualism norm since they employed code-switching in their Instagram conversations. In (04), the code-switching of the local language (Madurese) and Indonesian is used. Then, the code-switching of the local language (Pattinjo) and Indonesian is embedded in (05). Meanwhile, in (06), it is discovered that youth utilizes the code-switching of Indonesian and English. Even though young generations applied monolingual and double monolingual, this norm is mainly used by young generations to convey their ideas (Nortier, 2018). Another evidence of legacy and modernity in language use is presented as in (07).

(06) *Iyaa see mbak. Aku kyake udh fix tpi kok kek ada yg ngganjal.*
'Yes, sis. I feel certain, but it still feels like something is stuck.'

There is only one respondent who was indicated to apply poly-lingualism norm in WhatsApp communication. This conversation embedded more than two languages, namely Indonesian, Javanese, and English. In such a case, young people who use English indicate that they are already proficient in English. Even so, they still use Indonesian as their national language and Javanese as their local language. It is in line with McConvell and Florey (2005), which stated that various languages are used in language contact. In short, the youths tend to mix two languages in social media, which is considered an integrated bi- or multilingualism norm.

3.2 *Youth's reasons for using the language*

Apart from presenting the data related to the evidence of legacy and modernity in language use, it is significant to ask the selected respondents for the underlying reasons for using local, national, and English languages on social media. Their responses can be used to reflect their attitude towards Javanese, Sundanese, Madurese, and Pattinjo language as their ethnicity, Indonesian as the national language, and English as a symbol of modernity. Each of the data related to the reasons for employing the language can be seen below.

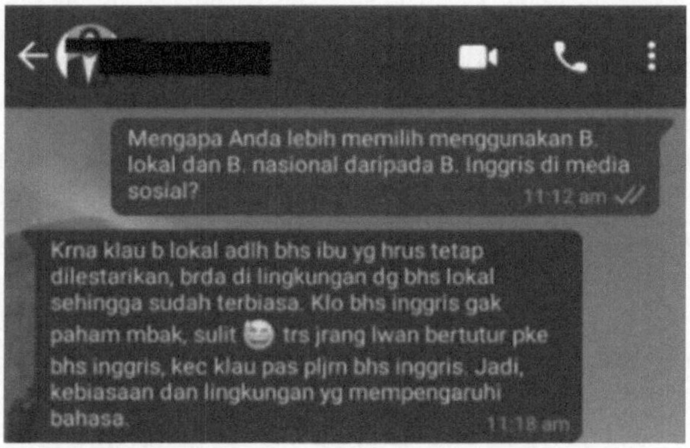

Figure 2. Social and traditional factors.

Figure 2 indicates the social and traditional factors affecting the preference of using the local language. Apart from being used to, another reason for using the local language is to maintain its language. As stated by Sagimin (2020), a minority of youth could maintain their local language. The data obtained shows that she has a positive attitude towards the local language (Javanese). Even though the local language gains a positive attitude, but it is discovered that most respondents have a negative attitude towards the local language (Figure 4). Besides, the negative attitude towards English is detected through the language-internal system, including grammar, vocabulary, and pronunciation. She added that English is considered difficult and rarely used to practice conversation within her social community. Another reason for using the language is displayed as follows.

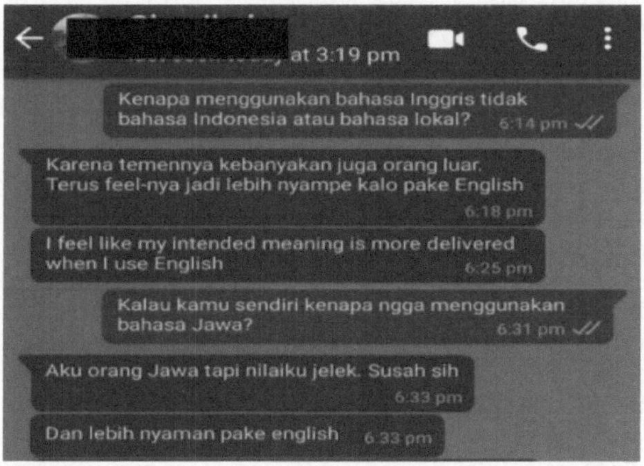

Figure 3. Prestige and power of the language.

Apart from social and traditional factors, it is found that prestige and power of the language also affect the youth's selection of language use. As in Figure 3, the youth preferred to use English due to the power of the language. In connection with that, she could convey her implied meaning in English easier than using local or national languages. This statement is in harmony with Putri (2018), which stated that English has a more accessible language system. Thus, English is widely used by young generations on social media.

The selected respondent had a positive attitude towards English. Meanwhile, she expressed that Javanese is considered difficult, particularly *krama*. Most of the youths confessed that Javanese is hard to learn. Besides, to avoid the mistake of using the local language (Javanese), the youths were likely to use Indonesian and English (Figures 3 and 4). Youths tend to use Indonesian due to the lack of proficiency in their local language (McCarty, Romero-Little, Warhol, & Zepeda, 2009; Zentz, 2015). Therefore, it can be drawn that the local language (Javanese) attains a negative attitude from younger generations. As supported by Zentz (2015), Javanese is more complex than Indonesian. The last reason for using the language is provided below.

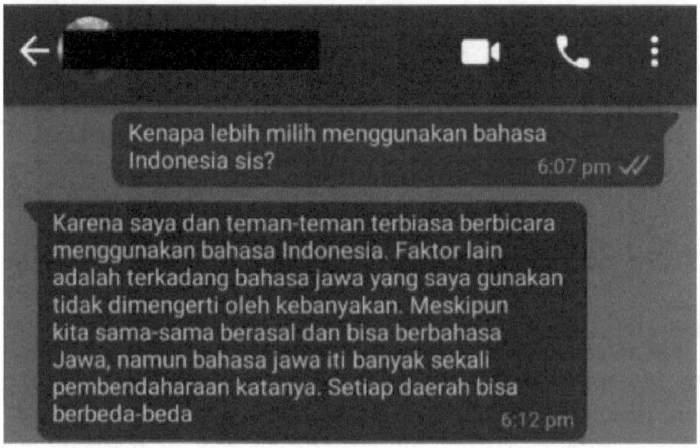

Figure 4. Language-internal system.

The last factor affecting the youth's language attitude is a language-internal system. It deals with sentence structure, pronunciation, grammar, and vocabulary (Jendra, 2010). In Figure 4, the respondent declared that Javanese is rarely used in her communication due to its rigid language system. There are three basic Javanese levels: *ngoko*' low level', *madya*' middle level,' and *krama*' high level' (Setiawan, 2013). The results also shed light that *krama* is not infrequently used for daily communication, especially the conversation between the younger and older generations. Moreover, Indonesian is the possible language use for youth to communicate with the elder.

With respect to Figure 4, the respondent reported that each region has different vocabularies that lead to misunderstanding while interacting with interlocutors. In spite of the fact that the respondent had difficulty utilizing Javanese, she further explained that she employed Indonesian as the primary language in the conversation within the community on social media. The respondent's response reveals that the Javanese language gains a negative attitude, while Indonesian is likely to get a positive attitude. A study by Zentz (2015) demonstrated that Javanese is less used due to the complexity of the internal system.

To sum up, the majority of the youth confessed that social and traditional factors influenced their attitude towards using language on their social media. Meanwhile, the rest of the youth claimed that the prestige and power of the language and language-internal system become factors in choosing language use. Thus, there are no respondents who stated that the historical background of a nation was one of the factors considered in using language on social media.

3.3 *Implication of using modernity through language*

To answer the third research question, the data could be detected through the youth's language attitude and the responses of the virtual interview in WhatsApp communication. The representation of the implication of using modernity through language is addressed below.

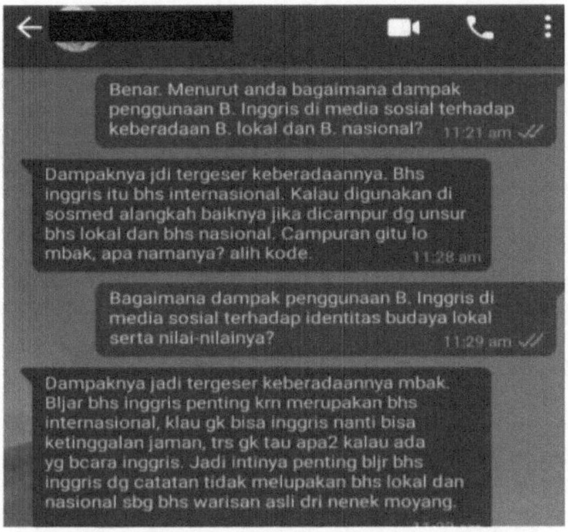

Figure 5. Implication of the language use.

From the data gained, most youth stated that the use of modernity in language could influence the emergence of local and national languages. As seen from Figure 5, the respondent suggested that the use of English should be combined with local and national languages. Fundamentally, English is regarded as the international language that requires young generations to use it. As a result, it is essential to learn and utilize English in conversation, but it is also noteworthy that local and national languages should be maintained as heritage languages (Figure 5).

Since the negative attitude towards the local language is discovered chiefly, language loss has progressively emerged. Language loss happens when the person's proficiency in their language decreases (Holmes & Wilson, 2017). Meanwhile, the usage of Indonesian in social media is shifted by English as a symbol of modernity. According to Pauwels (2016), language shift is when users ignore the local or national language and use other languages. In connection to the local language loss, there is also a loss of culture, identity, and value. Guardado (2002) stated that language loss is quite influential on changing the identity of people since there is a belief that language, identity, and culture are closely interrelated.

Concerning the data, it is detected that the local language is still used by a small number of the youth in social media. It is evidence that there is an effort to maintain the local language as a heritage. Based on the data, four local languages (Javanese, Madurese, Sundanese, and Pattinjo) still present their existence in social media (01, 02, 04, 05). As a result, the preliminary statement might be developed as a gap for future study.

4 CONCLUSION

Based on the documentation, youth are likely to combine two languages in social media, categorized as integrated bi- or multilingualism norm. Mostly, the youth admitted that social and traditional factors trigger them to choose the variety of languages. Other underlying reasons for embedding the language are due to the prestige and power of the language and the language-internal system. Thus, it becomes a reflection of their attitude towards language use on social media. It is sadly discovered that most of the youth considered Javanese to be a complex language to learn. This negative attitude threatens the existence of the local language on social media. As a result, it tends to influence the loss of culture, identity, and value. The minority of the youth still use the local languages on social media; it becomes an effort to maintain the local-ethnic languages. Therefore, there should be preservation efforts towards the indigenous languages in Indonesia.

REFERENCES

Arua, A. E., & Magocha, K. (2002). Patterns of language use and language preference of some children and their parents in Botswana. *Journal of Multilingual and Multicultural Development*, *23*(6), 449–461. https://doi.org/10.1080/01434630208666479

Blommaert, J. (2010). *The sociolinguistics of globalization* (1st ed.). Cambridge University Press.

Creswell, J. W., & Creswell, J. D. (2018). *Research design: Qualitative, quantitative, and mixed methods approaches* (5th ed). Los Angeles: SAGE.

Crystal, D. (2006). *Language and the internet* (2nd ed.). Cambridge University Press.

Darwati, D., & Santoso, B. W. J. (2017). Pilihan kode pada wacana konsultasi siswa kepada guru di SMK Ma'arif 4 Kebumen. *Seloka*, *6*(1), 93–99.

David, M. K., Tien, W. Y. M., Meng, N. Y., & Hui, G. K. (2009). Language choice and code switching of the elderly and the youth. *International Journal of the Sociology of Language*, *200*, 49–74. https://doi.org/10.1515/IJSL.2009.044

Greenfield, P. M., & Subrahmanyam, K. (2003). Online discourse in a teen chatroom: New codes and new modes of coherence in a visual medium. *Journal of Applied Developmental Psychology*, *24*, 713–738. https://doi.org/10.1016/j.appdev.2003.09.005

Guardado, M. (2002). Loss and maintenance of first language skills: Case studies of hispanic families in Vancouver. *Canadian Modern Language Review*, *58*(3). https://doi.org/10.3138/cmlr.58.3.341

Holmes, J., & Wilson, N. (2017). *An introduction to sociolinguistics* (5th ed.). Routledge Taylor & Francis Group.

Jendra, M. I. I. (2010). *Sociolinguistics: The study of societies' languages*. Yogyakarta: Graha Ilmu.

Jorgensen, J. N. (2012). Ideologies and norms in language and education policies in Europe and their relationship with everyday language behaviours. *Language, Culture and Curriculum*, *25*(1), 57–71. https://doi.org/10.1080/07908318.2011.653058

Leppanen, S., Pitkanen-Huhta, A., Piirainen-Marsh, A., Nikula, T., & Peuronen, S. (2009). Young people's translocal new media uses: A multiperspective analysis of language choice and heteroglossia. *Journal of Computer-Mediated Communication*, *14*, 1080–1107. https://doi.org/10.1111/j.1083-6101.2009.01482.x

Maghfur, A. 'Ali, Masruhan, & Indah, R. N. (2018). Language used in social media and its impact toward teens language acquisition. *ICONQUHAS & ICONIST*. https://doi.org/10.4108/eai.2-10-2018.2295456

McCarty, T. L., Romero-Little, M. E., Warhol, L., & Zepeda, O. (2009). Indigenous youth as language policy makers. *Journal of Language, Identity, and Education*, *8*, 291–306. https://doi.org/10.1080/15348450903305098

McConvell, P., & Florey, M. (2005). Introduction: Language shift, code-mixing and variation. *Australian Journal of Linguistics*, *25*(1), 1–7. https://doi.org/10.1080/07268600500110449

Musgrave, S. (2014). Language shift and language maintenance in Indonesia. *Language, Education and Nation-Building*, 1–17.

Nortier, J. (2018). Language and identity practices among multilingual Western European youths. *Language and Linguistics Compass*, *12*, 1–24. https://doi.org/10.1111/lnc3.12278

Pauwels, A. (2016). *Language maintenance and shift*. Cambridge University Press.

Putri, A. (2018). *Language attitude of English department students towards English at University of Sumatera Utara*. University of Sumatera Utara.

Sagimin, E. M. (2020). Language shift and heritage language maintenance among Indonesian young generations: a case study of Pamulang University. *Journal of Language, Literary, and Cultural Studies*, *4*(1), 21–37.

Setiawan, S. (2013). *Children's language in a bilingual community in East Java*. Scholars' Press.

Setiawan, S. (2020). Shifting from an ethnic language among younger generation in a metropolitan city in Indonesia. *Asian ESP Journal*, *16*(2.1), 110–129.

Setiawan, S., & Rahayu Nita, F. (2018). The trend of language use among netizens in Instagram "aslisuroboyo" account. *2nd Social Sciences, Humanities, and Education Conference (SoSHEC 2018)*, *222*, 384–389. https://doi.org/10.2991/soshec-18.2018.82

Sevinc, Y. (2016). Language maintenance and shift under pressure: Three generations of the Turkish immigrant community in the Netherlands. *IJSL*, *242*, 81–117. https://doi.org/10.1515/ijsl-2016-0034

Smith-Hefner, N. J. (2007). Youth language, gaul sociability, and the new Indonesian middle class. *Journal of Linguistic Anthropology*, *17*(2), 184–203. https://doi.org/10.4324/9781315675824-19

Zentz, L. (2015). "Love" the local, "use" the national, "study" the foreign: Shifting Javanese language ecologies in (post-)modernity, postcoloniality, and globalization. *Journal of Linguistic Anthropology*, *24*(3), 339–359. https://doi.org/10.1111/jola.12062.O

Innovation on Education and Social Sciences – Maureen et al (Eds)

Conversation topic of Indonesian speakers in first meeting contact situations Japanese conversation

S. Sanjaya
Osaka University, Osaka, Japan

ABSTRACT: Conversation topics are important for a conversation to take place. The choice of conversation topic also needs to be considered, especially if the interlocutor is a speaker with a different cultural and linguistic background. This study aims to identify what conversation topics are raised by Indonesian-speaking Japanese learners in conversations with first contact situations. The data was taken through roleplay conversation, with the participants being 20 Indonesian-speaking students and 20 Japanese-speaking students. Role-play conversations were carried out in two different settings, namely the setting of cultural exchange events at universities in Indonesia and the setting of conversations on trains in Indonesia. From the results of data analysis, it was known that in both settings, the conversation topic category that most native Indonesian speakers raise is the category of information related to the personality of the speaker and the interlocutor. Judging by the occurrences of the conversation topic, the category is seen the most in all conversations. From these results, it can be inferred that native Indonesian speakers tend to hope for a deeper relationship with native Japanese speakers.

1 INTRODUCTION

The topic of a conversation is an important thing in the occurrence of a conversation. Without a topic of conversation, the speaker cannot start the conversation, maintain the continuity of the conversation, and end the conversation with the interlocutor. In addition, conversation topics play a role in building relationships between speakers and interlocutors. If the conversation topics chosen are appropriate and in accordance with the interlocutor's situation, place, time, personal, and culture. On the other hand, if the chosen topic of conversation is not appropriate, building a relationship with the interlocutor is difficult or even impossible.

In conversations between speakers of different nationalities and languages (contact situations), such as native Indonesian speakers learning Japanese with native Japanese speakers, many things need to be considered in addition to the speaker's desire to get to know the interlocutor more deeply through conversation. As an example when a Japanese learner who speaks Indonesian meets a native Japanese speaker. He does not only need to have the courage to start a conversation but also needs to pay attention to what situation they meet in (first meeting or not), with whom he talks to (male or female, students, lecturers, tourists, or others), on what events (context of the meeting), and other aspects. However, the researcher frequently meets Japanese language learners who have high aspirations to improve their Japanese language skills verbally and hopes to establish communication and relationships with native Japanese speakers at the first meeting. On the other hand, in situations where there are potential problems in conversation in these contact situations, such as the emergence of misunderstandings and disruption of the native speaker's privacy. Therefore, the researcher found a need to analyze the topics of conversation between Japanese language learners (native Indonesian speakers) and native Japanese speakers to identify which topics are appropriate/inappropriate and safe/unsafe in the conversation of the contact situation at the first meeting. In addition, the development of conversation topics needs to be analyzed. By observing the

 DOI 10.1201/9781003265061-22

development of conversation topics, the researcher analyzed the responses of Indonesian speakers and Japanese speakers in discussing a conversation topic.

Research on the topic of conversation in conversations between Indonesian and Japanese speakers has been investigated by Sanjaya and Kusnendar (2018) and Sanjaya (2021). Sanjaya and Kusnendar (2018) analyzed the differences in the choice of conversation topics raised by native Indonesian speakers and native Japanese speakers in their conversations with people who speak the same language at the first meeting. The selected respondents were Indonesian and Japanese speakers who are already in the world of work (*shakaijin*). The researchers found that Japanese language learners are more likely to prefer or seek relationships with the same Japanese speakers (students) or the same age. It showed the need for research or studies that focus on choosing conversation topics for native Indonesian speakers who are Japanese language learners to native Japanese speakers or students. Sanjaya (2021) analyzed what issues Indonesian-speaking Japanese learners will adopt (topic selection) when having conversations with native Japanese speakers at the first meeting. However, the two studies' data (conversation topics) were obtained through surveys in which the answers were still in the form of respondents' consciousness of questions or statements about a conversational situation. If the respondent is in a real conversation situation, the topics raised by Japanese language students are not necessarily in accordance with the answers in the questionnaire. Seeing this, the researcher considered the need for research that focuses on the topic of conversation in the contact situation at the first meeting through observation of conversations between Indonesian-speaking Japanese learners and native Japanese speakers.

2 LITERATURE REVIEW

Communication between speakers of different cultures and languages requires intercultural understanding. This is because communication and culture are closely related. It is in line with Ishii and Kume (2013) statement, which states that culture and communication are closely related. Culture influences communication and the existence of communication contributes to the formation of culture. Thus, without understanding culture, speakers will find it difficult to understand how to communicate well. In the communication (conversation) that takes place between speakers of different languages, the topic of conversation becomes an important aspect of paying attention to. The conversation topics raised also need to consider the aspects of gender, sexual orientation identity, ethnic identity, citizenship, and even the status between the speaker and the interlocutor. If the topic of conversation is in accordance with these aspects, the potential for successful communication will be high (Martin & Nakayama, 2004). In addition to these aspects, another aspect that needs to be considered in the selection of conversation topics. In some situations, the conversation topics discussed might be offensive to the privacy of Japanese speakers. Sometimes, Japanese speakers find Indonesian speakers who suddenly ask questions or raise conversation topics that are considered normal by Indonesian speakers, not Japanese speakers, especially when conversations occur with someone they are meeting for the first time.

Burgoon et al. (1989) stated that privacy has a role in controlling access to whether information owned by the speaker (speaker) will be conveyed or not. Speakers control access to information by considering what situations and to whom protected information such as confidential and private matters can be conveyed. In addition, Moore (2008) states that privacy is cultural. The concept of privacy or the scope of privacy and how to control access differ depending on the kind of society in which one lives. The fact that the privacy concept and the scope of privacy vary based on society, the topics that fall under privacy also differ. Therefore, it is necessary to understand the culture and topics of conversation related to the privacy of the listener.

The ineffectiveness of the topic of conversation and a breach of privacy on the interlocutor's part can be a trigger for a failure to communicate, especially if the conversation occurs between two speakers of different languages. It is in line with Darmastuti (2013) opinion that differences in cultural backgrounds can cause problems if one speaker uses a symbol that is different from the symbol of the interlocutor, showing a difference in the meaning of the symbol. Meanwhile,

Shihabudin (2013) and Sukmono and Junaedi (2014) also stated that the communication processes between people with different cultural backgrounds often experience mistakes. The more diverse the cultural backgrounds, the greater the possibility of misunderstanding between the two speakers when communicating. Suppose the analogy is that the topic of conversation is a symbol, for communication between Indonesian speakers and Japanese speakers to run smoothly. In that case, both speakers need to understand the socio-culture of the two nations. Based on the findings of these experts, it can be concluded that in communication, understanding the topic of conversation that is raised or chosen is very important.

As the explanation above reveals, conversation is an important thing to consider in carrying out a conversation, especially if the conversation takes place between two speakers of different languages, such as a conversation between an Indonesian-speaking Japanese learner and a native Japanese speaker. Therefore, there is a need to study what conversation topics can be raised and what conversation topics should be avoided, especially if the conversation is made together with someone who the speakers are meeting for the first time. However, the researcher limited this research to the conversation topics raised by Indonesian-speaking Japanese students with the aim of identifying what topics they really bring up when having conversations with native Japanese speakers at the first meeting so that it can be seen whether the conversation topics raised close to or equal to the results of previous research conducted by Sanjaya (2021).

Research by Mimaki (2018) and Sekizaki (2016) examined the topic of conversation at the first meeting. Mimaki (2018) analyzed conversation topics between speakers of the same language (Japanese), while Sekizaki (2016) analyzed conversation topics in contact situations between two speakers of different mother tongues. The results of the analysis of [10] grouped conversation topics in native situations in the categories of campus life, affiliation, place of residence, similarities, regional origin, major or vocational, career path, and college entrance exams. Of these categories, the topic that was most widely raised was the topic category of campus life. Sekizaki (2016) analyzed the conversation topics between native Japanese speakers and international students from various countries (non-native Japanese speakers). From all the conversation groups, it is known that the chosen topics are the area of origin of foreign students, living in Japan, majors or vocational (especially areas of expertise), campus life. In conversational contact situations, topics related to living in Japan, climate, language lessons, friendships, food, and hobbies are many. Sekizaki (2016) also analyzes what topics are not raised, and the results of the analysis show that foreign students have a tendency similar to consciousness in native situation conversations. In other words, these foreign students are aware of choosing conversation topics close to native Japanese speakers when having conversations with native Japanese speakers.

This study differs slightly from the research conducted by Mimaki (2018) and Sekizaki (2016), as this study analyzed the conversation topics raised by native Indonesian speakers who are Japanese learners. Sekizaki (2016) analyzed the topics raised by foreign students studying in Japan. This research also uses speakers of foreign languages (not Japanese) as research subjects but is limited to native Indonesian speakers studying Japanese in Indonesia. However, the researches of both Mimaki (2018) and Sekizaki (2016) are used as a reference.

3 METHODS

3.1 *Aim of the research*

This study aimed to determine the conversation topic raised in the first meeting in terms of conversational analysis. This study also aimed to compare the results of the analysis with the results of previous research conducted by [2], which analyzed the topics raised by native Indonesian speakers in contact situation conversations through consciousness surveys. This study focused on the topic appointed by native Indonesian speakers without considering the topics raised by Japanese speakers. The topics raised in this study are topics that are initiated or asked by native Indonesian speakers to native Japanese speakers. Thus, this research can confirm whether the topics raised in the conversation are actually close to or the same topics that appeared in the survey [2].

3.2 *Data collection technique*

The data of this study were taken through roleplay conversations carried out by native Indonesian speakers who are Japanese learners and native Japanese speakers. The conversation participants consisted of 20 native Indonesian speakers (students of Sekolah Tinggi Bahasa Asing YAPARI-ABA Bandung and Universitas Muhammadiyah Yogyakarta) and 20 native Japanese speakers (students of Osaka University, Aichi Prefectural University, Kinjo Gakuin University, and Kobe University). All conversations were carried out in groups, with each group consisting of two people (one native Indonesian speaker and one Japanese native speaker). In the survey by Sanjaya (2021), data were collected through questionnaires and follow-up interviews. Through the questionnaire, the respondents were asked what topics would be raised in the conversation at the first meeting with contact situations with native Japanese speakers in five situations: cultural exchanges held at universities in Indonesia, tourist attractions in Indonesia, Japanese education exhibitions in Indonesia, Japanese culture festivals in Indonesia, and public transportation in Indonesia. Then, follow-up interviews were conducted to explore the answers of the respondents to the questionnaire. However, in this study, the conversation situation is only limited to two situations that have very different freedom of communication, namely cultural exchange at universities in Indonesia and conversations at the first meeting on a train in Indonesia. Thus, the data of this research is in the form of 20 conversations for each situation (setting).

As explained above, this study only analyzes conversations with two conversation situations through roleplay conversations. The reason for choosing these two settings is that the speakers' consciousness of the two situations is different. In situations of cultural exchange, both native Indonesian speakers and native Japanese speakers previously had an obligation to talk to each other and have the possibility to prepare conversation topics. On the other hand, at the first meeting conversation on the train, a native Indonesian speaker spontaneously invited a native Japanese speaker to have a conversation with the assumption that being able to have a conversation in that place is a rare opportunity, then the conversation will start if the native Japanese speaker accepts the invitation to the conversation. Conversation in this situation is sudden, and there may not be any preparation to determine the topic of conversation.

The data gathering process of this research was carried out using the non-participation observation technique (Zaim, 2014). Non-participation observation was used to observe the conversations made by native Indonesian speakers and native Japanese speakers, and the researcher was not involved in the conversation. Conversations were conducted online via the Zoom software. The conversations were recorded, and after the recording session, the researcher observed and listened to the conversation and analyzed the conversation topics that emerged. Kushida, Hiramoto, and Hayashi (2019) and Takagi, Hosoda, and Morita (2018) argued that the conversational analysis data comes from natural conversations, but Kato and Takiura (2016) argued that conversational data from roleplay conversations could be used as conversational analysis data. The researcher considers that the data for this study is difficult to extract from natural conversations, so by referring to the opinion of Kato et al., the conversation was carried out by participants by playing a role in two different situations (settings). The conversations in each situation were 15 minutes long. The researcher did not participate in the conversation but recorded the participant's conversation while conducting the non-participant observation. The recording results were used to record what topics were raised by native Indonesian speakers in each conversation.

3.3 *Data analysis technique*

After recording the conversation, the researcher categorized the topics raised in each conversation situation. The technique is slightly different from the topic categorization in Sanjaya (2021). In this study, there are no general situation topic categories consisting of groups of topics that speakers and interlocutors may discuss and topics that are only discussed by speakers and interlocutors speakers, and topic categories in particular situations. This is because this study only focuses on topics raised by native Indonesian speakers. If you look at the data collected in the two conversation situations, it cannot be confirmed that there are topics in specific situations. Thus, the categorization of topics in this study is grouped as in Figure 1.

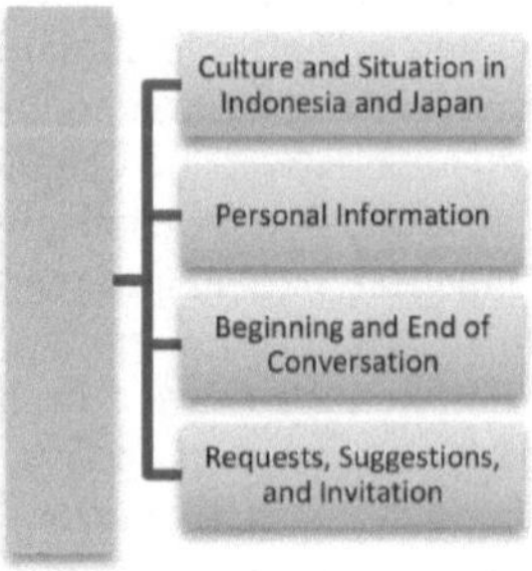

Figure 1. The categorization of topics.

After categorization, the researcher counted the occurrences of conversation topics raised by native Indonesian speakers (the number of times the topic appeared in each conversation situation) so which topics that were most discussed in conversation situations can be known and what Indonesian native speakers expect in contact situation conversations can be studied.

4 RESULTS AND DISCUSSION

As explained in the previous chapters, this study analyzed conversations in the setting of two conversation situations, namely the situation of cultural exchanges held at universities in Indonesia and conversations at the first meeting on a train in Indonesia. The results of the analysis are described as follows.

4.1 *First setting: conversations in cultural exchange events*

Based on data collected from the roleplay conversations, the conversation topics could be divided into four categories: culture and situation in Indonesia and Japan, personal information, beginning and end of the conversation, and requests, suggestions, and invitations, as shown in Table 1. In the culture and situation in Indonesia and Japan categories, there are culture, situation, and education subcategories, while in the personal information category, there are IT-based personal information, residence or location, plans, reasons, assessments, or experiences, skills, and competencies, and profiles subcategories.

Looking at the number of topic occurrences in all conversations in situation 1, the topic that is raised the most by native Indonesian speakers from the four categories is the personal information category (103 times). Moreover, it is known that the topics that fall into the profile subcategory (57 times) are the ones that appear the most in this category. The second most common topic that emerged was the topic in the culture and situation in Indonesia and Japan category (85 times). The difference of occurrences between these two topics can be seen. From this result, the researcher considers that native Indonesian speakers build interpersonal relationships opening up (self-disclosure) about themselves and exploring information about native Japanese speakers as their interlocutors.

This result is not very different from the results of the survey conducted by Sanjaya (2021) through questionnaires and follow-up interviews. The results show that the topics in the topic category of culture and situation in Indonesia and Japan and the personal information category are the topics that appear the most or are raised the most by native Indonesian speakers. However, in the results of the questionnaire, topics belonging to the category of culture and situation in Indonesia and Japan appeared 262 times (81.6%) out of 321 occurrences of all conversation topics, while topics belonging to the personal information category appeared 45 times (14%) of all occurrences of topics. However, the results of the follow-up interviews showed that topics that were included in the category of culture and situation in Indonesia and Japan appeared 23 times (41.4%) of the 56 occurrences of all topics, and topics that were included in the personal information category appeared 33 times (58.9%). Regardless, it can be concluded that the consciousness of the topic selection of native Indonesian speakers known or revealed in Sanjaya (2021) is not very different from the actual conversation conditions as seen from the data of this study. Let's look at the results of the analysis. It can be concluded that native Indonesian speakers choose topics in the categories described above to explore interpersonal relationships with native Japanese speakers.

Table 1. Category of conversation topics at first meeting: cultural exchange event.

Category	Subcategory	Conversation Topic	Number of Occurrences
Culture and Situation in Indonesia and Japan	Culture	Cuisine or food, artist, tourist attraction, or recommended place, Indonesian or Japanese culture, celebration in Japan, clothing, anime, Japanese drama (soap opera), Japanese kanji, festival, sports, Tokyo Olympics, dialect or regional language, Japanese, anime soundtrack Number of Topics: 15	45
	Situation	The situation of Indonesia or Japan, temperature, season, Japanese security, famous places, Momiji (red leaves in autumn), cities in Indonesia, the situation of the covid 19 pandemic, places in Japan, the presence of Indonesian restaurants in Japan Number of Topics: 10	29
	Education	Lectures in college, how to study, about majoring in Indonesian, student life in Japan, the school where you first learned English, the education system in Japan, teaching Japanese in universities Number of Topics: 7	11
		Total	85
Personal Information	IT-based personal information	Source of information of the speaking partner (interlocutor), social media Number of Topics: 2	5
	Residence or Location	Where the interlocutor lives, the speaker's residence, the city where the interlocutor is located. Number of Topics: 3	4
	Plans, Reasons, Assessments, or Experiences	The experience of coming to Indonesia, plans to go to Indonesia, reasons for not going to Japan, preparation of speakers to learn Japanese, plans after graduating from college, the experience of speaking partners to *onsen* (hot spring), reasons for speakers learning Japanese, the experience of speakers talking to Japanese people, desires or the speaker's plans to go to Japan, things he likes about Japan, experiences in Japan, things the interlocutor likes from Indonesia, Southeast Asian countries that the interlocutor wants to visit, is the interlocutor speaking to Indonesian speakers for the first time, impressions of Indonesia, lectures that are considered difficult Number of Topics: 17	28

(continued)

Table 1. Continued.

Category	Subcategory	Conversation Topic	Number of Occurrences
	Skills and Competencies	Speakers' Japanese language skills, foreign language learning, Japanese language difficulties, can the interlocutor speak Indonesian Number of Topics: 4	9
	Profile	Name, profession or status, age, place of the part-time interlocutor, hobbies, lecture major, activity at university, lecture level (level 1, 2, etc.), interlocutor thesis, regional origin, occupation of the speaker's parents, speaker's family, faculty, extracurricular or student activities of the interlocutor, university affiliation, research of the interlocutor, is the interlocutor an otaku Number of Topics: 17	57
		Total	103
Beginning and End of Conversation	–	Greetings, expressions of gratitude Number of Topics: 2	4
Requests, Suggestions, and Invitation	–	Invitation to recreation, invitation to meet again Number of Topics:: 2	3

4.2 *Second setting: spontaneous conversation on the train in Indonesia*

As shown in Table 2, in situation 2, the topics that native Indonesian speakers mostly raise are the same as conversations in situation 1, namely topics that are included in the category of personal information. However, the subcategory that appears the most in this category is not the profile subcategory (39 times) but the plans, reasons, assessments, or experiences subcategory (57 times). In addition, in this situation, compared to the profile subcategory in the personal information category, more topics in the culture subcategory in the culture and situation in Indonesia and Japan category were raised. In contrast to situation 1 in this situation, the speaker seems to want to build interpersonal relationships with the interlocutor, but it seems unlikely that there will be continued communication. As such, it appears that native Indonesian speakers raise topics with the aim of recognizing the culture and situation of the two countries by slightly avoiding topics related to profiles and focusing more on topics related to the experience of the interlocutor visiting Indonesia and plans, reasons, assessments, or experiences related to the current journey. However, when viewed from the appearance of conversation topics in all conversation groups, topics in the personal information category are the ones that appear the most.

The Sanjaya's survey (2021) also shows that the topics that are most often raised in this situation are topics that are included in personal information. From the results of the questionnaire, it is known that topics in the culture and situation in Indonesia and Japan category appeared 51 times (27 times) out of 184 occurrences of all topics. Topics in the personal information category appeared 60 times (32.6%) out of 184 occurrences of all conversation topics. In contrast, the follow-up results in the interview showed that the topic in the personal information category appeared 26 times (89.7%) of the 29 occurrences of all topics and topics in the culture and situation category in Indonesia and Japan appeared three times (10.3%) from the appearance of all conversation topics. In other words, the results of the questionnaire and follow-up interviews show that the most likely topic to be raised by native Indonesian speakers is the topic in the personal information category.

From the results of this analysis, it is known that in both situations, native Indonesian speakers often self-disclose their information and explore the information of their interlocutors with the aim of building more relationships with their interlocutors. However, in contrast to situation 1, in situation 2, a sensitive topic subcategory was found in the category of culture and situation in Indonesia and Japan, but the frequency of its occurrence was very small in all conversation groups that the researcher observed. In addition, in situation 2, there are also private topics such as the religion of the speaker and her family and the topic of remarriage of the speakers' parents, but only a few native Indonesian speakers talk about these types of topics. Seeing this, the researcher assessed that the reason why speakers raised sensitive topics and very private topics was that the possibility of the next meeting was low (the conversation was not oriented to long-term relationships).

Table 2. Category of conversation topics at first meeting: spontaneous conversation on the train.

Category	Subcategory	Conversation Topic	Number of Occurrences
Culture and Situation in Indonesia and Japan	Culture	Tourist attraction or recommended destinations, cuisine or food, anime, drama, train, etiquette on the train, how to ride a train, comics, accessories, religion in Indonesia, movies, ghosts in Indonesia, public transportation, anime soundtrack, Japanese kanji, Indonesian or Japanese culture, recommended souvenirs, Indonesian music, similarities between Indonesians and Myanmar people, Indonesian language. Number of Topics: 20	50
	Situation	Season, family restaurant prices in japan, weather, train ticket prices, famous place, prefectural information in japan, foreigners in Japan, temperature, snow in japan, covid 19 pandemic situation, train tickets in Indonesia, coldest place in japan, supermarkets in Japan, earthquakes in Japan, cities in Indonesia, people of Jakarta Number of Topics: 19	29
	Education	Number of colleges, college situation, how to study, Japanese grammar, education system Number of Topics: 5	5
	Sensitive Topics	Politics, relations between Indonesia and the People's Republic of China Number of Topics: 2	2
		Total	86
Personal Information	IT-based personal information	Social media, telephone number of the interlocutor, contact address Number of Topics: 3	5
	Residence or Location	The interlocutor' residence, the distance from the speaker's house to his/her high school Number of Topics: 2	4

(continued)

Table 3. Continued.

Category	Subcategory	Conversation Topic	Number of Occurrences
	Plans, Reasons, Assessments, or Experiences	Destination of the interlocutor, destination to Indonesia, currently traveling alone or with someone, when they arrived in Indonesia, the impression of Indonesia, with whom they came to Indonesia, length of stay in Indonesia, experience to Japan, destination of the interlocutor to destination, the experience of riding the shinkansen train, impressions on Indonesians, wishes or plans to go to Japan, reasons for taking the train, plans to go to South Korea, things the interlocutor wants, the experience of the interlocutor abroad, the reason the speaker rarely takes the train, the reason the interlocutor learns Mongolian, the experience of the interlocutor to Hokkaido, the reason for going to a destination, have you been to Jakarta or not, the reason for going to Indonesia, the interlocutor's interest in the sea in Jakarta, the impression of the train being taken, foreigners who like Japan, speakers experience while in Jakarta Number of Topics: 26	57
	Profile	The speaker's nationality, name, regional origin, occupation or status, the speaker's family religion, remarriage of the speaker's parents, hobbies, college affiliation, level of study, whether or not the interlocutor is the speaker's first Japanese friend Number of Topics: 10	39
	Skills and Competencies	Japanese vocabulary that is not understood, foreign language learning, spoken Japanese ability, whether or not the interlocutor can speak English Number of Topics: 4	5
		Total	110
Beginning and End of Conversation	–	Greetings, gratitude, requests for permission to converse Number of Topics: 3	12
Requests, Suggestions, and Invitation	–	An invitation to recreation, an invitation to eat together, an invitation to meet again, an invitation to be close friends Number of Topics: 4	7

From the data that have been gathered in this study, it can be seen that the research results are not so different from the results of the research by Sanjaya (2021). The topic raised the most is the topic related to personal information. From this, it can be judged that native Indonesian speakers expect deeper relationship building when having conversations with native Japanese speakers at the first meeting in both situation 1 and situation 2. However, Adams, Murata, and Orito (2009) stated that the further apart a native Japanese speaker is from the group, the less information (including personal information) he will provide. Therefore, if many topics related to personal information are raised to native Japanese speakers at the first meeting, it is likely to intrude on the privacy of native Japanese speakers. In order to be able to explore information related to native Japanese speakers who are their interlocutors, speakers must enter the *uchi* group (the inner circle) of native Japanese speakers.

5 CONCLUSION

From the explanation above, it can be seen that the conversation topics that native Indonesian speakers mostly raise are topics related to the speaker (him/herself) and the interlocutor (personal information). However, whether or not native Japanese speakers accept all topics raised in the whole conversation and whether from the existing topics whether there are topics that offend the privacy of native Japanese speakers has not been revealed. In addition, this research only analyzes the topics raised by native Indonesian speakers only, while how the topics raised flow in conversation have not been analyzed. Therefore, as a continuation of this research, the researcher assesses the need for an analysis of topic development in contact situations. After that, it should be continued with an analysis of the responses of native Japanese speakers to topics related to privacy raised by native Indonesian speakers. Thus, it is possible to study effective communication strategies to build relationships between native Indonesian speakers and native Japanese speakers who are both linguistically and culturally different.

REFERENCES

Adams, A. A., Murata, K. & Orito, Y. (2009). The Japanese sense of information privacy. AI & Society, 24 (4):324–341. DOI: https://doi.org/10.1007/s00146-009-0228-z

Burgoon, J. K., Parrott, R., Le Poire, B. A., Kelley, D. L., Walther, J. B., & Perry, D. (1989). Maintaining and restoring privacy through communication in different types of relationships. *Journal of Social and Personal Relationships*, 6(2), 131–158. DOI: https://doi.org/10.1177/026540758900600201

Darmastuti, R. (2013). *Mindfulness dalam Komunikasi Antarbudaya*. Yogyakarta: Buku Litera.

Ishii, S., & Kume, T. (2013). Intercultural Communication Research Methods: from Topic Selection to Paper Writing. Tokyo: Yuhikaku.

Kato, S., & Takiura, M. (2016). *A Guidebook for Research in Pragmatics*. Tokyo: Hituzi Shobo.

Kushida, S.., Hiramoto, T.., & Hayashi, M. (2019). *An Introduction to Conversation Analysis*. Tokyo: Keiso Shobo.

Martin, J. N., & Nakayama, T. K. (2004). *Intercultural Communication in Context*. New York: McGraw-Hill.

Mimaki, Y. (2018). *Discourse Analysis on Politeness-How People Interact at First Encounters*. Tokyo: Kuroshio Shuppan.

Moore, A. (2008). Defining Privacy. *Journal of Sociophilosophy*, 39(3): 411–428.

Sanjaya, S. (2021). 'Conversation Topics on Contact Situation in Introductory Conversations Between Indonesian and Native Japanese Language Speakers: A Survey Study on Indonesian Japanese-Language Speakers'. *Social, Humanity, and Education (ICoSIHESS 2020)*. Yogyakarta, 23-24 October 2020. DOI: https://doi.org/10.2991/assehr.k.210120.148

Sanjaya, S., & Kusnendar, M. (2018). Deskripsi Topik Pembicaraan yang Tidak menyinggung Privasi Bagi Penutur Bahasa Jepang dan Penutur Bahasa Indonesia. *Journal of Japanese Language Education and Linguistics*, 2(1) 199–212. DOI: https://doi.org/10.18196/jjlel.2112

Sihabudin, A. (2013). *Komunikasi Antarbudaya Satu Perspektif Multidimensi*. Jakarta: Bumi Aksara.?

Sekizaki, H. (2016). Topic Schema at First Meeting of Contact Situation: Implications from Conversations between Japanese Students and Foreign Students at a Japanese University. *Nihongo Kyoiku Ronshu* (31) 17–32. DOI:https://doi.org/10.15068/00137464

Sukmono, F. G., & Junaedi, F. (2014). Komunikasi Multikultur Melihat Multikulturalisme dalam Genggaman Media. Yogyakarta: Buku Litera.

Takagi, T., Hosoda, Y., & Morita, E. (2018). *Basics of Conversation Analysis*. Tokyo: Hituzi Shobo.

Zaim, M. (2014). *Metode Penelitian Bahasa: Pendekatan Struktural*. Padang: FBS UNP Press.

Innovation on Education and Social Sciences – Maureen et al (Eds)

The pragmatic marker '*ya*' in Indonesian Chinese: Perspective of language contact

X. Renfei
Central China Normal University, Wuhan, Hubei, China

ABSTRACT: Indonesia is a multilingual country, and languages always have an impact on each other in the process of language contact. The present work intends to investigate the influence of Indonesian on Indonesian Chinese based on the functional variation of 'ya' in Indonesian Chinese. Different from Mandarin Putonghua, Indonesian Chinese 'ya' serves as a pragmatic marker that has the functions of positive response, as a tag question or a discourse marker. It plays a role in repairing discourse, summarizing discourse, and responding to discourse, and the most important function is to establish unity which can be explained from the language contact. Indonesian social and cultural values and the status of Chinese phonology and lexicology are another aspect of reasons.

1 INTRODUCTION

1.1 *Language contact*

Language contact has been studied extensively in different languages, such as English (e.g., Alexandra, Fleißner, Kim, & Newerkla , 2020; Anna, 2021), Germany (Lenz, Fleißner, Kim, & Newerkla, 2020), Japanese (e.g., Reinier, 2021), Hebrew (e.g., Hovav, Reshef, & Taube, 2019), Chinese (Shi & Zhu, 1999; Wu, 2004), and so on. The language contact arises when some new words will incorporate with each other across the languages, and sometimes the spread of new sounds and sentence structures, etc. (Thomason, 2001).

As we know, for a long while, the academic circles have always been concentrating efforts on the structualist paradigm during the study of language contact, especially the language borrowing (see also, Andersen, 2014; Unuabonah & Daniel, 2020, etc.), which mainly counts and classifies the types of loanwords according to the adaptability of the form and pronunciation (Zener & Kristiansen, 2013). The vocabulary borrowing has ever been become a pronoun in the study of language borrowing to a certain extent. However, there has been a pragmatic turn in the study of language borrowing in recent years (Andersen, Cristiano, & Ilić, 2017).

1.2 *Indonesian Chinese*

Chinese, like many other languages in the world, has produced a variety of regional variants in the process of overseas dissemination and inheritance due to the influence of economic, geographical, cultural, and social environment, and the development of the language itself. And in the process of long-term contact with the local language, Chinese naturally was affected by the local language, which is distributed at all levels from pronunciation to pragmatics.

Numerous studies have proved that China and Indonesia have a long history of cultural exchanges (Kong, 1999). Studies have shown that the history of Chinese emigration to Southeast Asia can be traced back to the Han Dynasty. From the Han Dynasty to the Ming and Qing Dynasties, whenever the new and old regimes changed, the people who could not bear the war and the declining dignitaries would emigrate one after another. Southeast Asia became the first choice for Chinese immigrants because it was easy to reach.

 DOI 10.1201/9781003265061-23

The arrival and rooting of Chinese immigrants not only led to the birth of a new race, but also the natural import of the language. In the process of overseas immigration, Chinese—their mother tongue—would also be brought over with the flow of immigrants and continued to pass on to their future generations.

In this significance, the Indonesian Chinese has two bedding of meanings: one is to represent an ethnic group of Indonesian Chinese, maybe better to be called Chinese Indonesian, and the other is to represent a language of Indonesian Chinese. As the concept of a language, we contend that Indonesian Chinese is a regional variant of Chinese derived in Indonesia.

1.3 *Language contact in Indonesian Chinese*

With the trade, cultural exchange, and mixed residence with the local ethnics, with the increase in frequent communication between different languages, the language contact has had natural results where the most common manifestation of one language being influenced by another language is borrowing.

With the long-term cultural integrating, the language borrowing comes automatically, with the result that Chinese loan words can be seen everywhere in Indonesian, Javanese, and many local languages, such as *kucai, lentkeng, teh, tahu, bacang, lumpiah, cawan, kemoceng, barongsai, capgomeh, angpau,* and so on (Kong, 1989). Vice versa, Chinese is also being influenced by Indonesian, Javanese, and other local languages. For example, Juta is not a monetary unit in Chinese, and it is also troublesome to change it into a Chinese monetary unit. Therefore, the Chinese Indonesian people created a unique unit of measurement *Tiao* which is equivalent to Juta in Indonesian. In addition, *Tempe, sate, rendang* have also entered into the system of Chinese vocabulary.

But anyhow, little attention has been paid to grammatical borrowing in the past studies, and the discussion of pragmatic borrowing is rare, as is the discussion of overseas Chinese of their spoken language.

2 METHODS

Therefore, we take the functional variation of '*ya*' in Indonesian Chinese as an example to explore the influence of Indonesian on Chinese.

Considering the different situations where some speakers use Chinese as their mother tongue or heritage language and some speakers speak Chinese as their second language, we collected the data from the balanced samples: 2 local Indonesians studied Chinese as a second language, 2 Chinese Indonesians studied Chinese as a second languages and 2 Chinese Indonesians who speak Chinese as their mother tongue and whose age is over 70. All of them are fluent in Chinese communication.

The corpus is collected by natural collection method, which collects the natural conversation in the process of natural conversation or a phone call with them. And for more comprehensive and effective purposes, we also use some heuristic data and collect some data in the form of interviews and finally a corpus with a dimension of 101,065 characters has been formed.

3 RESULTS AND DISCUSSION

3.1 *Results*

With the investigation, it is found that some usages of '*ya*' in Indonesian Chinese natural spoken language are different from those in Mandarin Putonghua. In addition to the common usages, there are the following functional variation usages, which can't be explained by the mutation or the change of the pronunciations.

3.1.1 *Interjection*

3.1.1.1 Affirmation

As an exclamation, several usages of '*ya*' are the same as Mandarin, but also have two special functions. The first function is to show promise and affirmation. For example:

(1)

A:	Nǐ	dào	le	ma?
	you	arrive	(tense particle)	(modal particle)
B:	→ ya,	wǒ	gāng	dào!
	(interjection)	I	just	arrive

A: Have you arrived yet?
B: Ah, I just arrived!

(2)

A:	Wǒ	yǒu	jiàn	shì	xiǎng	máfan	nín,
	I	have	(measure)	matter	want	disturb	you
	wǒ	xiǎng	qǐngwèn	jīnnián	de		
	I	want	ask(polite)	this year	(particle)		
	Xīnchūnliánhuānhuì		wǒmen	réng			
	Spring Festival's Party		we	still			
	yìqǐ	jǔbàn	kéyǐ	ma?			
	together	hold	can	(particle)			
B:	→ ya,	hǎo	ā!	Méi	wèntí	ya.	
	(interjection)	OK	(particle)	no	problem	(particle)	

A: Can we still hold the spring festival party together?
B: Ah, no problem.

(3)

A:	Lǎoshī	zuìjìn	hǎo	ma?	
	teacher	recently	good	(particle)	
B:	→ Ya,	chàbùduō,	hái	kéyǐ	lā.
	(interjection)	almost	rather	fine	(particle)

A: How are you recently, dr.?
B: Ah, almost, it's OK.

In the above three examples, '*ya*' is used in the response sentence to express an affirmative response or promise to the speaker's question.

3.1.1.2 Confirmation

The second type is used to confirm the speaker's views, requirements in the response turns.

(4)

A:	Nǐ	de	Hànyǔ	zhēn	hǎo!		
	your	aux.	Chinese	really	good		
B:	→Ya,	wǒ	de	lǎoshī	yí	zhème	shuō.
	(interjection)	my	aux.	teacher	also	so	speak

A: Your Chinese is really good!
B: That's true. So did my teacher.

(5)

A:	Nǐ	zhuǎnlù	yìxiē gēn nǐ	qīzi	de		
		you	transcribe	some	with	your	wife aux.
		duìhuà	huòzhě	qítā	rén	de	dōu kéyǐ.
		dialogue	or	other	people	aux.	all fine
B:	→ya…ya…,	míngbai.					
	(interjection)	understand					

A: You can transcribe some conversations with your wife or others.
B: Ah, ah, I see.

The above '*ya*(s)' is read in a descending tone, where the sound is like [ia] or [iya]. They can be combined with *en, um, hmm* and other confirmative exclamations and there is often co-occurance with the ones like *Ok, good, right* and *no problem* indicating positive confirmation.

3.1.2 *Tag question*

Tag question is a question (such as "*isn't it*" in "*It's fine, isn't it?*") added to a statement or command. Usually the tag question is served by the brief question in a form of short sentence which comes at the end of a statement to ask if the statement is correct, or to gain the assent of or challenge the person addressed. Nevertheless, some scholars, such as Holmes (1983), Wouk (2001), and Nordenstam (1992) also regard *ya, yeah, uh-huh, right* and so on as tag questions, because of their additional interrogative functions.

The '*ya*' in Indonesian Chinese also has the function of additional interrogation, which is located at the end of the sentence to indicate a slight query. For example:

(6) → Wǒmen xiān shàngkè, ya?
we first have class tag question
Shall we have class first?

(7) → Shuā le yá jiù shuìjiào ya?!
brush (tense) teeth then sleep tag question
Brush your teeth and go to bed.

(8) →Wǒmen míngtiān xiàwǔ qù kéyǐ ma? Peter ya?
we tomorrow afternoon go OK (particle) (name)tag question.
Can we go tomorrow afternoon? Peter?

(9) A: →Dìdi huílái le ya?
brother go back (particle) tag question
B: Ya, zài xízǎo ne.
(interjection) be doing take a bath (particle)
A: is your brother back?
B: Ah, he is taking a bath now.

(10) A: →Ni xiàzhōu qù Zhōngguó ya?
you next week go China tag question
B: Ya.
(interjection)
A: Are you going to China next week?
B: yes.

Different from the previous exclamation, the one as a tag question we read it in an ascending intonation. Secondly, '*ya*' in this usage looks similar to the modal particle "ma" in Mandarin Putonghua, but the difference is that we can make a pause before '*ya*', which can be separated by a comma in writing. In the sense, it is equivalent to "is it OK?", "is it fine?", "is it right?", etc., and all of these are not the characteristics of the modal particle "ma". Third, we know that "ma" is a typical modal particle to construct an yes-no interrogative sentence, comparing with this, the additional interrogative sentence, which is composed of '*ya*', is basically similar to soliciting the other party's opinions or making their own information confirmed by the other party where the degree of query isn't as intense as the yes-no interrogative sentence.

3.1.3 *Discourse marker*

Discourse markers refer to those linguistic components that do not express truth value semantic relations in discourse, but convey procedural meaning, which are syntactically optional and indispensable in communication (refer to Blackmore, 2000; Fraser, 2009, 1999; Schiffrin, 1987; and so on).

The Indonesian Chinese '*ya*' sometimes plays the part of a discourse marker in discourse, which helps the progress of communication go smoothly in interaction.

According to its location in the turn, the '*ya*(s)' that plays the function of discourse marker can be divided into three categories.

3.1.3.1. *Initial*

(11) A: Ni hái chī diǎnr ba?
you still eat a little (particle)
B: →ya, méi shìr, yǐjīng gòu le.
no problem already enough(particle.)
A: I think you'd better eat a little more.
B: Yah, no problem, it is already enough.

(12) A: èn, (qìchē) wǒ xiǎng yě shì
(interjection) car I think also
mài diào bǐjiào hǎo.
sell rather better
B: → ya, búguò, nǐ zhīdào, xiànzài Yìnní yìqíng ....
however you know now Indonesia pandemic
A: I also think we'd better sell the car.
B: Hmmm, however, you know, the pandemic in Indonesia now...

The '*ya*' in the first example is to give a positive response to the speaker's invitation which is polite and appreciative, indicating the gratitude to the other party. At the same time it makes the preparation for the subsequent negative reply, so that even if the hearer refuses the invitation, it won't hurt the other party's face. '*ya*' in the second example is also a discourse marker, which indicates a positive response to the speaker's discourse, and simultaneously gives the speaker time to think deliberately. And it is also used in combination with other discourse markers just as *you know*, *however*, *how shall I put it*, and so on.

3.1.3.2. *Internal*

(13) Suīrán tā shì gēn wǒmen yìqǐ qù de,
although he is with we together go (particle)
dànshì,
but
→ya
yah
nǐ zhīdào,
you know
tā zhǐshì yí gè sījī...
he. is only one (measure word) driver
Although he went with us, but you know, he's just a driver.

(14) Xiànzài qù Bromo bǐjiào wēixiǎn,
now go (name of a volcano) rather dangerous
huǒshān róngyì huápō,
volcano easy slide
suóyǐ,
so
→ya,
yah
háishì qù Batu,
we'd better. go (name of a scenary)

nàlǐ huánjìng hǎo,
there environment good
yě bǐjiào hǎowán.
also rather interesting
It's dangerous to go Bromo now, the volcano is easy to slide. Therefore, we'd better go Batu, where the environment is good and it's also fun.

It's dangerous to go Bromo now, the volcano is easy to slide. Therefore, we'd better go Batu, where the environment is good and it's also fun.

'*Ya*' in the above two cases seems the same as in Mandarin Chinese for they can all be put after the conjunctions, but there are differences in phonetics, rhythm, and intonation. First of all, '*ya*' in Mandarin Chinese if connects with the previous discourse markers, it should immediately be at the heel of them, such as "*suóyi ya*" and "*dànshì ya*", and the discourse marker and '*ya*' can't be separated by a pause, while Indonesian Chinese '*ya*' and the conjunction can be separated by a full pause (a comma in writing), and phonetically in a metrical foot, which is not alike with Mandarin Chinese in length, intensity, pitch, and mute time in the pronunciation. Secondly, the intonation is also different, where '*ya*' in Mandarin Chinese is read in a descending intonation and '*ya*' in Indonesian Chinese is read in an ascending intonation.

3.1.3.3. *Final*

(15)	A:	Wǒ	qùnián	hái	zài	nà	jiā	
		I	last year	still	in	that	(measure word)	
		→gōngsī	gōngzuò	ya				
		company	work	yah				
	B:	èn,	nǐ	gēn	wǒ	shuō	guò ...	
		Ah	you	with	me	tell	(particle)	
	A:	ya,	jīnnián,	ni kàn,	méi	jǐ	gè	yuè,
		yah.	this year	you see.	no	several	(measure word)	month
		jiù	daobì	le...				
		then	go bankrupt	(particle)				

A: You know, I worked in that company last year...
B: I see, you ever told me that...
A: Hmm, but you see this year, it went bankrupt in a few months...

In the above example, the speaker makes the application of '*ya*' at the end of his turn, the goal is to intend to interact with the listener and hope that his utterances can be responded or recognized by the hearer. And occasionally the speaker will use another '*ya*' in the next turn, so as to echo with the utterance he just made and confirm that the listener has correctly and completely received his information. The difference is that '*ya*', the one interacting with the listener, reads in an ascending intonation, while '*ya*', the one reconfirming, reads in descending way.

3.2 *Discussion*

3.2.1 *Perspective of Language Contact*

Then, is the emergence of Indonesian Chinese '*ya*', function variant the result of the change of the language itself or the influence of other languages?

Searching for the sandhi of the phonetic change of '*ya*', and even the grammaticalization of Mandarin '*ya*', it is difficult to find the explanations of the changes of the above functions of '*ya*' in Indonesian Chinese from the Chinese itself. However, the study of Indonesian '*ya*' can provide enlightening ideas for our interpretations.

Wouk (2001) examined the use of the two allomorphs of the Indonesian word for yes, '*ya*' and '*iya*', in Indonesian conversational data. She found that both are frequently used responsively, as an affirmative marker or a conversational continuer. They can also be used as a tag question to initiate an exchange or make a request that has a low social cost, and as a discourse marker. Let's give an example:

(16)	Y: jadi	dari	rumah	P	tu	ya,
	so	from	house	[name]	that	yeah
	R: ya					
	yeah					
	Y: itu	agak	kebelakang-nye	tu	ya,	
	that	rather	to:the:rear-gen	that	yeah	
	mau	arah	ke:(.75)	apa	tu,(.50)	
	want.	direction	to	what	that	
	[click]					
	R: arah	kapling,				
	direction	plot:of:land				
	Y: iya					
	yes					

Y: so from your house huh,
R: yeah
Y: it's kind of behind it huh, you go towards: (.75) whtchamacallit, (.50) [click]
R: toward the field
Y: yes

As can be seen from the above example, the pragmatic marker '*ya*' in Indonesian is completely corresponding to the functional variant in Indonesian Chinese where the first '*ya*' in the dialogue is used as a tag question, and the second one is used as an affirmative marker or a confirmative marker, and the third one is used as a conversational continuer.

And as Wouk (2001) investigated, 88% of the occurrences of '*iya*' are shown to comprise a turn or are turn-initial, while 74% of the '*ya*(s)' are intonation units final or turn-final. So is in Indonesian Chinese, the difference is their pronunciation distinct are mainly reflected in intonation. The proportion of 53.6% of the occurrences of '*ya*', which read in ascending are shown to be the final of a sentence (it is also called intonation unit final) or comprise turn final, and the proportion of 44.5% which read in descending intonation are shown to comprise a turn or are turn-initial, and only around 2% of the '*ya*(s)' occur in the sentence internal.

And as Wouk (2001) suggested, as a discourse marker, '*ya*' has a number of different functions, occurring in repairs, in conclusion drawing, in echoing, most frequently, however, they are used as a means of building solidarity through the creation of fictive common ground.

And the function of pragmatic markers '*ya*' in Indonesian Chinese can basically make the same generalization and classification. Wouk (2001) explained the high frequency of this type of usage in Indonesian conversation can be seen as a reflection of Indonesian cultural values, which place a premium on maintaining the appearance of cooperative behavior.

In addition to these, we consider that the monosyllabic characteristics of Chinese characters and the position where modal particles in Chinese can be located are also the reasons for the variation of the function.

4 CONCLUSION

Chinese not only belongs to China, but also belongs to the world. With the spread of Chinese overseas, over time a set of different characteristics will be produced, for which being influenced by the local languages is one of the main reasons. For the variation of Indonesian Chinese '*ya*', language contact is only the most appropriate explanation and it can be explained properly.

Language is the reflection of culture. If Wouk's inference is correct, i.e., in Indonesia there is a strong orientation towards solidarity and social harmony holds a high value in Indonesian traditional culture. Then, the Chinese here means the Chinese Indonesian people, who need to integrate into the main ethnic groups, where they must be influenced by Indonesian cultural orientation, and necessarily accept the Indonesian cultural values.

The Javanese term *gotong royong* (meaning collaborative work) will be invoked regularly. The Chinese who is willing to live in harmony with Indonesia's local ethnics, has to learn the social style and be appreciated and adapt their value orientation. Conversely, the solidarity in Indonesian laying stress on the harmony interaction will also get others' favorable impression, so that they will all be vying to imitate following.

REFERENCES

Alexandra, N. L., Fleißner, F., Kim, A. & Newerkla, S. M. (2020). *give* as a put verb in German–A case of German-Czech language contact?, *Journal of Linguistic Geography* 2020, 8(2):1–15.

Andersen, G. (2014). Pragmatic borrowing, *Journal of Pragmatics*, 67: 17–33.

Andersen, G., Cristiano, F. & Ilić, B. M. (2017). The pragmatic turn in studies of linguistic borrowing, *Journal of Pragmatics*, 113: 71-76.

Fraser, B. (1999). What are discourse markers. *Journal of Pragmatics*, 31: 931–952.

Fraser, B. (2009). Topic orientation markers. *Journal of Pragmatics*, 41: 892–898.

Hovav, M., Reshef, Y., & Taube, M. (2019). *Language Contact, Continuity and Change in the Genesis of Modern Hebrew*, Amsterdam: John Benjamins.

Kong, Y. Z. (1999). *Cultural Exchange between China and Indonesia*, Peking: Peking University Press.

Reinier, S. (2021). The Dutch Language in Japan (1600-1900): A Cultural and Sociolinguistic Study of Dutch as a Contact Language in Tokugawa and Meiji Japan, *Dutch Crossing*, 45(2):211-214.

Schiffrin, D. (1987). *Discourse Markers*. Cambridge: Cambridge University Press.

Shi, D. X., & Zhu, Z. Y., (1999). The Influence of English on the Syntax of Written Chinese in Hong Kong-Language Changes Caused by Language Contact. *Foreign languages*, (4):2–11.

Thomason, S. G. (2001). *Language Contact: An Introduction*. Washington, D.C.: Georgetown University Press, 2001.

Unuabonah, F. O. & Daniel, F. O. (2020). *Haba*! Bilingual interjections in Nigerian English: A corpus-based study, *Journal of Pragmatics*, 163: 66-77.

Wouk, F. (2001). Solidarity in Indonesian conversation: The discourse marker *ya*, *Journal of Pragmatics*, 33(2):171–191.

Wu, A. Q. (2004). The Influence of Language Contact on Language Evolution. *Minority Languages*, (1):1–9.

Innovation on Education and Social Sciences – Maureen et al (Eds)

Indonesian female migrant domestic workers and the alternative narratives of "The heroes of remittance" and "part of the family"

D. Irawaty
Department of Anthropology, State University of New York (SUNY), Binghamton, NY, USA

ABSTRACT: Indonesian transnational female domestic workers have been labeled as "the heroes of remittance." The label has been approached to present pseudo heroization of migrant workers by both the state and society. These female migrant domestic workers also deal with the narrative of "part of the family" that also becomes mainstream. This paper analyzes the impacts of both narratives on the social and work conditions of female migrant domestic workers. I observe how their responses toward the dominant narrative reflect their agentive capacity in challenging the mainstream and hegemonic narratives as well as in providing the alternative ones. I argue that personal story plays a crucial role in expressing personhood, identities, and personal position among Indonesian female domestic workers in Singapore and Hong Kong vis-a-vis the dominant narratives, in this case, of "the heroes of remittance" and "part of the family."

1 INTRODUCTION

Indonesian migrant female domestic workers appear to be a driving force of the national economy. However, the government does not pay serious attention to the personal lives of migrant workers. I analyze the personal narratives of Indonesian female migrants who work as domestic workers (IFMDWs) in Singapore and Hong Kong. This paper discusses two narratives of "the hero of remittance" (*pahlawan devisa*) and "part of the family" (*bagian keluarga*) dominant within Indonesian society. My goal is to give these IFMDWs rooms to talk about their life stories and use their voices regarding the above narratives. Female migrant domestic workers are a central actor in the discourse in the narratives.

In this paper, I offer, firstly, alternative narratives produced by female domestic workers to indicate a capacity to challenge the dominant narratives and show their resistance. Their personal narratives exhibit their positionality concerning the state, mass media, and society and their position within the larger political and economic structure. Secondly, the narrative of "part of the family" is created as a fantasy to benefit employers in reducing power, lowering wages, maximizing loyalty and service, and expecting more respect.

Finally, the narrative of "national heroes of remittance" is re(produced) to conceal the manipulation of informal economy excluded from labor censuses and unprotected by labor legislation; it masks the government's irresponsibility in handling abuses against migrant workers.

1.1 *Discourse analysis on dominant narrative as public discourse*

Van Dijk (2008) argued that discourse analysis needs to pay attention to power and domination. Foucault (1980) addressed how powerful regimes produce knowledge to control public discourse. Blommaert (2005) discussed the effects of power and power relations in a discourse in creating inequality and the importance of knowing the regimes of power. In the Foucauldian sense, thinking and talking connect to certain ideologies to maintain and circulate power in society. The narratives about female migrant domestic workers are a public discourse to influence general ways of thinking.

 DOI 10.1201/9781003265061-24

I refer to Gumperz's (1982) contextualization cues to see the pronoun choices, non-linguistic devices or signs, intonation, code-switching, etc., to contextualize the storytelling and the storytelling event.

Feminist critical discourse analysis can be used to see how "gender ideology and gendered relations of power (re)produced, negotiated and contested in representations of social practices, in social relations between people, in people's social and personal identities in texts and talk" (Lazar, 2005, p. 11). Gender intersects with other social-political identities such as sexuality, class, race, geography, age, education, nationality, religion, etc. This is a crucial concept to see how female transnational domestic workers have been treated as the second citizen, second sex, second class, the undervalued workers, and *pembantu* (literally helper/servant). Ladegaard (2020) showed that Indonesian migrant domestic workers in Hong Kong were constructed as incompetent and stupid just because they cannot speak Mandarin. Lan (2006) found that domestic workers in Taiwan have been desexualized because their female employers are worried that their husbands will be attracted to their domestic workers. Intersectionality analysis is needed to comprehensively capture and make sense of the whole picture and look at the problem.

IFMDWs have "the reflexive project of the self" (Giddens, 1991) as an ongoing negotiation between individual and social identity. Following Bakhtin's (1935) heteroglossia, there is no dominant narrative; what IFMDWs have done is heteroglossia because narratives are the continuity of texts or discourses that never end. The concepts of resistance and agency are used to discuss the IFMDWs' capacity of constructing narrative and producing story. Bakhtin's (1935) idea of metamorphosis (human transformation) is useful to analyze the agency of IFMDWs in deconstructing stories. Ochs and Capps (1996) and De Fina (2003) discussed resistance by speakers against the dominant narrative when they produce alternative narratives.

1.2 *Work relation, narratives, and identity*

Paid domestic work is an informal economy that blurs professional relationships. Sotelo (2001) differentiated maternalism and paternalism. Personal relations in domestic work appear to be a mechanism of oppression and control. Maternalism requires rituals of deference by employees that lead to unequal relationships. Maternalism obscures paid work and unpaid services. Tappert and Dobner (2015), Lan (2006), and Sotelo (2001) agreed that maternalism is a one-way relationship that claims loyalty and reinforces inequalities. Personalism is a two-way relationship based on mutual respect with the recognition of the domestic worker as an individual and as a person. However, the relationship remains asymmetrical and maintains distance. Both maternalism and personalism could be used to conceal low wages and poor working conditions. Tappert and Dobner (2015) addressed the term "part of the family" as a fictive construction through the integration of workers into a family. Treating domestic workers as "daughters" produces "superordinate-subordinate relationships" between domestic workers and their employers (Anderson, 2000).

Rollins (1990) found the forms of deference in terms of deferential occupation in three types of the most common linguistic deference: (1) Most domestics are called by their first names and expected to call their employers by their last names; (2) employers expect their domestic workers call them by such deferential names such as: "Ma'am"; and (3) domestic workers are called "girls" regardless of age. This reflects the inferiority of domestic workers' whole identity by which they are expected to accept the demeaning treatment.

Maternalism conceptualizes domestic workers as childlike, a characteristic of servitude throughout history and replication of feudalism. According to Rollins (1985, p. 178), maternalism has a complex dimension; it is about "love, economic exploitation, respect and disrespect, mutual dependency, intense self-interest, intimacy without genuine communication, mutual protection." Through maternalism, employers exercise control over their domestic workers (Bakan, Stasiulis, & Stasiulis, 1997). When domestic workers are considered "one of the family," employers do not have the responsibility of protecting their rights (Romero, 2016).

Lan (2003) addressed personalism which turns professional relationships into a family personal intimacy, but the structure is unequal based on different statuses. She used "boundary work" to explain the socio-spatial boundaries in domestic and private relationships in which employers reproduce social inequality. In contrast, domestic workers try to negotiate social distance and find alternative boundaries. Anderson (2000) pointed out that domestic workers sell their labor and "personhood" as their employers demand obedience and loyalty. The result is the commodification of domestic workers' labor power and personhood. Positioning domestic workers as part of the family would blur work relations and manipulate emotional relationships that obscure domestic workers' exploitative work conditions. Bakan and Stasiulis (1997) argued that domestic workers often accept their devalued status in such a circumstance.

2 METHODS

This study focuses on IFMDWs working in Singapore and Hong Kong. I conducted interviews with twenty respondents through WhatsApp. I took ten narratives of ten participants as a case study for deeper discussion and analysis. According to Sibai (2015), the validity and quality of the narrative does not come from the quantity of persons we interview because personal experience is different from one person to another I applied a snowball sampling technique by assigning my networks as the key informants to approach other participants. It was a semi-structured open interview. For some participants, I did multiple interviews to gain more data.

Most participants came from Java. The participants' ages range from 35 to 50. The participants are described as domestic workers. Their tasks include household work, babysitting and taking care of the elderly of their employers' family. They live in their employers' houses.

3 RESULTS AND DISCUSSION

I provided the Indonesian version of the excerpts of a case study of ten respondents. The Indonesian version is the original language used in the interviews to anticipate the lost in translation. There are some words and expressions that I cannot find the translation for in English because those words are too cultural. I even cannot find it in an Indonesian dictionary because there are no specific meanings of the words; they are only expressions to emphasize some words. I wrote the Indonesian language of the transcription before I translated it into English for this paper.

3.1 *On "the Heroes of Remittance"*

In Indonesia, the government applies the politics of economic developmentalism by pushing women to participate in the labor market as a cheap laborer. The women are expected to be stereotypically docile, detailed, and desirable whether as factory workers or migrant domestic. Some of the ubiquitous discourses on migrant workers as pahlawan devisa appeared in mass media including social media such as Facebook. One medium reported how the government calls them pahlawan devisa, while five media wrote about the ironic and traumatic experiences of these pahlawan devisa.

http://www.bbc.com/indonesia/laporan_khusus/2012/04/120423_nasibtki,
http://regional.kompas.com/read/2010/11/19/08161095/Nasib.Pahlawan.Devisa.di.Negeri.Citra,
http://harian.analisadaily.com/opini/news/kisah-pilu-pahlawan-devisa-kita/902/2014/01/23,
http://ekspresionline.com/2016/02/02/negara-menjulukinya-pahlawan-devisa/,
http://www.dw.com/id/ironi-pahlawan-devisa/a-19111231,
https://www.facebook.com/beritapahlawandevisa/

Twenty participants showed their resistance through their narratives. They made the stories available as first-hand narratives from persons working as migrant domestic workers. De Fina (2003) and De Fina and Georgakopoulou (2012) mentioned that narrative presents the self in relation to social experiences and makes sense of those experiences. Using a participation framework, we can put their narratives into the framework of taking a stance and showing alignment and mis(dis)alignment with respect to the issue and the omnipresent dominant narrative. Five participants showed their footings (Goffman, 1979, 1981) in performing their own stance, identities, and affiliation or (dis)alignment.

Tuti, Yati, Ranti, and Ningsih agreed that they are heroes, but not national heroes of remittance; they are heroes to their own families which they helped to survive. They felt that because they worked hard abroad, they deserved to be called "pahlawan devisa." Ranti said that the label is hyperbole produced by the state and perpetuated by mass media. Santi contended that the term does not benefit domestic workers because it keeps domestic workers from being exploited and abused by their employers. The participants disagreed whether the term was empty and did not follow up with any action to protect them from abusive employers. As De Fina and Georgakopoulou (2012) argued, narrative can be used as an argumentative device to counter-arguments against a certain position, such as proffering, backing, or disputing. Storytelling is not only about cooperation but also about conflict and confrontation. Some of the women were aware of how the government benefits economically from those working abroad but does not care about the lives of domestic workers.

The respondents claim that the failure to protect their rights as citizens to be free from violence denies the grand narrative of "hero of remittance" and their claim is not without reason. The government does very little, indeed almost nothing, when migrant domestic workers experience violence. The situation is in sharp contrast to the Philippines, where the "national heroes" narrative is popularly applied to their migrant domestic workers. In one case, Flor Contemplacion, a female domestic worker in Singapore faced a death sentence, and the people of the Philippines showed "people's power" in the streets, insisting the government lobby the Singaporean authority in support of their hero (Rodriguez, 2002).

Clegg (1993) pointed out that narrative can be either concealment or revelation. For the government, mass media, and other institutions as well as employers, the narrative of "part of the family" and "heroes of remittance" is a strategy of concealment of the real situations, manipulation, abusive relationships and the creation and reproduction of myth to take advantage of domestic worker's labors and services. For domestic workers, narrative is revelation because they voice their opinions, feelings, perspectives, and stances. The narrative is deployed to maintain and perpetuate the myth, defray their resistance, and keep them docile.

All participants thought that the government has to care for migrant domestic workers and fight for their rights. The government tends to ignore cases of violence faced by domestic workers in the workplace. It cares only about the money these workers send through remittances and how it benefits from the recruitment, training, and departure, etc. Ranti said clearly that domestic workers like herself are not heroes. All participants found a contradiction in pahlawan devisa because the government treats domestic workers as second-class citizens rather than as heroes.

The respondents highlighted the government's role in protecting those working abroad from violence and abuse as well as fighting for their rights. All participants showed resistance toward the government and the ambiguous narrative it has created. We can see how they positioned themselves with respect to violence against domestic workers, the role of the government and the term pahlawan devisa. They expressed their feelings and took a stance that disagreed with the ignorant government, and they dis-aligned from the government's attitude of not assisting workers experiencing violence and abuse. They criticized the government's contradictory attitude and demanded government concern on their behalf if it considers them heroes of remittance.

Santi and Ranti had mixed feelings when they explained their feelings and whether they agreed or disagreed. They are happy and proud of it; it is good; and they agreed about it. On the other hand, they did not agree because they were the only hero of their own family. Santi even said that she was no longer called pahlawan devisa when she came back to her village one day. They agreed that the term is hyperbolic in which they do not get benefit at all because there are many IFMDWs who still have difficulties in their workplace and the government does not really care about them. All twenty participants agreed that the Indonesian government does not pay attention to them, so the term does not have any meaning and benefit for them.

I analyze contextualization cues, such a pronoun choice in expressing either personal or collective identity as domestic workers. The women used the first-person pronoun I in their narratives.

According to Kerby (1991), the use of I in narrative generates a sense of personal identity and creates individual meaning in their particular life stories for themselves and shows who the self is. The use of I differentiates them from other domestic workers and emphasizes their own personal experience, story, and perspective. Though there are many domestic workers, each participant pointed to herself as the subject and not as a representative of others. They used they and them when referring to other domestic workers. Santi, Tuti, and Ningsih used them and they when talking about some domestic workers facing problems from their employers, such as physical abuse, being unpaid, etc. Tuti and Ningsih said perhaps they had these problems because they were illegal. They differentiated between documented and undocumented domestic workers; they believed undocumented migrant workers will experience problems the government will not care about. We see the exclusion of others by blaming undocumented migrant domestic workers.

The tone or intonation is another key part of the narratives. Yati, Ranti, and Ningsih used them and when narrating about the government they viewed as unwilling to fight for their rights while caring only about their money. Ranti used we and us to draw a collective identity among domestic workers to emphasize the oppositional position between them (the government) and we (domestic workers) to express that the government does not care about domestic workers. Domestic workers showed resistance to and dis-alignment with the government. When talking about the ignorant government, they raised their intonation and emphasized their speech to show the importance and disappointment with the government. At the same time, they hoped the government might actually treat domestic workers as "real heroes" and be helpful to them. Tuti and Yanti used "our" when both mentioned the Indonesian government. In my understanding, the use "our government" was meant to build an affiliation with the interviewer because we belong to the same country to show their disagreement and invite me to have an alignment with them that the Indonesian government is not responsible and does not pay much attention to the problems that IFMDWs have.

3.2 *On the narrative of "Part of the Family" (Bagian Keluarga)*

Like "heroes of remittances," "part of the family" is a dominant narrative about domestic workers. It refers to a work relation in which domestic workers work and live in their employers' houses with their employers. The concept comes from employers and the agency to give a good impression and lure prospective domestic workers. Most domestic workers do want to be treated as part of their employers' families. How does this concept really work? How do IFMDWs understand this narrative? How do they feel? Do they have their own interpretations and expectations of this narrative? Research organized in 2006 by Rumpun Gema Perempuan in which I was involved discovered how the narrative of "part of the family" disguised the nature of employee-employer relationships and blurred the work relationship. By being positioned as "part of the family," domestic workers' status as workers was never clear (Irawaty, 2008).

All participants realized that the narrative of "part of the family" is a lip service, metaphor, and hyperbolic. The respondents face an ambiguous circumstance of a personal-emotional relationship. Ayu's expression was clear about what personalism is in her response to "part of the family." As a domestic worker, she has a right to enjoy free time when she has finished her tasks. She does not want to sell her "personhood" (Anderson, 2000). In a feudalistic relationship, employers "own" the personhood of their domestic workers as the basis of loyalty and obedience (Irawaty, 2008). This relationship still continues in modern capitalism. Zarembka (2003) viewed that migrant domestic workers endure a form of contemporary slavery. Rena, one of the respondents, realized that if she is regarded as part of her employer's family, she will not be paid for doing household chores and taking care of an elder family member.

The analytical framework is based on the literature on "part of the family," "like one of the family" (Harris & Childress, 1986), fictive kinship (Tappert & Dobner, 2015), fake family, "not a member of the family," "not one of the family" (Anderson, 2000, 2002;; Catherine, 2008; Lan, 2006), and "like a daughter" (Young, 1987), maternalistic and personalistic relationships, parent-children relationships, master-servant relationships, pseudo families, quasi-families, imagined communities, surrogate family, etc. Um (2015) mentioned, "one of the family" leads to the rejection of the legal status of care workers as workers and allows employers to pay their domestic workers inadequate wages and to insist upon flexible labor while at the same time denying the responsibility of protecting the basic rights of workers.

Domestic workers are aware of the physical and psychological boundaries between them and their employers. They feel they are not "like one of the family" and sometimes they feel they are. As Anderson (2000) pointed out, the disadvantages of a domestic worker being incorporated into a family circle far outweigh the advantages, such as lower wages and hiding the most discreet forms of exploitation. Jackson (2017) underlined the complicated relationships to live and love in the racial and social hierarchy of domestic work. The phrase used to elucidate affection of employers for domestic workers blurs and conceals the labor conditions and unprotected work.

A respondent, Shelly, thought that her employer treats her as part of the family. However, she contradicted herself by saying "I wish I could be part of their family" which is in my opinion ambiguous because the linguistic term "pengennya sih" (wish) she expressed is an informal language to refer to a conditional sentence. She talked about a hypothetical situation that did not happen. Another informant, Fitri, expected that all employers are supposed to treat their domestic workers as part of their families, but she evaluated and contradicted herself at the end of her sentence saying they also want to be treated as workers and be paid adequately. Fitri said that being considered part of the family means that they hoped to have free time and freedom after finishing their work. This implies that when they work, they want to be treated as workers; when they are done working, they want to be treated as part of the family. What they said does not happen yet; that is their hope and dream. They asked for freedom, recognition and appreciation of their work and needed "me time" as other family members have.

In positioning themselves within the narrative of "part of the family," the respondents tried to make sense of their stories and experiences through narrative production as a way of negotiation. De Fina (2003) argued, through narrative, people can create and negotiate understanding of social realities; narrative can be a site of resistance and transformation to show agentive capacity. The informants express their agreement and disagreement with the narrative and voice their own notions. De Fina and Georgakopoulou (2012) called this process negotiations and shifting (mis)alignments. They shifted their alignment and (mis/dis)alignments in their narratives.

Following Bakhtin (1935), I argue that participants use "double voicing" within the dialogue themselves. They had two different voices that indicate that they seemingly have two personalities with two different voices but, in my view, they applied a dialogical approach within themselves in terms of taking a stance and showing affiliations and alignments. They have mixed feelings when they explain their feelings and whether they agreed or disagreed. They addressed that they felt happy to be acknowledged as part of the family. However, they did not think and feel that they were treated as part of their employers' families. They agreed that the phrase is only a term created with a specific purpose to make them feel good about the relationship, but the "part of the family" never happened and will not happen since they have the boundary with their employers in terms of social and economic status, class, race/ethnicity, religion, and nationality.

I discuss the contextualization cues, such as linguistic preferences, intonation and code switching as a way of indexing a person by using pronouns as indexing group identity (Koven, 2015) or as indexing in-group and out-group member (Modan & Shuman, 2011) and shifting in and shifting out of collective identities (Van De Mieroop, 2015). Looking at the intonation, I sometimes found the informants internalized an inferiority syndrome through whispers or lowered voices. Rena did so when she called herself a "pembantu" (literally, helper) who works in her employer's house. When she said she wanted to be treated as part of the family, she raised her voice as if she expected her employer would do that for her.

Shelly did code switching from Indonesian to English when she was copying what her employer told her. She switched to Javanese, in the middle of her conversation throughout the end of her story. Four participants did code-switching from Indonesian to Javanese only for some words but for less than half of their conversation. Since all of the participants are from Java, they are bilinguals or multilingual. Some respondents from Java code-switched several times during our conversations. Most of them spoke Indonesian with a Javanese accent. Two respondents come from Lampung, but they are Javanese. Shelly switched from Indonesian to English when she said it was impossible to express an emphasis; she raised her voice because, in her view, it is not possible for employers to treat their domestic workers as part of the family. She used English to express "impossible" that domestic workers will be considered as part of their employers' families.

Shelly and Ayu switched to Javanese in the last half of the interview. They did not ask me where I come from and assumed that I understood Javanese. Both invited me to join her for having a collective identity as a Javanese person. I was honored that they trusted me and developed a friendly interaction by using their ethnic language. People do not use a local language unless they are familiar with their speaking partners. Though I was not born and raised in Java, I understand Javanese because my parents are Javanese. They positioned themselves with respect to me as an interviewer and invited me to be in her conversation. I found it interesting that Raisya used a repetition to say the same word three times, "lip service" to emphasize that "part of the family" is a lip service. I argued that she tried to align and build affiliation with me by switching code through one of the local languages that she spoke well.

I analyze the use of linguistic pronouns, such as we and I, us and them. All participants used I and switched to we when they talked about domestic workers as a whole; they used I when talking about their experiences. Rena used I during almost the whole conversation to talk about her employers and employers in general. Shelly, Raisya, Ayu, and Fitri used we when they discussed all domestic workers who hope their employers would treat them as part of the family and emphasized the loyal character of domestic workers. Shelly used them when commenting about new domestic workers who did not dare to report violence committed against them by their employers. They had to keep silent to keep working so they could pay their debts to their brokers. Shelly shifted out in terms of the collectivity among domestic workers because she was not part of the new domestic workers she was talking about. She showed that domestic workers do not all have and share the same problems. Collective identity is about inclusion and exclusion, belonging and not belonging, so shifting in and out is common when people tell a story or narrative.

4 CONCLUSION

Personal story plays a crucial role as a medium of expressing personhood and personal position in the lives of Indonesian female migrants who work as domestic workers in Singapore and Hong Kong vis-a-vis the dominant narratives of, in this case example, "the heroes of remittance" and "part of the family." They have an agentive capacity to produce alternative narratives or alternative interpretations about the dominant narrative. They assigned the alternative narrative to show critical responses against their employers, the government and the mainstream media that maintain the popular narrative. The process of producing the narrative through the interviews in this research became a medium by which they could express their voices and show their positionality among the dominant narratives. In the specific context of socio-linguistics, these female domestic workers were able to express certain words or phrases, such as the use of "I" and "they" or "I" and "we," in their narrative that indicates their positionality within the dominant narratives showing the socio-emotional situations of being included and excluded.

REFERENCES

Anderson, B. (2000). *Doing the dirty work?: The global politics of domestic labour*. Palgrave Macmillan.

Anderson, B. (2002). Just another job? The commodification of domestic labor. In B. Ehrenreich & A. R. Hochschild (Eds.), *Global Woman. Nannies, Maids and Sex Workers in the New Economy* (pp. 31–55). New York: London: Granta Publications.

Bakan, A. B., Stasiulis, D., & Stasiulis, D. K. (1997). *Not one of the family: Foreign domestic workers in Canada*. University of Toronto Press.

Bakhtin, M. (1935). The Dialogic Imagination. Austin: University of Texas. *First Published.*

Blommaert, J. (2005). Key topics in sociolinguistics. Cambridge: Cambridge University Press.

Catherine, C. H. (2008). *Work and Occupations*, 35, 507–508.

Constable, N.(2007). Maid to Order in Hong Kong: Stories of Migrant Workers. Ithaca, NY: Cornell University Press

Clegg, S. R. (1993). Narrative, power and social theory. *Narrative and Social Control: Critical Perspectives*, 21, 15–45.

De Fina, A. (2003). Identity in narrative: a discourse approach. In *Identity in Narrative: A Study of Immigrant Discourse. Amsterdam: John Benjamins Publishing Company* (pp. 11–30). Philadelphia: Joh.

Fina, A. de, & Georgakopoulou. (2012). *Analyzing Narrative: Discourse and Sociolinguistic Perspectives*. New York: Cambridge University Press.

Foucault, M. (1980). *Power/knowledge: Selected interviews and other writings, 1972-1977*. Vintage.

Giddens, A. (1991). *Modernity and self-identity: Self and society in the late modern age*. Stanford university press.

Goffman, E. (1979). Footing.

Goffman, E. (1981). *Forms of talk*. University of Pennsylvania Press.

Gumperz, J. J. (1982). *Discourse strategies*. Cambridge University Press.

Harris, T., & Childress, A. (1986). Like One of the Family: Conversations from a Domestic's Life.

Irawaty, D. (2008). Redefining Cultural Practices and Reconstructing a Colonial Legacy: Foundations for Female Household Workers' Empowerment in Indonesia. Brandeis University.

Jackson, K. C. (2017). "She Was a Member of the Family" Ethel Phillips, Domestic Labor, and Employer Perceptions. *Women's Studies Quarterly*, 45(3/4), 160–173.

Kerby, A. P. (1991). *Narrative and the self*. Indiana University Press.

Koven, M. (2015). Narrative and cultural identities. *Handbook of Narrative Analysis*, 388–407.

Ladegaard, H. J. (2020). Language competence, identity construction and discursive boundary-making: Distancing and alignment in domestic migrant worker narratives. *International Journal of the Sociology of Language*, 2020(262), 97–122.

Lan, P.-C. (2003). Negotiating social boundaries and private zones: The micropolitics of employing migrant domestic workers. *Social Problems*, 50(4), 525–549.

Lan, P.-C. (2006). *Global Cinderellas: Migrant domestic and newly rich employers in Taiwan*. Durham and London: Duke University Press.

Lazar, M. M. (2005). Politicizing gender in discourse: Feminist critical discourse analysis as political perspective and praxis. In *Feminist critical discourse analysis* (pp. 1–28). Springer.

Modan, G., & Shuman, A. (2011). Positioning the interviewer: Strategic uses of embedded orientation in interview narratives. *Language in Society*, 40(1), 13–25.

Ochs, E., & Capps, L. (1996). Narrating the self. *Annual Review of Anthropology*, 25(1), 19–43.

Rodriguez, R. M. (2002). Migrant heroes: Nationalism, citizenship and the politics of Filipino migrant labor. *Citizenship Studies*, 6(3), 341–356.

Rollins, J. (1985). *Between women: Domestics and their employers* (Vol. 71). Temple University Press.

Rollins, J. (1990). Ideology and servitude. *At Work in Homes: Household Workers in World Perspective*, 74–88.

Romero, M. (2016). *Maid in the USA*. New York: Routledge.

Sibai, S. A. (2015). Narratives of Spanish Muslim women on the hijab as a tool to assert identity. In *Identity and migration in Europe: Multidisciplinary perspectives* (pp. 251–268). Springer.

Sotelo, P. H. (2001). Domestica: Immigrant Workers Cleaning and Caring in the Shadows of Affluence (Berkeley and London. University of California Press.

Tappert, S., & Dobner, M. (2015). Being a member of the family? Meanings and implications in paid migrant domestic and care work in Madrid. In *Migrant Domestic Workers and Family Life* (pp. 276–299). Springer.

Um, S. (2015). Struggling to Make Time for Family: Work and Family Life of Korean-Chinese Institutional Care Workers in South Korea. In *Migrant Domestic Workers and Family Life* (pp. 257–275). Springer.

Van De Mieroop, D. (2015). Social identity theory and the discursive analysis of collective identities in narratives. In De Fina, A. & Georgakopoulou, A. (Eds.). The handbook of narrative analysis (pp. 408-428). Malden, MA: Wiley-Blackwell.

Van Dijk, T. A. (2008). *Discourse and power*. New York: Macmillan International Higher Education.

Young, G. E. (1987). The Myth of Being" Like a Daughter." *Latin American Perspectives*, 14(3), 365–380.

Zarembka, J. M. (2003). America's dirty work: Migrant maids and modern modern-day. In B. Ehrenreich & A. Hochschild (Eds.), *Global Woman: Nannies, Maids and Sex Workers in the New Economy* (pp. 25–47). New York: Henry Holt.

Art and culture

Innovation on Education and Social Sciences – Maureen et al (Eds)

Archiving local culture through transnational popular culture research

E.M. Dukut
Soegijapranata Catholic University, Semarang, Central Java, Indonesia

ABSTRACT: Popular Culture works have been seen as products that are not worthy for academic research. A factor for this is that the research into everyday life's cultures from the low to the middle class have not been of interest to academics. Through the lens of Cultural Hybridity and Phenomena perspectives for the analysis of the Popular Culture products on *punklung*, Jamaican reggae, MacDonald's rice porridge, TikTok, and Vaseline hand body lotion, this article shares a research result that the popularity is maintained by mixing local-global and irrespectively global-local elements. In this way, Popular Culture products can continue to be transnational, while at the same time archiving local culture.

1 TRANSNATIONAL POPULAR CULTURE

Popular Culture research is varied. Not only can there be talk of popular box office movies, music, or celebrities, but there is also talk of trendy clothes or makeup advertisements, famous food, or whatever else is being phenomenal in particular social media apps. Most of the time, a cultural product is regarded as popular due to the mass audience or consumers enjoying the product. The mass criterion is usually the key to popularity because having products available in mass means easy access to the creation and having it cheap enough to buy. After all, popular cultural objects are texts and practices of the working classes as well as texts and practices produced and consumed on a mass scale. Thus, they are usually mass-marketed products that many people use (Schuck, 2020).

The United States of America has been known to produce numerous kinds of popular culture. One original reasons is that US television shows have been aired throughout the rest of the world since the 1920s. Furthermore, due to the coming of the internet nowadays, these shows are even more popular with their reproduction or promotion through live YouTube channels, indicating that most popular culture products originate from the USA still. The Americas Got Talent show is one form of popular culture that has globalized worldwide due to its franchise for British, Australian, Indonesian, and many other countries' Got Talents. However, with the entrance of Western popular culture in its development to the East, the South Korean culture is recently trending, enabling the entrance of the Eastern culture to the USA, Indonesia, and other parts of the world.

An article on the Jamaican Rastafarian movement is another example of how transnational and influential a popular culture product could become. As a movement, it made use of reggae music to popularize its contemporary socio-religious Black Messiah movement. Only some people have been aware that within the last three decades, the "global penetration of the pop music" (Savishinsky, 1994, p. 260) was made successful through the mass-produced audio and video cassette recording industries of reggae music, which later developed into live shows on the television and in the form of live-streams on YouTube. Initially, entertaining lower-class neighborhoods of the Caribbean islands, the upbeat rhythm of the reggae music has interested young generations to follow the loose and freestyle of the colorfully clad Bob Marley with his authentic solid music. In Zimbabwe, it was reported that young fans of Bob Marley had "began sporting dreadlocks, smoking ganja, speaking Iyaric and wearing Rasta colors" (Savishinsky, 1994, p. 268). Therefore, transnational popular culture could start researching a global product that locals enjoy, or it could be the other way round, i.e., an originally local cultural product that becomes a global culture.

DOI 10.1201/9781003265061-25

2 LOCAL-GLOBAL, GLOBAL-LOCAL PHENOMENON

Recently, in Indonesia, there is a local popular culture that may become a globally acclaimed phenomenon. Almost like reggae, the phenomenon is a musical performance that chooses to play at the corner of traffic lights. The location selected has been understood by some passersby as a form of protest to the Covid-19 condition, which limited people to be entertained in a big field for the massive crowd. However, other individuals see that the phenomenon is one way to appreciate an attempt to elevate a local culture into a global culture.

The musical instrument used in the performance is a traditional instrument known as an angklung. If an angklung set, made from bamboo, is usually played with a traditional percussion known as kendhang by drumming on it with the palm of someone's hands, this musical set of angklung is paired with a western drum and cymbals. Alongside it is a xylophone or kulintang-like instrument, which is also made from bamboo. One great instrument is a homemade 25-liter empty bucket covered with cow's hide, which functions as the bass drum. Some people call the people playing *Calung Malioboro*; meanwhile, Punklung players are the people who are dressed up like punks (Dukut & Nugroho, 2021). Though only playing for a few seconds, depending on how long the red traffic light is on until it turns into green, it is obvious that it has attracted car and motor vehicle drivers to give their attention and smile for the quick one-stoplight entertainment. Interestingly, this phenomenon is found in cities on Java Island. With some of the performances uploaded to YouTube, it will likely be a local turned into a global popular culture.

Another phenomenon enjoyed by the masses during the Covid-19 condition, which has impacted society to work mostly at home, is the innovation of TikTok in social media. TikTok is a video platform application from China launched in 2016 (Yu, 2019). It allows a user to upload a 15–60 second video showing the user doing something with her/his current hobby. The hobby could be cooking something, singing a favorite song, making popular dance moves, and many more. Ever since March 2020, Indonesian TikTok users have increased by 20 percent (Massie, 2020). When users upload their TikTok video, they are considered successful when followed by other TikTok users. Although originating from China, a 16-year-old girl named Charli D'Amelio has been making choreographies and viral songs since June 2019; she managed to be successfully followed by 90.2 million followers in a year (Leskin & Haasch, 2020). It is proven that a local culture can eventually become a global popular culture indeed.

Global–local culture is evident in the US MacDonald's burger, which has been transnationalized in many countries. Starting with its branches in international airports, it quickly makes its way to building up MacDonald's restaurants in many cities worldwide. The popularity of the restaurant's menu, however, experiences a transformation. For example, in attracting locals to consume a McDonald's product continually, the franchise owners do not only sell the usual kind of menu found in the USA, but they also include some local menus which are packaged following the McDonald tradition. For example, with some Asians being more comfortable with rice than bread as their carbohydrate source, some Indonesian MacDonald restaurants provide rice porridge or rice rather than the usual bun, sausage, and omelette. Indeed, popular culture products do find ways to keep themselves continually popular.

3 METHODS

Most people take popular culture products for granted. Because it is something they deal with in their everyday lives, they never thought it could be valuable data for research. Using my popular culture class to make my students realize the rich data are in front of them, I encourage them to archive their local culture to show how popular culture is continually developing its transnationality.

As a process, my students are initially given an understanding of the definition and characteristics of popular culture. After this, they are given readings to study and determine why they point to a popular culture product. Next, they are asked to observe their surroundings and find a popular culture product to discuss in class. As a final exercise, after receiving input on what is right or wrong with the local popular culture, which can employ cultural hybridity and phenomena theory to support their explanation, they write it up as an academic paper.

In one class term, there are around thirty students. However, only two things will be discussed for this article because their class academic paper was then developed into their undergraduate thesis. One is about the TikTok phenomena that originated from China. The other is about the popularity of Vaseline hand body cream, which originated from the USA and became a favorite for Indonesians due to its whitening formula. As the major sponsor for their undergraduate thesis, I work closely with them to support my bigger research on popular culture, which adopts the cultural hybridity and phenomenon perspectives. The results of their study are discussed here.

4 RESULTS AND DISCUSSION

4.1 *TikTok phenomenon*

Catharina Maria Nora Praviana is a student that became interested in researching TikTok phenomena. In addition to informing that TikTok has been popularly watched and used by many Indonesians since 2017, she highlighted the condition that TikTok not only helped users archive personal videos, but it also made users narcist in a positive way (Praviana, 2021). As a popular culture, TikTok manages to fulfil the criteria of (1) creating a sense of pleasure for the users, (2) being easy to obtain because it is mass-produced, (3) having a low price, (4) manipulating in promoting itself, (5) creating a global-local lifestyle, (6) reflecting people's daily life, (7) reflecting a specific phenomenon, (8) temporariness, (9) practicality, (10) having a profit-making aim (Dukut, 2020). In providing continual popularity, Praviana (2021) also finds that Indonesian users used hybrid cultural elements in their TikTok uploads. One of those is the selfie uploading batik apparels, such as trousers, culotte, cardigan, sweaters, jumpsuits, and the already conventional long dress or skirt. Another form of cultural hybridity is how Erica Banks sang Buss It with a remix of Sundanese musical instruments. It is similar to the research above on mixing eastern and western musical instruments for *Punklung* music, which I did with a junior lecturer, who is under my auspices.

In seeing narcist as a positive value for TikTok users, Praviana (2021) considers that it not only entertains but also elevates self-confidence and positive interaction among other TikTok users. This is a positive fact as it increases the users' immune system to make them healthier during Covid-19 conditions. This narcist behavior is not a traditional Indonesian culture that was once known to be shy in expressing oneself and usually makes Indonesians have a low profile. With the entrance of TikTok, however, Praviana also finds that young people now have better self-esteem themselves. This is the reason why Indonesian TikTok users are increasing in number day by day.

4.2 *Vaseline hand body whitening and tanning*

Fei Bheola Putri Haryan's (2021) research compared two Indonesian Vaseline hand body advertisements and two USA Vaseline hand body advertisements. By using Roland Barthes' semiotics, she found that Indonesian advertisements have also used cultural hybridity elements to popularize Vaseline hand body advertisements among female consumers. Being a product originating from the USA, there has been the same branding in the font and colors used for the product name Vaseline. The apparels used by the models in the advertisement were also deliberately maintained in the navy blue or light blue color that Vaseline carries worldwide for its products. The model, however, is seemingly made local with Indonesian models who can be wearing a hijab and the other not. Both models, however, strive for the global Eastern beauty from Japan, South Korea, and the old-time Dutch colonialized time, as well as experienced by Javanese princesses in Indonesia, whereby the white complexion becomes the criteria for a standardized beauty. This is why the advertisements often contain texts or visuals that suggest how important are the elements of whitening someone's skin.

By contrast, however, in the USA, the standard of beauty has developed from white porcelain beauty into a tanned beauty. Due to multicultural ethnic values and the four weather seasons experienced in the USA, the white porcelain beauty has been regarded as a sign of paleness or illness. Tanned beauty is now considered sexier because it entails that women are more healthy-looking by showing how much more time a person can be doing sports under the sun. This becomes the reason why tanning salons are popular in the USA, in addition to a hand body lotion that would give a tanning tone to their skin.

5 CONCLUSION

Some kinds of research can be discussed to show how popular culture products can be researched academically. However, it could not become an interesting academic finding if there is no opportunity to train a junior lecturer and some students on how to use the criteria of popular culture with perspectives of cultural hybridity and phenomena for the analysis. As shown in the article, a talk about *Punklung*, Jamaican reggae, MacDonald's rice porridge, TikTok, and Vaseline hand body lotion are ways to show that possibility.

ACKNOWLEDGMENT

An acknowledgment is given to the English Department, Faculty of Language and Arts of Soegijapranata Catholic University for allowing me to research Popular Culture with my auspices, consisting of a junior lecturer and two undergraduate thesis students.

REFERENCES

Dukut. E.M. (2020). Jajanan Jalan Tol-Budaya Populer Makanan Cepat Saji. In *Dampak Jalan Tol Terhadap Pulau Jawa* (pp. 132-159). Semarang: Universitas Katolik Soegijapranata. Retrieved from: https://scholar.google.com/citations?view_op=view_citation&hl=en&user=7doKc8EAAAAJ&citation_for_view=7doKc8EAAAAJ:bKqednn6t2AC

Dukut, E.M. & Nugroho, Y.Y.T. (2021). History and Transformation of Calung to Punklung. *Makna: Jurnal Kajian Komunikasi, Bahasa, dan Budaya, 8*(1), 51–69. https://doi.org/10.33558/makna.v8i1.2468

Haryan, F.B.P. (2021). *A Popular Culture Analysis of Vaseline Hand Body Ads using Roland Barthes Semiotics* (Unpublished undergraduate thesis). Soegijapranata Catholic University, Semarang.

Leskin, P. & Haasch, P. (2020, Dec 25). Charli D'Amelio has taken over as TikTok's biggest star. These are the 40 most popular creators on the viral video app. *Insider*. Retrieved from: https://www.businessinsider.com/tiktok-most-popular-stars-gen-z-influencers-social-media-app-2019-6?r=US&IR=T

Massie, A. (2020). Kehadiran TikTok di Masa Pandemi (The Presence of TikTok in the Pandemic Era). *SSRN Electronic Journal*. DOI: https://doi.org/10.2139/ssrn.3633854

Praviana, C.M.N. (2021). *The Influence of Cultural Hybrid TikTok Application on Narcissism* (Unpublished undergraduate thesis). Soegijapranata Catholic University, Semarang.

Savishinsky. N.J. (1994). Transnational Popular Culture and the Global Spread of the Jamaican Rastafarian Movement. *New West Indian Guide/ Nieuwe West-Indische Gids*,68(3-4), 259-281. DOI: http://dx.doi.org/10.1163/13822373-90002653

Schuck, R. (2020). *Introduction to Popular Culture*. Retrieved from: https://tophat.com/marketplace/arts-&-humanities/communication- & -media/textbooks/introduction-to-popular-culture-raymond-schuck/809/25108/

Yu, J.X. (2019). Research on TikTok APP Based on User-Centric Theory. *Applied Science and Innovative Research, 3*(1). DOI: https://doi.org/10.22158/asir.v3n1p28

Innovation on Education and Social Sciences – Maureen et al (Eds)

Displaying power and solidarity through the pronoun use in "The Voice Kids Indonesia Season 4"

R.J. Firdaus & S. Setiawan
Universitas Negeri Surabaya, Surabaya, East Java, Indonesia

S. Weda
Universitas Negeri Makassar, Makassar, South Sulawesi, Indonesia

ABSTRACT: Pronouns can be used to determine the relationship between the speaker and the listener during a conversation. This case is applied to *tu* and *vous* theory, which discusses the formal and informal context. The study examines the theory of power and solidarity through the pronoun used by the speaker, such as to the coaches, participants, and others, to determine their social relationship. The study also adds the power theory to determine why the speaker can be said to have power. As a result, the study finds that four of six categories exist while the last two do not exist because the event includes an informal event. The coach tried to make the participants join their team and used casual words. The power is also found in coaches backed by music institutions, including music experts and famous singers, as well as in their suggestions and compliments that the participants listen to.

1 INTRODUCTION

Language is one of the important things used by people to communicate. Language has various types such as written, spoken, and sign language. The first type people learn in their early years is spoken language. Abdul-Wahab's study (2016) claimed that speech greatly impacts the social life either of individuals or of communities. Thus, many researchers use this chance to analyze what people say in their daily interactions and linguistic choices to gain communicative goals. Wardhaugh (2006) added that someone chooses some decisions when he or she is speaking, such as the topic, way, and manner from words and sentences. He also mentioned that linguistic choices determine the social relationship between the speaker and the listener or the listeners in each case. Language cannot only be used to communicate with others, but it is also used to indicate personal identity either in individuals or in communities in the political, economic, and social context (Agung, 2016).

In society, some cases are sometimes found where the lower classes use formal language to the upper classes. At the same time, the upper classes use informal or casual ways but give formal ways too. According to Brown and Gilman (1968), the cases are found in *tu* and *vous* theory in which the analysis is taken from the way, words, and sentence of the conversation between speaker and listener. *Tu* and *vou*s do not only exist in the French language but also in other languages with other names such as English (thou/you), German (*du/sie*), Swedish (*du/ni*), etc. The pronoun is also one of the factors that can indicate the practice of *tu* (casual way) or *vous* (formal way) within the conversation. Pronouns are words that appear in other words such as nouns, noun phrases, or other pronouns to avoid repetition (Fajar, 2019) and address someone.

Speech can have power inside and it depends on the people who say it. Coultas (2003) demonstrated that the person could have language and power if the people are backed by an institution, such as police, law, government, etc. For instance, a motorcycle or car user normally follows the police's instructions, indicating the power in language when the police give a warning. Coultas

DOI 10.1201/9781003265061-26

(2003) did not only mention that the people backed by an institution can be said to have power in language, but other aspects may also happen to be said to have power in the language, such as social position, age, gender, etc. Power can also be applied in solidarity theory by looking at the pronouns of the conversation (Brown & Gilman, 1968).

A previous study by Susanti and Indarti (2016) analyzed the linguistic strategies in reflecting power and solidarity through divergent movies. They used an English movie as their informant and used a linguistic strategy to mark the solidarity. In Brata's study (2013), he analyzed the power and solidarity semantics in the Bible in the translation of cultural address terms. In Steinhauer's study (2010), he analyzed Indonesian gender words and Indonesian pronouns. The things that make the present study different are that the current research uses power and solidarity, intended by Brown and Gilman (1968), and power, intended by Coultas (2003), to show why the speaker can be said to have power. The present study uses a television programs, The Voice Kids Indonesia Season 4, on the GTV channel as the informant, which has not been a research object in any prior studies.

The present study aims to examine the pronoun use and the power in The Voice Kids Indonesia Season 4 on the GTV channel on television. Television is one of the media that mostly evolves in the technology and information context (Candra, 2010). Mustika's study (2012) added that television programs almost always consist of entertainment because people mostly like watching entertainment programs on television. The present study focuses on the pronouns used by coaches, participants, and other speakers and on the power within the conversation. This study is important because it reveals the role of pronouns in determining the power-solidarity relationship among participants. To ease the analysis, this study has created three questions outlined to focus on the specific object and not to be broad. The questions are: (1) What pronouns do the speaker use to the listener and vice versa? (2) What is the theory shown in the conversation? (3) What aspects influence the conversation in indicating power?

1.1 *The pronouns of power and solidarity*

According to Brown and Gilman's study (1968), the relationship between participants in a given conversation can be decided through the use of pronouns. They used *tu-vous* theory applied in power and solidarity terms in which the pronoun is taken from the power and solidarity view. The *tu-vous* theory is a theory that discusses the way the speaker is using the language "Do they use the casual or familiar language or formal language?" This theory is firstly found in French in which *tu* addresses a single person and consists of someone who is familiar while *vous* addresses strangers or someone with respect and conveys formality (Ismail, Aladdin, & Ramli, 2014). Wardhaugh (2006) demonstrated that the relationship could be looked at through the use of such address terms as first name, last name, title, nickname, etc.

Ismail et al. (2014) also added detail about *tu-vous* theory by giving examples in social life. *Tu* is used when a child speaks to his or her mother or a friend talks to other friends, while *vous* is used when a student talks to his or her teacher. In some cases, a child can also use *vous* to his or her mother to give respect. The pronouns here are mister (Mr), Miss (Ms), professor, doctor, and other calls. Susanti and Indarti's study (2016) revealed that power is the asymmetrical relationship between both speaker and listener or listeners where one is subordinate to another, while solidarity is the symmetrical relationship where both speaker and listener or listeners indicate equality. Parents to children and teachers to students are examples of power and friends to best friends are examples of solidarity.

Brown and Gilman (1968) created six categories to showing the relationship between the speaker and the listener, presented in Figure 1.

Superior is when people with high status communicate with those who have high status; for example, a wife talks to a husband. Inferior is when the people who have low status communicate with those who have low status; for example, children talk to other children. Equal is when the people with high status communicate with those who have low status and vice versa, for example, parents talk to their children. Solidarity, here, is when both speaker and listener or listeners use

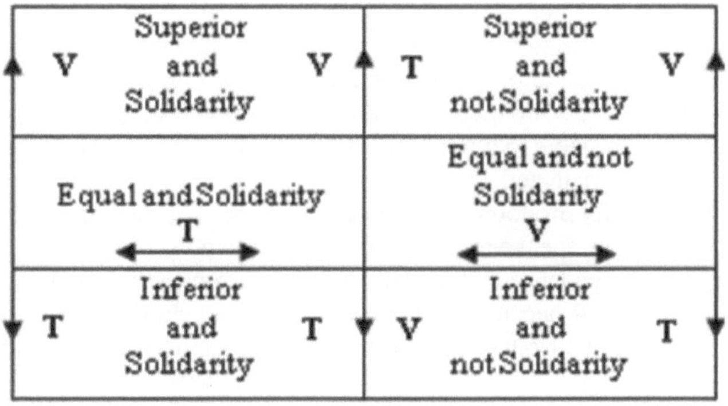

Figure 1. Six categories in Brown and Gilman's theory.

casual and friendly words to address each other, while not solidarity is when the speaker uses casual and friendly words but still the listener uses formal words to address each other and vice versa.

1.2 *Power*

Susanti and Indarti (2016) stated that power relates to authority, control, independence, and an extension of involvement. Coultas (2003) mentioned that power can be influenced by aspects such as people's institution, age, gender, situation or circumstance, occupation, authority or domination, etc. This theory is applied in the analysis to add more information on why people can have power in a conversation.

Coultas (2003) explained in more detail "what does the mean of language and power?," particularly in the English language, by giving such clues as: (1) age can be a factor when someone is considering as having power – a young and an old person can be considered as powerless within society because they sometimes feel that they are not being heard; (2) some researchers claim that in society men have more power within communication; (3) social position can confer or deny the power; (4) the people who have skills in the use of language can be considered powerful because they can manipulate others to do what they say. The advertisement and media are the instances; (5) occupations can have more power because they have particular vocabularies than those who do not know the terminology; (6) some researchers state that those who have strong regional or ethnic language are less powerful because they are unable to access the education system and excluded from certain types of jobs or institutions; (7) the people can be considered as powerful if they are backed by an institution that is socially recognized as having power such as police, law, etc.; (8) people can have power when they can take control or talk over in a certain situation; and (9) English speakers from ethnic minorities often claim that they are less powerful because of the degree of access and they need to follow the standard form of language.

2 METHODS

This study uses qualitative analysis as the outcome of qualitative research in the form of sentences (Creswell, 2007). Creswell (2007), as cited in Mair's (1969) study, said that language is included in ethnographic research in which ethnography is a qualitative design that explains and analyzes the shared and learned pattern of values, behaviors, beliefs, and language of the culture-sharing group. This present study provides the English version since the conversation is Indonesian, as in Martini's study (2018).

This study analyzes The Voice Kids Indonesia Season 4 program on the GTV channel from a power and solidarity perspective but the data is taken from the YouTube channel to allow the study to watch the videos repeatedly. The reason why the present study takes this object is that the program is a music competition where it gives different classes either for the judges (coaches) or the participants, meanwhile the program was designed for children around 7-15 years old. It can be a chance to analyze the solidarity since the judges should capture the participants to join their teams. However, it is possible that non-solidarity appears. The competition also gives a lot of communications that can be analyzed.

This study is limited to only pronoun use in data collection. The data are in the form of conversation and looking for why the speaker can be said to have the power within the conversation from other aspects, such as age, gender, etc. The data are taken from the Blind Audition round in The Voice Kids Indonesia Season 4 because in this round the speakers do a lot of communication to give feedback to the participants from coaches either in terms of their performances or their vocals.

3 RESULTS AND DISCUSSION

This study takes some videos to prove the power and solidarity theory. To allow the analysis, this section is segmented into six categories: (1) superior and solidarity, (2) equal and solidarity, (3) equal and not solidarity, and (4) inferior and solidarity.

3.1 *Superior and solidarity*

Superior and solidarity are conditions where the speaker and the listener have the same position as the upper classes. They address each other by using their first name or different names they created to determine the intimacy or solidarity between them. The evidence can be seen in (1) between Coach Isyana (CI) and Coach Yura (CY).

(1)

CY	:	We have the same time in turning, don't we?
CI	:	I think yes
CY	:	Hey!!!!, Allahu akbar, Isyana Sarasvati!!!
CI	:	What?, do I (*aku*) block you?
CY	:	Who's then? It's impossible if Marcell does it.
CI	:	Because I (*aku*) really want her.

In (1), the conversation indicates superior and solidarity because the coaches use the first name to address each other. When coach Yura calls coach Isyana by her complete name, the situation consists of joking about capturing the participant because coach Yura and Iky were blocked by coach Isyana. Then coach Isyana replies and uses the word "*aku*." Based on Murtisari, Fabrian, Lolyta, Lukitasari, and Rahardjono's (2019) study, the word "*aku*" indicates closeness. The power, stated by Coultas (2003), is found within the conversation when coach Yura argued with coach Isyana but coach Marcell did not join the debate. It indicates that older people are powerless since coach Marcell is older than other coaches. Powerless, in this context, is because coach Marcell does not want to involve himself in their debate. The coaches are upper class because Isyana (Hafidha, 2020) and Yura (Pamugarwati, 2020) are musicians.

3.2 *Equal and solidarity*

Equal and solidarity is a condition where the speaker and the listener have different classes but the speaker is equal to the listener and vice versa. In this category, the listener also accepts the equal by using the speaker's nickname or first name and putting himself as equal as the speaker. For example, when judges talk to participants, judges sometimes address a participant by his first name then the participant replies to the call by title. In some cases, the participants addressed the judges by an informal pronoun. The reply of an informal pronoun shows solidarity if the judges also use participants' first names or nicknames, as seen in (2) between Coach Isyana (CI), Coach Yura (CY), Coach Marcell (CM), and Faith (F).

(2)

CI	:	When I (*aku*) was listening to Faith's singing, especially when I (*aku*) turned, your (*kamu*) voice is very effortless. You (*kamu*) sing in your way. You (kamu) are not like the person who tries to change your (*kamu*) timbre, you (*kamu*) are just being you (*kamu*) ….
CM	:	The most touching is that I (coach) can feel your comfort. You (*kamu*) success conveying how comfortable you (*kamu*) are with either your (*kamu*) instrument, or your (*kamu*) voice, or your (*kamu*) performance. Those are very touching.
CY	:	…well we do not think about anything but just the feeling you (*kamu*) deliver to us through the song. But when you (*kamu*) were singing, whom do you (*kamu*) want to ….
F	:	When I (*aku*) am doing music, I (*aku*) always remember my (*aku*) grandfather because my grandfather made me learn piano well that is my (*aku*) first time to start falling in love with music.

In (2), the conversation indicates equality and solidarity because the coaches use the word "*kamu*" in which they tried to be equal as the participant but the coaches also address her by her first name. The participant also addresses herself using the word "*aku*." Murtisari et al. (2019) explained that the word "*aku*" is used when the speaker speaks with close people and is used in an informal context. Because the participant, Faith, used the word "*aku*," it means she also tries to put herself in the same position as the coaches to get the intimacy. The power, stated by Coultas (2003), is also found when the coaches give their suggestions and compliments to the participant because they are experts in music and know a lot about music. They are also backed by a music institution. Faith listened to the suggestions to improve her performance in the next round. The upper class is the coaches and the lower class is the participants.

3.3 *Equal and not solidarity*

Equal and not solidarity is a condition where the speaker and the listener have different classes but the speaker puts himself as equal as the listener and vice versa. In this category, the listener rejects to place himself as equal as the speaker by using the speaker's title and last name. For example, when a teacher talks to his students, a teacher talks to his student by student's first name or probably nickname and the student replies by the teacher's title, like "Mr" or "Ms." The similar case can be found in (3) between Coach Isyana (CI) and Ghaftan Rifqi (GR).

(3)

CI	:	Eh, you (*kamu*) see who turns into you (*kamu*) at the first time, right?. That means I (*aku*) want you to be part of my (*aku*) team. What is your (*kamu*) name?
GR	:	Ghatfaan Rifqi
CI	:	Age?
GR	:	13 years old
CI	:	If I (*aku*) sing the song "pamer bojo" with you (*kamu*), do you (*kamu*) want to join my (*aku*) team?
GR	:	Yes, *Mbak*

In (3), it can be seen that the relationship in the conversation between the coach and the participant is equal and not solidarity. In the conversation, the coach tries to put her position on the same level as the participant by using the pronoun "*kamu*" but the participant replies to the coach's call by using "*mbak*," which is addressed to the person who is older than the speaker in Javanese language. In (4), the power, as stated by Coultas (2003), exists because many words related to music context are used by the coach, such as stage act and performance. The participant also listened to the coach because the coach is a famous singer and expert in music experience (Hafidha, 2020), even though she has a team in producing music or we can say that she is backed by a music institution.

(4)
CI : If you (*kamu*) join my team, you (*kamu*) can get a lot of courses, such as how to do stage act is, how performance is to get the better performance in the future. If you (*kamu*) have succeeded, I (*aku*) promise you (*kamu*) will be directed by kak Iky

3.4 *Inferior and solidarity*

Inferior and solidarity is a condition where the speaker and the listener have the same position as lower classes. They address each other by using their first name or different names they created to determine the intimacy or solidarity between them. For example, when a person or a friend talks to another friend, they commonly address each other by their first names or probably unique names they created. By addressing their first names or unique names, it shows the solidarity between them. This condition can be seen in (5) between Mirai (M), Kak Omesh (KO), and Yoya (Y).

(5)
M : Hello Yoya.
KO : What do you (*kamu*) want to say to your (*kamu*) brother?
M : Yoya, pray for Mbak Aziel to be part of the grand final here.
Y : Never quit Kakak Mirai.
M : Thank you, Yoya.

In (5), the conversation indicates inferiority and solidarity. Both speakers, Mirai (M) and Yoya (Y), are inferior because they are a participant and a child (inferior). The solidarity is shown when Mirai calls Yoya by his first name and Yoya calls Mirai by her first name but adds "kakak" to indicate respect. Even though Yoya adds the title "kakak," it still shows solidarity because they are brother and sister. Yoya also addressed her by her first name. In this conversation, the power is found from the age aspect. The power, in this context, is shown in Mirai because Mirai is Yoya's sister or Mirai is older than Yoya.

Four of six categories are found in this study while two of six are not found: superior and not solidarity, and inferior and not solidarity. Two of the six categories in this study are not found because The Voice Kids Indonesia Season 4 is an informal event. The study also finds that the context can influence the relationship as in (5). The system of this competition is that the coaches should build their team by choosing participants. To get the participants, the coaches try to take friendly or casual words so that the participants do not get intimidated. Using casual and friendly words shows that the coaches try to be equal or put their position as the same as the participant. The power is also found when the coaches give suggestions and compliments to the participant. The participants listened to what the coaches gave. It shows that the coaches, Isyana (Hafidha, 2020), Yura (Pamugarwati, 2020), and Marcell (Marhendri, 2020), have power because they are famous singers, music experts, and backed by music institutions.

4 CONCLUSION

In society, people sometimes use formal or casual words and sentences to deliver their speech to listeners. This case creates a theory, called *tu* and *vous* theory in which one of the objects is the pronoun. Pronouns can be used to determine the relationship within a conversation. Coultas (2003) explained that power can be found within a conversation and influenced by some aspects, such as age, gender, situation, occupation, etc.

This study found only four categories in The Voice Kids Indonesia Season 4. Two categories superior and not solidarity, and inferior and not solidarity do not exist. This is because the program is an informal event. Moreover, the system is that coaches should build their team by choosing the participants. It means that they should speak casually to capture the participants to join their team. This study also found that all of the coaches have power because they are music experts, are backed by music institutions and famous singers, so all of their compliments and suggestions are listened to by the participants. The present study leads future researchers to prove the superior and not solidarity and inferior and not solidarity categories in different objects since those categories have not been proved in the present object.

REFERENCES

Abdul-Wahab, A. (2016). Power and Solidarity in Social Interactions: A Review of Selected Studies. *Journal of Language and Communication*, *3*(1), 33–34. Retrieved from: https://journalfbmk.upm.edu.my/ojs3/index.php/jlc/article/view/77

Agung, W. K. S. (2016). Solidarity Expressions Emerged in Addressing System as Linguistic Signal. *Jurnal Bahasa Lingua Scientia*, *8*(1), 21–30. DOI: https://doi.org/10.21274/ls.v8i1.293

Brata, F. I. M. (2013). Exploring Power and Solidarity Semantic in Translation of Culture Terms of Address in the Bible. *Linguistik Indonesia*, *31*(2), 187–205. Retrieved from: https://garuda.ristekbrin.go.id/documents/detail/950158

Brown, R., & Gilman, A. (1968). The Pronouns of Power and Solidarity. In J. A. Fishman (Eds.), *Readings in the Sociology of Language* (pp. 252–275). The Hague: Mouton & Co. N.V. Publishers. DOI: https://doi.org/10.1515/9783110805376.252

Candra, N. R. A. (2010). Perkembangan Media Penyiaran Televisi Menjadikan Televisi Sebagai Kebudayaan Masyarakat. *Jurnal Seni Media Rekam*, *1*(2), 187–198. DOI: https://doi.org/10.33153/capture.v1i2.497

Coultas, A. (2003). *Language and Social Context: Power*. London & New York: Routledge.

Creswell, J. W. (2007). *Qualitative Inquiry and Research Design: Choosing among Five Approaches* (2nd ed.). Sage Publications, Inc.

Fajar, M. (2019). Power and Solidarity Reflected in the Use of Pronouns Portrayed in the Hybrid Political Discourse. *Journal of English Education Linguistics and Literature*, *5*(2), 57–72. https://doi.org/10.32682/jeel.v4i2.991

Hafidha, S. I. (2020). Biodata Isyana Sarasvati, Penyanyi Multitalenta dengan Segudang Prestasi. *Liputan6*. Retrieved from: https://hot.liputan6.com/read/4375236/biodata-isyana-sarasvati-penyanyi-multitalenta-dengan-segudang-prestasi

Ismail, I. R. S., Aladdin, A., & Ramli, S. (2014). Vous ou tu?: Towards Understanding the Politeness Concept in French. *Procedia – Social and Behavioral Sciences*, *118*, 184–189. DOI: https://doi.org/10.1016/j.sbspro.2014.02.025

Mair, L. (1969). [Review of *The Rise of Anthropological Theory: A History of Theories of Culture.*, by M. Harris]. *Man*, *4*(1), 144–145. DOI: https://doi.org/10.2307/2799288

Marhendri, D. (2020). Tak Banyak yang Tahu, Marcell Siahaan Dulunya Ternyata Drummer Band Underground. *Merdeka.Com*. Retrieved from: https://www.merdeka.com/jabar/fakta-marcell-siahaan-dahulunya-adalah-drummer-band-underground-tak-banyak-orang-tah.html?page=all

Martini, A. (2018). Conversational Implicature of Indonesian Students in Daily Conversation. *Indonesian EFL Journal*, *4*(1), 93. DOI: 10.25134/ieflj.v4i1.889

Murtisari, E. T., Fabrian, D. D., Lolyta, R. D., Lukitasari, D. R., & Rahardjono, V. C. (2019). The Use of Indonesian First-Singular-Pronouns by Students Interacting with Teachers: Saya or Aku?. *Kajian Linguistik Dan Sastra*, *4*(1), 79–90. Retrieved from: https://journals.ums.ac.id/index.php/KLS/article/view/7811

Mustika, R. (2012). Budaya Penyiaran Televisi di Indonesia. *Jurnal Penelitian Teknologi Informasi Dan Komunikasi*, *3*(1), 1–6. Retrieved from: https://mti.kominfo.go.id/index.php/mti/article/view/35

Pamugarwati, A. (2020). Profil Yura Yunita, Pelantun Lagu Cinta dan Rahasia. *Kompas*. Retrieved from: https://entertainment.kompas.com/read/2020/03/12/160700766/profil-yura-yunita-pelantun-lagu-cinta- dan-rahasia-?page=all

Steinhauer, H. (2010). Gender and the Indonesian pronouns. *Wacana Journal of the Humanities of Indonesia*, *12*(2), 295. DOI:10.17510/wjhi.v12i2.119

Susanti, Y. E., & Indarti, Y. (2016). Reflecting Power and Solidarity through the Relativity of Linguistic Strategies in Divergent Movie. *Journal of Anglicist*, *5*(2), 145–151. Retrieved from: http://journal.unair.ac.id/download-fullpapers-anglicist74028b924ffull.pdf

Wardhaugh, R. (2006). Solidarity and Politeness. In *An Introduction to Sociolinguistics* (5th Ed.) (pp. 260–283). Blackwell Publishing Ltd. Retrieved from: http://staffnew.uny.ac.id/upload/132107096/pendidikan/Book+ for+Sociolinguistics.pdf

Innovation on Education and Social Sciences – Maureen et al (Eds)

National pride provoking heroes and anti-heroes in Ajidarma's "Eyewitness"

A. Salam
Universitas Gadjah Mada, Yogyakarta, Special Region of Yogyakarta, Yogyakarta, Indonesia

S. Ningtyas
Universitas Dian Nusantara, Jakarta, Special Capital Region of Jakarta, Jakarta, Indonesia

ABSTRACT: Seno Gumira Ajidarma's short story collection "Eyewitness" positions the conflict between Indonesia and East Timor as the narrative's core. Heroes and anti-heroes were born among East Timorese through the Act and political withdrawal. Both are non-ideological reactions motivated by national pride, but in different ways; the former is by taking action, and the latter is by refusing to take action. This article aimed to identify the two reactions and explain the national pride hidden behind the characters' subconscious. This qualitative research used a textual analysis method with a psychoanalytic approach and integrated Žižek's political theory and Smith's theory of national pride. The investigation results on sixteen stories showed that national pride leads subjects with different social, political, and cultural backgrounds to maintain territorial unity. In "Eyewitness," the national pride that underlies the Act and the political withdrawal both have significance in the independence history of Timor-Leste.

1 INTRODUCTION

Historically, East Timor has been an integral part of the unitary state of Indonesia, precisely when Timor was included in the territorial territory of the Sriwijaya and Majapahit Kingdoms (Strating, 2014). Not long after East Timor proclaimed its independence from Portugal in 1975, the Indonesian government invaded the territory by including it in the list of provinces in the state—although the sovereignty with the de facto government was never recognized by the United Nations (Nevins, 2002). This claim has an argumentative basis of improving the situation after hundreds of years of the territory being under colonial rule and especially because the majority of East Timorese are considered to have strong kinship relations with the Indonesian people.

During the two decades under Indonesian occupation, a conflict expanded between the pro-independence group who fought for liberty and the pro-integration group who desired East Timor to remain part of Indonesia. Because of the dispute, a quarter of the population of East Timor died as a result of physical fighting, starvation, and disease (Hill, 2001). The pro-independence party at that time, which was under Fretilin (Frente Revolucionária de Timor-Leste Independente), was the largest party that brought East Timor's independence from Portugal (Candio & Bleiker, 2001), and dreamed of self-determination. On the other hand, the separation of East Timor from Indonesian territory was feared to disrupt the unity of the Unitary Republic of Indonesia, given that Indonesia is an archipelagic state.

Political and physical tensions between the Indonesian government and those fighting for the independence of East Timor led to a massacre—which is even mentioned as a genocidal practice (Harper, 2005; Robinson, 2011), especially by independence authorities of East Timor, human rights activists, and reporters who spread the news from the field (Saul, 2001). One of the Indonesian journalists who witnessed the conflict, Seno Gumira Ajidarma, recounted what he witnessed in

DOI 10.1201/9781003265061-27

a collection of short stories entitled "Saksi Mata" (Ajidarma, 2002). The book, which was later translated into English with the title "Eyewitness" (Ajidarma, 2015), illustrates the position of the East Timorese, who were victims of this humanitarian conflict and the pro-Indonesian militias who played the role of power holders.

"Eyewitness" summarizes the roles of privileged subjects and oppressed subjects in the 1999 East Timorese crisis. By applying the concept of the Act and political withdrawal introduced by Žižek, this study revealed how subjects who experience oppression from the treatment of the dominant subject show a tendency to choose two different reactions: take radical action through the Act or withdraw through political withdrawal. Besides, national pride hidden behind the character's subconscious is also analyzed in this study.

Based on the researchers' investigation, several studies have been carried out previously related to the object of this research. Relating to the material object, Ferdinal (2013) explored several essential aspects of modern Indonesian literature, including censorship, resistance, and transformation. Short stories entitled "Saksi Mata," "Pelajaran Sejarah," and "Misteri Kota Ningi"—the three of them are included in the short story collection of "Saksi Mata"—were the main object of Ferdinal's research, which are considered to be included in the resistance literature. Ajidarma is considered to have used the power of allegory to fight the oppression in East Timor. Several other studies, such as those conducted by Hartono (2015) and Arifin (2019) provided an analysis from the perspective of hegemonic theory and political and aesthetic dissensus, while Wahyuni (2019) compared the role of journalism and literature in the literary work.

In addition to the research above, many studies focusing on literature review have been conducted regarding the formal object of this study. A thorough overview of Žižekian's practice of action is described in a study by Stejskalová (2011) in sequence, starting with Žižek's background of philosophical thinking, the emergence of the Act, to political withdrawal as an alternative to radical action. In contrast to the tone of criticism in Stejskalová's study, several other studies were appointed to show the important role of Žižek's practice of action (Bojesen & Allen, 2019; Kang, 2021; Schaap, 2020). The subject's radical action is seen as a tool to get out of the limits of capitalism (Robinson & Tormey, 2005), while political withdrawal is seen as a modern resistance that positions the subject outside the hegemonic symbolic order.

Based on the explanation above, it can be seen that this study presents a novelty by breaking down the negating practice of actions that leads to the emergence of heroes and anti-heroes in the short story collection "Eyewitness." Furthermore, this study draws a red line behind the two practices of actions that come from the same subconscious. As far as the researchers have explored, no studies have attempted to establish an interdisciplinary relationship between Žižek's theory of action and Smith's theory of national pride. Therefore, this study reveals the birth of heroes and anti-heroes in the short story "Eyewitness" by Seno Gumira Ajidarma, which manifests in two different non-ideological practices of actions.

1.1 *The act*

Departing from Lacanian psychoanalysis, Žižek makes a derivation of the conception of subjectivity, which states that the subject is always at a point where reality is inconsistent. This inconsistent reality—the barred Other, to borrow Lacan's term—automatically leads to the barred subject. Subjects who always feel lack will continue to strive to fill the void themselves by reaching the Real, which is anti-symbolic (Sidle, 2006; Zizek, 2008).

Against the background of inconsistent realities, Žižek argued that the ruling social order would not be able to shackle the subject through ideological interpellation continuously. However, there is a negative dimension that provides a way for the subject to withdraw themselves. The nature of the subject can lead them to freedom of subjectivity, where the subject has the potential to release the shackles of their social environment (Zizek, 2000).

Based on Žižek's perspective, the subject's attempts to withdraw from their natural circle are disruptive. These efforts put the subject in a condition of radical possibilities that open up new opportunities for change. The impulse that lives in the subject goes beyond their biological life and the pleasure principle. The impulse that Lacan adopted from Freud's concept of death drive then formed the subject. At this point, all the possibilities that arise will form radical actions that are revolutionary.

1.2 *Political withdrawal*

The radical action proposed by Žižek drew criticism from many thinkers because it was considered to open up opportunities for the emergence of leftist radical thought and practice. As an alternative to radical action, Žižek offered the idea of the politics of withdrawal. By adopting the concept of *Versagung* (denial of the symbolic order) introduced by Lacan, these politics indoctrinate the subject to refuse to take action. As a negation of radical actions, the subject in this phase is free from all dominant ideologies in the social order. Even further, the subject can move beyond the magnetism of hegemony.

Žižek explains political withdrawal through an analogy from Herman Melville's short story entitled "Bartleby, the Scrivener." This story presents the main character Bartleby, a worker who initially shows high dedication at work. However, the monotonous routine made him finally reach the culmination point where he refused everything his boss ordered. His statement, "I would prefer not to," becomes a legendary rejection statement and a sign of withdrawal (Stejskalová, 2011).

Žižek reformulated the reluctance to take such action as political withdrawal, which is also known as Bartleby politics. This rejection is a form of modern resistance that positions the subject outside the hegemonic symbolic order. While it may seem like a conscious move at first, Žižek insists that this withdrawal is a subconscious expression of the death drive (Sidle, 2006).

1.3 *National pride*

National pride is understood as a form of ownership, love, and loyalty of a group of individuals from the same historical background to a particular territorial unit (Smith, 1991). In simple terms, national pride is the emotion individuals perceive towards their country and nation (Smith & Kim, 2006). National pride, a collective sentiment, is based on individual identification and assessment of a nation based on the perception of the experience encountered by the majority of the group's members (Vlachová, 2019).

National pride—which is the emotional attachment between a particular individual and their country—is conceptually different from national identity, which is a cognitive and non-affective factor that determines the normative limits of one's involvement in a country (Morrison, Tay, & Diener, 2011; Reeskens & Hooghe, 2010). On the one hand, national pride can be based on positive feelings towards specific achievements of a state. On the other hand, national pride can also represent a person's belief in the superiority of their state, which is based on positive feelings about the characteristics of that state.

Individuals belonging to large ethnic groups tend to express higher levels of national pride than individuals belonging to small ethnic groups. In addition, the lack of political power by specific individuals is closely related to the deficient level of national pride. Individuals belonging to discriminated groups have the lowest level of national pride, followed by individuals belonging to vulnerable groups, junior partners, and self-excluded groups, respectively. In this case, political inequality is also strongly related to the lowest national pride (Ray, 2018).

2 METHODS

This qualitative research used a textual analysis method, one of the methods in literary research that involves close encounters with the works being explored and explanations related to detailed matters based on specific approaches (McKee, 2003). Since this study referred to Žižek's perspectives (especially those rooted in Lacanian psychoanalysis), the researchers applied a psychoanalytic approach to explore the meanings presented in the text.

The data sources of this study were taken from sixteen short stories by Seno Gumira Ajidarma, which are bound in a short story collection entitled "Eyewitness" (Ajidarma, 2015). The book is an English translation version by Jan Lingard and John H McGlynn of "Saksi Mata" (Ajidarma, 2002), which the author originally wrote in Indonesian. The research sample used consisted of quotations of words, phrases, clauses, sentences, and paragraphs taken from the translated work to facilitate the analysis process.

By referring to McKee (2003), the data collection technique in this study included several steps: (1) reading the short story collection of "Saksi Mata" (Ajidarma, 2002) repeatedly and thoroughly to obtain an initial understanding as an interpretation material; (2) reading the English translated version entitled "Eyewitness" (Ajidarma, 2015), which was the data source, repeatedly and thoroughly; (3) marking words, phrases, clauses, sentences, and paragraphs that have the potential to become research samples; and (4) categorizing the research sample by classifying it as data of the Act or political withdrawal with reference to the theory used.

After selecting and reducing the sample based on the relevance of the data to the topic raised in the study, the researchers conducted data analysis. Textual analysis was carried out by interpreting data related to the Act and political withdrawal that gave birth to heroes and anti-heroes and revealing how national pride is a subconscious driving factor for both actions. Finally, the conclusions of the research were drawn, and some suggestions were recommended for further research.

3 RESULTS AND DISCUSSION

3.1 *The Act versus political withdrawal as subjects' reactions in "Eyewitness"*

The struggle for independence that involves two parties with different views often ends in bloody clashes that reap many victims. This was also experienced by two opposing parties in the 1999 East Timorese crisis. In this case, East Timor was the side that desired independence, while on the other hand, Indonesia held the mission of maintaining the unity of the Republic of Indonesia. With these contrasting goals, the two parties were engaged in a dispute that could not be solved without the intervention of international parties (Ball, 2001; Burchill, 2000).

As Ajidarma later retold in "Eyewitness," the subjects involved in the conflict seem to be separated into privileged subjects and oppressed subjects. Indeed, it is undeniable that there will be parties who hold power and others who are disadvantaged in friction, either materially or spiritually. Regrettably, the privileged ones are often those who hold authority, especially the ruling government and the military personnel, while the oppressed ones are those of the community.

As in the case of "Eyewitness," the privileged subjects are military personnel who serve on behalf of the state. They are people in uniform who scooped out the eyewitness' eyes ("Eyewitness"), soldiers who chopped the ears of the rebels ("Ears"), an intelligence officer who arrests Manuel ("Manuel"), military escorts who string up Salvador's corpse over the gateway of the town ("Salvador"), potentiometer operators who give Januario's body electric shocks during his investigations ("Electricity"), and other authorities who hold control over subjugated victims in the 1999 East Timorese crisis.

When privileged subjects are confronted with a different reality, they tend to show active ignorance by excluding and marginalizing the experiences of others who are in a different position from them. At this point, social inequality emerged between the two groups. While active ignorance allows dominant subjects to feel comfortable in the inequality of the social structure that benefits them, epistemic friction arises with interruptions in the structure that continues to be reproduced (Schaap, 2020). In other words, epistemic friction is generated when marginalized groups expose their social perspectives to the public sphere.

The oppressed subjects demonstrated different reactions in responding to their perceived unfair treatment. As depicted by Ajidarma in his sixteen stories, some characters resist revolutionary actions, while others prefer to do nothing. In Žižekian's view, these subjects are both indicated to carry out the non-ideological practice of act. One side tends to confirm authentic political agency

(Stejskalová, 2011), while the other prefers to refuse any action in particular by withdrawing politically (Kang, 2021).

In the first short story with the same title as the short story compilation book, the eyewitness seeks justice by testifying in court. The story tells that a group of people dressed as ninjas scoop out the eyewitness's eyes while he is sleeping. The entire eye-scooping episode only happens in a dream, but his eyes are gone when he wakes up. However, even though he no longer has any eyes, he comes to the court to testify about a massacre he has witnessed.

"Remember it all well, because although there were many eyewitnesses, not one of them is prepared to testify in court except you."

"Yes sir"

"Once again, are you still willing to testify?"

"I am sir."

"Why?"

"For the sake of justice and truth sir." (Ajidarma, 2015, p. 5-6)

An excerpt of dialogue in the hearing above shows how the eyewitness sacrifices his safety by testifying in court in front of many people, the judge, the prosecutor, and the defense lawyer. It is even ironic when the one who is prepared to testify in court is a person who has lost a pair of eyes. What he has done is something revolutionary that can drive the mass to question justice and truth.

At the end of the story, the eyewitness has another dream, in which several men in ninja uniforms tear out his tongue. Because the hearing is adjourned and to be resumed the next day, the plot twist indicates that the eyewitness will testify in the next hearing even though he no longer has any tongue. As a representative of the massacre victims, the eyewitness becomes a hero by voicing the truth, even without any tongue to speak.

Still, in the same vein, the main character in "Clandestine" also incites the spirit of resistance. The story is about an underground movement built by people in the dark sewer tunnel of a city. Their goal is to take revenge, destroy the foundation of the city above them, and replace the old system with a better, more correct, and stronger one. The unnamed character is initially confused about the purpose of the people in the underground city. However, as he is seeking freedom of thought, he finally accepts the offer of the rebels' spies. The evidence can be seen in these excerpts of the following dialogue:

Once again my guide thrust the drill at me.

"This is our leader," he said. "This is our ideology." I still didn't understand, but I took the drill anyway.

"Resistance," he said again passionately, "our ideology is resistance. We don't care about winning or losing, we just keep resisting."

"Yes, but…"

"Our ideology does not condone questions. It can't be otherwise, our ideology is always right, perfect and faultless. Only by fanatical and militant adherence to it can we destroy our enemies. Don't keep asking questions. Do something. Prove your resistance with actions." (Ajidarma, 2015, p. 56)

As always underlies post-structuralism thinking, ideology is one of the factors beyond human subjects' control which always determines their social structures, leaving them with only little expectation for the political agency (Stejskalová, 2011). Proposing something novel as an alternative to these political deadlocks thus, Žižek suggests a concept of the Act. In this idea, as the one done by the leading character in "Clandestine" who keeps on drilling and crawling in the ground, the subject takes authentic revolutionary acts. He creates a new individual identity that is free from the symbolic order.

The following short story is told in a slightly more abstract way. Unlike the hero in the first short story who dares to appear alone as a radical subject, the epistemic heroes in "Ears" emerge as voices that incite rebellion. These heroes who seek justice by demonstrating an open resistance to the authorities fight the soldiers on the battlefield. The soldiers initially cut off people's ears as a warning of the risks they face if they incite rebellion. However, since the soldiers never know precisely who their natural enemies are, they eventually cut off the heads of anyone under suspicion.

All the soldiers were busy because everyone was resisting. Everyone was an enemy, and everyone was under suspicion. Rebellions flared up in every corner. The rebels whispered the spirit of struggle even into the ears of babies still in the womb (Ajidarma, 2015, p. 10).

On the other side, reactions that are completely different from those shown by the previous characters are also demonstrated by another group of characters in this book. By taking the opposite direction to the Act, marginalized subjects may also withdraw from the active ignorance perpetuated by the dominants. Not only by taking a subversive act, but the subject can also achieve self-preservation through political withdrawal.

In essence, characters with different missions can appear simultaneously in one short story. For example, in "Salvador," Salvador and Carlos Santana are both prominent models of epistemic heroes. Salvador is the leader of the people who forever exhorts rebellion, while Carlos Santana is the next leader who bravely takes control over late Salvador. Meanwhile, the character *I* in the story resembles an epistemic anti-hero since he refuses to take any action after seeing Carlos Santana's message written in blood on the walls.

On the walls of the gateway, he read words written in blood that was still wet and tricking to the ground.

I took the body of Salvador
our brave leader
I, Carlos Santana
Now lead the struggle

The wanderer turned his donkey around and headed away from the town, not going in after all.

He said to his donkey, "Let's be off, brother, I don't want to get involved."

The donkey nodded his head in agreement. (Ajidarma, 2015, p. 27)

Instead of joining Carlos Santana, who risks his life by being on the opposite side of the ruling government, the character *I* prefer not to get involved in any way. His passiveness is a sign of political withdrawal—a Žižekian response also known as Bartleby politics. Bartleby becomes a political role model in the modern form of resistance. Through his statement, 'I would prefer not to,' Bartleby appears to juxtapose executive power, or particularly the sort of power embodied by modern liberal cultures (Bojesen & Allen, 2019). By following in the footsteps of Bartleby, the character *I* in "Salvador" chooses to become an epistemic anti-hero.

The most radical form of political subjectivity can be observed in the response demonstrated by Januario. The story tells the life of a prisoner who is hit repeatedly by an electric current of 110 volts during an interrogation with the authorities. Instead of fighting back, he resignedly accepted every electric shock given along with each question asked to him by the interrogation officers. His resignation indicates a political withdrawal. As Bartleby does in Menville's story, the subject does not take any revolutionary action at all. He shows no reaction to the active indifference that continues to be socially produced and makes the withdrawal become his only form of political response.

3.2 *The national pride hidden behind the subjects' subconscious*

Based on the findings from the investigation in the previous section, it can be seen that the "Eye-witness" portrays two oppositely different groups of subjects. Epistemic heroes and anti-heroes are born differently from the many victims of violence in the 1999 East Timorese crisis. The former takes extremist action by doing the Act, while the latter refuses to take any action by demonstrating political withdrawal.

In Žižek's view, the marginalized subjects who do the Act have the potential to release the shackles of their social environment. Meanwhile, political withdrawal can also have the same impact by disrupting active ignorance from within its field of possibilities. Both are non-ideological. When subjects decide to take one of the practices of action, either to take action or to refuse to take action, they are not influenced by any particular ideology.

Through entirely different reactions, subjects with different social, political, and cultural backgrounds have the same purpose: to maintain the same territorial unity. Their response is motivated by national pride, a sense of belonging, affection, and loyalty shared by a group of people who have a standard historical connection to a particular territorial unit (Smith, 1991). Take an example from Manuel, the main character of a short story with the same title. A narrative fragment of the story is provided below:

> Our peaceful town was now full of foreign troops, with lots of spies everywhere and always suspicious of us. We went to school but weren't allowed to think our own way. We didn't speak our own language or study our own history and it was impossible for us to express our own ideas and aspirations because every time that happened, someone would always be arrested, tortured and thrown into jail without trial. (Ajidarma, 2015, p. 15)

Manuel tells his experience living in his country to a man whose identity is revealed as an intelligent officer at the end of the story. He is unfamiliar with everything in his own country, despite being born and raised there. Everybody is living under suspicion. Despite living amid the invasion, he still survives in his country even though he lives in the bush until seventeen. He eventually gets back to his town even though he feels that he is not his own master.

At the time of the dispute, official records are questionable. As depicted in "The Mystery of the Town of Ningi (or The Invisible Christmas)," the archives save invalid data. The figures cannot tell the correct information. Even worse, the history books do not somewhat tell the historical event to the next generation, as portrayed in "The History Lessons." The local community is as if being blinded by reality. However, with all these limitations, people keep struggling to build their nation. What the people have done is fundamentally based on the same motivation, which is national pride.

Junior is a representative of epistemic heroes in "Eyewitness." As the son of a guerrilla family, Junior has been raised and educated by Sister Tania since he was very young. Once he grows up, he has a great determination to help people who are struggling for independence. As shown in the excerpt of the dialogue below, he even promises to give his soul as a form of contribution to his homeland.

"Stay with us, Junior. Now that you've gotten an education, we'll look for work for you, something from which you can live. You'll be able to contribute to your homeland."

"I will give it my soul, Sister Tania. Plus, I want to find my mother." (Ajidarma, 2015, p. 74)

National pride has certain levels, as revealed in research conducted by Ray (2018). He argued that individuals with a low national pride generally come from discriminated groups, vulnerable groups, junior partners, and self-excluded groups. As an absolute part of these groups, Junior certainly does not have great national pride for the land of East Timor. Nevertheless, Junior desires to support his country with his whole heart despite his lack of national pride.

In the struggle for independence, particularly in the 1999 East Timorese crisis, in this case, the ordinary people are not the only ones showing their national pride. The Indonesian soldiers also came to the battlefield with a mission of national unity. Both parties held their national pride for their states. This can be observed in this narrative taken from "The Rosary" below:

> Soldiers weren't the only ones with the courage to die. Anyone had the courage to die to defend his life. Especially a life of independence and self-respect. (Ajidarma, 2015, p. 30)

The narrative above underlines that anyone, both the soldiers and ordinary people, dares to die. Ironically, they dare to die to secure their lives, lives explicitly of independence and self-respect.

The General also depicts sacrifice in the "Blood is Red, General." In the story, he is the commander of the soldiers in the dispute. He even nearly died when his head was hit by shrapnel on the battlefield. Behind the whole thing, he is highly proud of his battle scars. He considers his occupation as a soldier to be the most honorable among all professions, as soldiers devote their entire life to serving the state. The evidence is shown below:

> Not everyone could wear a wound with pride. The retired general was proud of his battle scars. He often thought there was no profession more noble than being a soldier. He based this opinion on his belief that once you became a soldier you were laying your life on the line. This was more than just a profession. (Ajidarma, 2015, p. 60)

The General's pride is motivated by national pride. Referring to the theory of national pride proposed by Ha and Jang (2015) and Vlachová (2019), there are mainly two factors underlying national pride. Authentic pride is commonly perceived by those who have a positive feeling of a state. However, since this pride is generally connected to rapid economic development, a low level of corruption—and most importantly—ideal government, it cannot be linked to the pride perceived by the General in the story. The General is more affected by hubristic pride that is closely related to characteristics of the state (e.g., its culture and traditions). With the position of Indonesia at that time having more decisive points over Timor-Leste, which was still struggling as a new country, it is not surprising that the General has great hubristic pride for Indonesia.

4 CONCLUSION

Ajidarma's "Eyewitness" represents the roles of privileged and oppressed groups from an actual historical event of the 1999 East Timorese crisis. The sixteen stories compiled in the book portray the victims' reactions through two contrastive practices of action: the Act and political withdrawal. The radical subjects who take revolutionary actions are born to be epistemic heroes, while the others who prefer to withdraw politically are epistemic anti-heroes. Nevertheless, these non-ideological practices are stimulated by the same energy source of national pride hidden behind the characters' subconscious.

REFERENCES

Ajidarma, S. G. (2002). *Saksi Mata.* Sleman: Bentang Pustaka.

Ajidarma, S. G. (2015). *Eyewitness* (Transl.: J. Lingard & J. H. McGlynn). Jakarta: The Lontar Foundation.

Arifin, Moch. Z. (2019). Menim(b)ang disensus: Politik dan estetika Seno Gumira Aji Darma dalam cerpen saksi mata. *Atavisme, 22*(1), 47–60.

Ball, D. (2001). Silent witness: Australian intelligence and East Timor. *Pacific Review, 14*(1), 35–62.

Bojesen, E., & Allen, A. (2019). BARTLEBY IS DEAD: Inverting common readings of Melville's Bartleby, the Scrivener. *Angelaki – Journal of the Theoretical Humanities, 24*(5), 61–72.

Burchill, S. (2000). East Timor, Australia, and Indonesia. *Bulletin of Concerned Asian Scholars, 32*(1–2), 59–65.

Candio, P., & Bleiker, R. (2001). Peacebuilding in East Timor. *Pacific Review, 14*(1), 63–84.

Ferdinal, F. (2013). Censorship, resistance and transformation in modern Indonesian literature. *Mediterranean Journal of Social Sciences, 4*(1), 269–272.

Ha, S. E., & Jang, S. J. (2015). National identity, national pride, and happiness: The case of South Korea. *Social Indicators Research, 121*(2), 471–482.

Harper, E. (2005). Delivering justice in the wake of mass violence: New approaches to transitional justice. *Journal of Conflict and Security Law, 10*(2), 149–185.

Hartono. (2015). Cerpen Saksi Mata karya Seno Gumira Ajidarma sebuah analisis dengan teori hegemoni. *Diksi, 16*(6), 97–110.

Hill, H. (2001). Tiny, poor and war-torn: Development policy challenges for East Timor. *World Development, 29*(7), 1137–1156.

Kang, W. (2021). I would prefer not not-to: Critical theory after Bartleby. *Interventions, 23*(3), 356–367.

McKee, A. (2003). *Textual analysis: A beginner's guide.* London: SAGE Publications.

Morrison, M., Tay, L., & Diener, E. (2011). Subjective well-being and national satisfaction: Findings from a worldwide survey. *Psychological Science, 22*(2), 166–171.

Nevins, J. (2002). The making of "ground zero" in East Timor in 1999: An analysis of international complicity in Indonesia's crimes. *Asian Survey, 42*(4), 623–641.

Ray, S. (2018). Ethnic inequality and national pride. *Political Psychology, 39*(2), 263–280.

Reeskens, T., & Hooghe, M. (2010). Beyond the civic-ethnic dichotomy: Investigating the structure of citizenship concepts across thirty-three countries. *Nations and Nationalism, 16*(4), 579–597.

Robinson, A., & Tormey, S. (2005). A ticklish subject? Žižek and the future of left radicalism. *Thesis Eleven, 80*(1), 94–107.

Robinson, G. (2011). East Timor ten years on: Legacies of violence. *Journal of Asian Studies*, *70*(4), 1007–1021.
Saul, B. (2001). Was the conflict in East Timor 'genocide' and why does it matter? *Melbourne Journal of International Law*, *2*(2), 477–522.
Schaap, A. (2020). Do you not see the reason for yourself? Political withdrawal and the experience of epistemic friction. *Political Studies*, *68*(3), 565–581.
Sidle, S. D. (2006). Resisting the urge to do nothing. *International Journal of Zizek Studies*, *11*(1), 115–118.
Smith, A. (1991). *National identity*. Reno, NV: University of Nevada Press.
Smith, T. W., & Kim, S. (2006). National pride in comparative perspective: 1995/96 and 2003/04. *International Journal of Public Opinion Research*, *18*(1), 127–136.
Stejskalová, T. (2011). Žižek's act and the literary example. *Moravian Journal of Literature and Film*, *2*(2), 55–74.
Strating, R. (2014). Contested self-determination: Indonesia and East Timor's battle over borders, international law and ethnic identity. *Journal of Pacific History*, *49*(4), 469–494.
Vlachová, K. (2019). This country is not for anyone: Explanations of low national pride in the Czech Republic. *Nationalities Papers*, *47*(6), 1000–1012.
Wahyuni, D. (2019). Pertarungan jurnalisme dan sastra dalam menguak kebenaran. *Paradigma: Jurnal Kajian Budaya*, *9*(3), 231.
Žižek, S. (2000). *A ticklish subject: The absent centre of political ontology*. London: Verso.
Žižek, S. (2008). *The sublime object of ideology*. London: Verso.

Innovation on Education and Social Sciences – Maureen et al (Eds)

Zine as alternative media: A case study of football-themed zines

B.M. Mahardika
Universitas Gadjah Mada, Yogyakarta, Special Region of Yogyakarta, Yogyakarta, Indonesia

P. Retnaningdyah
Universitas Negeri Surabaya, Surabaya, East Java, Indonesia

ABSTRACT: Popular culture has changed a lot from time to time in terms of form and function. This includes the use of popular culture as a medium to express the producers' voices. As a product of popular culture, zine is closely associated with stigmatized groups such as football fans. This paper aims to explain the spirit raised by zinester in producing a zine and the ideas they convey to the readers. Data were obtained through interviews with three sources, namely the authors of Sorak-Sorai Zine, Footballtropika, and Fortuna Zine. Those three sources are the authors of football-themed zines. The result of this paper is that among football fans, zines are used as a medium of expression because so far, through the mass media, football fans are often considered something that is given a bad stigma and through zines, football fans finally have unlimited space and do not need to rely on mainstream media. By producing zines, the author of Sorak-Sorai Zine, Footballtropika, and Fortuna Zine had shown the spirit of do-it-yourself culture.

1 INTRODUCTION

The existence of fast-spreading mass media such as newspapers and television has changed the order of ideas in society. The spread of ideas originating from other countries is also carried over to a country by the mass media presence. Things that are popular in society also cannot be separated from the role of the mass media, which continues to discuss and provide information related to the popularity that many people like. Popular culture has its characteristics and forms in different periods of time. Popular culture is also almost always associated with something popular that existed in the past. The commercialization of popular culture is getting easier to find and obtain information because of industrial and economic demands increasingly pressing the community. The definition of popular culture also varies; even one definition defines it as a culture that many people like, significantly depending on the calculation of sales of goods such as sales figures for CDs, DVDs, books, and the number of attendees at a concert or sports event. This definition is a quantitative definition based only on calculations. In other words, something that is popular can be proven by its popularity (Storey, 2014),

Do-it-yourself (henceforth, DIY) culture is one of the sub-cultures that emerged in response to pre-existing popularity. DIY culture can also be said to be a counter form to something that previously dominated a particular segment. The debate regarding whether popular culture can be categorized as a culture has occurred. Still, one thing that cannot be denied is that popular culture has become a habit that accompanies and even becomes a guideline for the daily life of certain people. The term zine is very closely related to magazines or magazines, but several aspects distinguish the two. By definition, zines are magazines published independently and with minimal capital to construct identity, create community, and explore social activism (Chidgey, 2020). This non-commercial nature is what distinguishes zines from magazines in general which are usually found

 DOI 10.1201/9781003265061-28

in bookstores. By category, 'zines' can be divided into 15 different genres: fanzines (related to music, television, sci-fi, etc.), political and activist zines, personal zines, underground scene zines, sex zines, travel zines, comics, literary zines, art zines, and others that are not included in typology (Duncombe, 2008) . Zine as a form of DIY culture is generally written by non-professionals and produced and distributed by the author (Spencer, 2005). In addition, Zine is a product that is not paid for and is usually only distributed to the closest people or within the same subculture. Interestingly, the people who write zines are not very money or profit-oriented, which means specific reasons cause the idea to be applied. Zine has been produced by many people from various regional and political backgrounds. What can be known more profoundly is whether in the modern era and capitalism as it is today, zinesters still apply the early characteristics of zines, such as not being profit-oriented and the author's form of expression.

Zines are not trivial products but are diverse voices from the underground world trying to show their identity amid capitalism and the shadows of the mass media (Duncombe, 2008). It can be seen that DIY culture has a solid community base but tends to be narrower because it has been focused on one subculture (Spencer, 2005). Janice Radway (2018) argues that zines are produced as a response to commodification of mainstream society. Considering its 'rebellious' characteristics (Ferris, 2001), zines do not really follow any specific formats in terms of content and design (Duncombe, 2008).

A number of studies have been conducted to investigate the various purposes of zines. In the area of literacy research, zines are examples of literacy as situated social practices that are constantly negotiated with people's roles and goals of the social groups (Barton & Hamilton, 2000) Previous literature has shown that there are at least two approaches to the study of zines: zines to show resistance and zines to tell one's life stories (Kempson, 2015). As DIY culture, zines have been used as alternative media for underrepresented groups to express the voices that show how girls use zines to express their voices as adolescence and develop social capital through network and support from their peer supporters (Schilt, 2003). While zines may not have an explicit intention to promote social change, some might have had cultural or political impacts by increasing public awareness of a particular issue (Radway, 2018). This is exemplified by Kempson's (2015) study of feminist zines, which reveals that zines also serve to voice instances of third wave feminism as a way to exercise feminist subjectivities. Studies on zines as life stories offer an alternative approach to zines not as a medium to challenge cultural and media hegemonies, but more as a channel to construct people's narratives which may not find a place to be distributed in mainstream media (Poletti, 2008). In this study on football zines, both approaches—zines as resistance and zines as personal life stories—can be traced through the contents and process of production. Carrying a similar characteristic of DIY culture, football-themes zines are argued to serve the purpose of offering different sides of Indonesian football which may not likely be covered in mainstream media. Moreover, the zines under study also inform the slices of life of people who are at the backstage of the football field. Borrowing Erving Goffman's (1959) dramaturgical concept of Frontstage/Backstage, this paper argues that football-themed zines demonstrate the zinersters' attempt to portray how people earning their living from football negotiate the spatial boundaries.

2 METHODS

Data collection was carried out directly by conducting interviews with three different sources to understand the broader and real context. These three resource persons are also people who have produced zines or commonly referred to as zinesters. Interviews were conducted with three sources at different times. Interviews at various times were conducted to avoid having answers that tend to follow other sources. The three interviewees were zinester Sorak-Sorai Zine on 5 June 2021, Footballtropika on 6 June 2021, and Fortuna Zine on 8 June 2021. The three sources were zinesters who wrote football-themed zines.

3 RESULTS AND DISCUSSION

3.1 *Zine as a form of expression*

The zine categorization covers different themes from personal affairs, art, and music to political elements. Zine as an alternative medium for expression was chosen by zinesters because there are no standard rules that require zine writers to exist and follow specific writing standards. In a written zine, of course, there are messages or ideas to be conveyed to the reader. When deciding to produce a zine, the initial purpose was for various reasons, and one of the main ones was actually to fight the domination of the mass media and build the mindset of the readers. The narratives constructed by the mainstream media are considered to have changed people's perspectives to become uncritical. Concerning football, the target community is the football fans themselves to think more sensitively and critically on events that involve the mainstream media and football fans. Sorak-Sorai's initial goal of producing a zine was to ignite and change the mindset of football fans in Indonesia who have always been the target of the mainstream media to become the "accused" for things that happened inside and outside the stadium.

> "*Basically to change people's mindset. How, from now on, we're aware of the existence of new sub-cultures among fellow supporters. Come on guys! Let's change our mindset that's still attached to the early 2000s. When supporters are associated with riots and chaos. This is what we've been trying to change. When you're stigmatized by the society, you know how to behave and react. Also, when the club you're proud of is having a situation, you know what to do to support.*"
>
> (Sorak-Sorai Zine, 2021)

Rebuilding the mindset, especially for football fans in Indonesia, is the main goal so that they are not easily perceived as a problem in the social life. In addition, efforts to build criticality towards clubs that are supported by supporters are also considered a foundation for maintaining a better relationship between fans and clubs. Lastly, zine as a product also symbolizes or has a purpose so that a zine is not considered trivial and lower than a book because, in the zine itself, the values of freedom of speech are solid.

> "*This zine is actually a medium that may be considered trivial, not worth mentioning in books or academic writing. I admit that zine is mostly based on opinions. But then the point is how the content is accepted by fellow supporters who come from various backgrounds.*"
>
> (Sorak-Sorai Zine, 2021)

Commercialized products such as books have indirectly created categorization in written works. This is what zinester Sorak-Sorai has been trying to change so far, that written results in the form of zines are also worthy of consideration and appreciation.

Zines are also an alternative media to express the author's opinion departs from the ideas and experiences that the author has done before. Zinester Footballtropika, who previously learned a lot from the street library movement, decided to start writing zines as a form of exhaustion for the football ecosystem in Indonesia.

> "*Ideally there should be writings related to football. Not just technical matters that show mismanagement in Indonesian football. Supporters' voices have never been represented.*"
>
> (Footballtropika, 2021)

Efforts to create a space that is not dependent on mainstream media is the goal of zinester Footballtropika so that Indonesian football fans have a place to voice and express their opinions. By doing this, the voices of football fans in Indonesia who previously had no space can finally speak their aspirations and views.

Likewise, zinester Fortuna Zine, who wrote about the world of football superstars, especially PSIM Yogyakarta, produced a zine. Around 2015, there was not much access to information and space owned by PSIM Yogyakarta supporters. For that reason, Fortuna Zine tries to build space so that supporters can have more space and even know information about PSIM Yogyakarta.

"To say it's about football...well, not really. It's more like a fanzine. In fact, it's a supporters' zine. So, lives of being supporters are what this zine is about. The sections in this zine talk about what happens outside the field. We talk to people who are involved in PSIM, but not technically about the game. There's also a section on players. But their lives not related to football. Such as what he does for a living apart from being a football player. And then we also write about rising stars. You know...new player who are on the spotlight. And then there's music closely identified with PSIM...and also the historical notes...with the help of bawahskor. And the last section is opinion section."

(Fortuna Zine, 2021)

As one of the active elements in football, football fans do not always have room for expression. On the other hand, many mainstream mass media reported more about riots involving football fans than reporting on creative activities carried out by football fans. Because the dominant mass media is more focused on portraying supporters on something that is not good, the efforts made by Fortuna Zine can be a solution so that the dominance of the mass media can be countered or even balanced. Zines are not trivial products but are diverse voices from the underground world trying to show their identity amid capitalism and the shadows of the mass media (Duncombe, 2008). Through the zine produced and the idea initiated by Sorak Sorai, Footballtropika, and Fortuna, it shows that as football fans who are often smothered by the culture of capitalism, such as the inner demand to buy original merchandise made by club officials and the bad news and stigma that is imaged by the mass media, they do not necessarily remain silent. This idea of a zine is something that is produced as an attitude to their use and judgment.

3.2 *No profit, just gain support*

Do-it-yourself culture is often associated with doing things independently. In terms of production, DIY culture relies on its capital (no other party to finance it). One important thing that needs to be built and maintained is communication with fellow DIY culture activists or, in this context, the zinesters. According to Spencer (2005), zinesters have an awareness that the strength of the zine community lies in communication between fellow zinesters by supporting each other to provide resources and share experiences with the aim that others are also able to produce their zines.

The independence of capital from other parties is an advantage that zinesters have to be freer in conveying ideas or propaganda to readers. Even when asked about the profits made from producing zines, Sorak-Sorai mentioned that the profits obtained so far were used again to produce future zines.

"We have never taken any profits. What we get, we use it for production. I suppose it's because our system is different from mainstream magazines. They are funded. This may be a unique characteristic of zine. But in fact there is a zine about a well-known group band in Indonesia. It's funded and can enter Gramedia bookstore. This created a controversy. There's also another interesting example. We have a fellow zine from Bandung. One whole page is open sponsorship. But then we cannot say it's an industry. They use a partnership system. Like...there is a friend who runs a garment business. They put it there."

(Sorak-Sorai Zine, 2021)

What Sorak-Sorai mentioned about the sales profits that are reused to produce zines and how the relationship with other parties, such as fellow zinesters or with other parties, supports what has been explained by Spencer (2005) about the significant role of communication between similar fellow communities. From this, it can be seen that DIY culture has a solid community base but tends to be narrower because it has been focused on one subculture.

Zine can also be produced from a person's point of view or a personal point of view. Footballtropika, a zine writer who uses his point of view, agreed with how relationships with fellow zinesters have a significant role to play in helping in making zines in the future. The same applies to the profit distribution of zines that are reused to produce future zines.

"From the beginning there has been a zero capital. We didn't even have tools for writing. So we borrowed from friends who had the equipment. Initially it was digitally distributed, but then I thought that this zine wouldn't receive appreciation if distributed for free. People may think that it's there to

be shared and uploaded. But not to be read. When the zine was printed and sold, it actually received more appreciation. We use the profit for production and we do research...gather data out of town. Once we distributed the zine in a digital format following the hard-copy one. It received quite an attention...until we had an issue with erroneous link. Having connections with fellow zinesters really helps us get more ideas and references.

(Footballtropika, 2021)

The fluid nature of zines makes it easy for anyone to contribute. There are no fixed rules or standards that make zines an alternative for people who feel they have no place to speak. According to Jesse (2010), zinesters and readers use zines as a medium to share and learn the kind of knowledge based on personal experience that can become a daily practice.

Meanwhile Zobl (2009) asserts that zines function as spaces for active participation and critical reflection. This is what Fortuna Zine applies in distributing zines. The selection of print zines aims to get a more comprehensive network of people interested in consuming their zines. In addition, Fortuna has their ideal idea of not getting paid for the zines they distribute because creating a zine in print format is for networking.

"It's free you know. Free so that we can meet new people personally when delivering it. When people want to pay, we don't accept it. From there...interesting conversations occur. That's how new networking is built. I think that's the main purpose of creating this zine in a print format.

(Fortuna Zine, 2021)

What Sorak-Sorai, Footballtropika, and Fortuna have done is a form of their attitude towards what has been received so far (in this context, it is a negative stereotype that is always associated with football fans). In addition, there are efforts to ward off and prevent homogenization that continues to be applied by the mass media to this day. According to Jesse (2010), zines are a struggle because zines are the product of action based on the zinesters' articulated opposition to media companies and their simplified representation and homogenization impact.

4 CONCLUSION

Through zines, groups, or people who were not previously exposed or even only exposed to the wrong parts are able to express their ideas freely. By not relying on any party, zinesters are able to prove that those whom the mass media may have poorly reported about actually have capacities to do positive activities. In the long term, if things like this are done, they can certainly fight the dominance of the mainstream mass media. By implementing a DIY culture in this zine, the zinesters of Sorak-Sorai, Footballtropika, and Fortuna desire to break away from the stigmas built by the mainstream mass media. Departing from the zine they created, football fans' independent mindset and attitude is expected to create a platform to express their voices so as to be independent from the mainstream mass media. By implementing and developing this DIY culture, football fans in Indonesia are on their way to create their own path to strengthen their identity.

REFERENCES

Barton, D., & Hamilton, M. (2000). *Literacy practices in situated literacies: Reading and writing in context.* New York: Routledge.

Chidgey, R. (2020). Zine culture. *The International Encyclopedia of Gender, Media, and Communication*, 1–5. https://doi.org/10.1002/9781119429128.iegmc052

Duncombe, S. R. (2008). Notes from the underground: Zines and the politics of underground culture. *ProQuest Dissertations and Theses*, pp. 394–394. Retrieved from http://navigator-iup.passhe.edu/login?url=http://search.proquest.com/docview/304276902?accountid=11652%5Cnhttp://fn9cr5xf4p.search.serialssolutions.com/?ctx_ver=Z39.88-2004&ctx_enc=info:ofi/enc:UTF-8&rfr_id=info:sid/ProQuest+Dissertations+%26+Theses+Globa

Ferris, M. A. (2001). Resisting mainstream media: Girls and the act of making zines. *Canadian Woman Studies*, *21*(1).
Goffman, E. (1959). The presentation of self. In *Life as Theater: A Dramaturgical Sourcebook*. https://doi.org/ 10.4324/9780203787120
Jesse, N. (2010). *Alternative literacies , resistance , and spatial representations in the do-it-yourself (DIY) Culture of Zine Publication in New Orleans*.
Kempson, M. (2015). 'My Version of Feminism': Subjectivity, DIY and the feminist zine. *Social Movement Studies*, *14*(4), 459–472. https://doi.org/ 10.1080/14742837.2014.945157
Poletti, A. (2008). *Intimate ephemera: Reading young lives in Australian zine culture*. Melbourne: Melbourne University Press.
Radway, J. (2018). Zines, Half-lives, and afterlives?: On the temporalities of social and political change. Modern Language Association Stable.? http://www. jstor.org/stable/41414086 the changing profession Afterlives?: On the. *Pmla*, *126*(1), 140–150.
Schilt, K. (2003). "I'll resist with every inch and every breath": Girls and zine making as a form of resistance. *Youth and Society*, *35*(1), 71–97. https://doi.org/10.1177/0044118X03254566
Spencer, A. (2005). *DIY: The rise of lo-fi culture*. Marion Boyars.
Storey, J. (2014). *From popular culture to everyday life*. New York: Routledge.
Zobl, E. (2009). Cultural production, transnational networking, and critical reflection in feminist zines. *Journal of Women in Culture and Society*, *35*(11), 28–37.

Innovation on Education and Social Sciences – Maureen et al (Eds)

The ideological resistance of Japan's *Kamikaze* pilots in a book entitled *Kike Wadatsumi no Koe*

S. Subandi, D. Nurhadi, L.P. Hartanti & M.R. Mael
Universitas Negeri Surabaya, Surabaya, East Java, Indonesia

ABSTRACT: The use of certain language structure patterns is based on certain motifs which represent the speaker's ideological representation. It can be found in *Kike Wadatsumi no Koe* book, a document collection written by *Kamikaze* pilots. This study aims to describe the forms and functions of the Japanese pilot's resistance ideology that constructed the text structure of *Kike Wadatsumi no Koe*'s discourse within the CDA framework. The three-dimensional concept of CDA Fairclough is used to examine the construction of the resistance ideology concept. The representation dimension presents a description of the author's psychological condition, the relation dimension presents the facts of the relationship between the author and the characters, and the identity dimension is indicated by the pronoun 'I' which refers to the author. Besides, the use of enclitic 'I' indicates the author's personal identity while the use of plural personal pronoun 'we' or 'our' is to attract the reader.

1 INTRODUCTION

How speakers choose and use the language is based on their ideology that builds a self-narrative belief to achieve language goals. The reason is that reality can be manipulated with language through diction and language style to create a particular language structure and realize the effectiveness and efficiency of language activities. It means that language style and diction formulate into a specific pattern which is a form that represents an ideology of the speaker's language. In a language, there is an ideational, which refers to a particular representation of what speakers want to convey in discourse text. It is in line with Fairclough (1993) and Eriyanto (2001), which states that speakers use language based on a certain ideology. While we realize that how speakers use language conveys certain ideological values, we need a comprehensive analysis that includes the history of language formation, social relations, and constructive social contexts to analyze the language.

The language symbol system used by humans is a manifestation of the ideology used to explain and organize social systems related to their role in preserving and changing social structures (Fairclough, 2003). Therefore, every language used is actually intertwined with the speakers' motive to be conveyed to the interlocutor, packaged in a certain language formulation that can lead the interlocutor to a certain ideology (Subandi & Mustofa, 2019; Thompson, 2007). It is also supported by Shukry's (2013) and Hua's (as cited in Halim, 2014) opinion, which state that conflicting perceptions can say something about the role of language in the spread of an ideology. Moreover, it is in line with Gendeng's view (2010), which denotes that an ideology (individuals believe) is to mobilize the intuition of a group and contribute to forming solidarity.

Socially and historically, language is a form of action in a dialectical relationship with social structures. It is supported by Wolfowitz (1991) which stated that a language variation is a form of intervention in social structures and situations into language forms and formulations. The Second World War had a significant influence on the social situation, particularly for the Japanese people. At the end of World War II, the Japanese government made policies and war strategies, namely the *tokubetsu kougekitai* (special attack troops), which the world viewed as a form of a brutal attack,

 DOI 10.1201/9781003265061-29

better known as the *Kamikaze* attack (Ojong, 2001). The impact of the social situation caused by the Second World War was felt in the social-psychological condition, which was actualized in the form of language style used in the 79 documents written by Japanese youths who were serving military service as *Kamikaze* pilots. The missions carried out by the pilots who carried out the *Kamikaze* attacks were interpreted as a form of fanaticism and blindness as well as a suicide attack, and the pilots who carried out the *Kamikaze* were seen as suicide pilots (Maynard in Kharismasari, 2006). However, Japan assumed that it is more as an embodiment of a patriotic spirit wrapped in a spiritual understanding that is nobler than just suicide (Richad, 2002). The present article examined how the construction of the resistance ideology concept owned by the pilots of *Kamikaze* actors based on wills, letters to family and/or lovers, and diaries that have been recorded in *Kike Wadatsumi no Koe*. It was reviewed based on the CDA Fairclough framework. In CDA, ideology is a central concept in critical analysis. Therefore, discourse is never separated from ideology, manipulating the reader/interlocutor towards an ideology.

Daeng (2016) explains that ideology is a system of thinking in dealing with problems, challenges, and obstacles formulated in a policy to win against something or achieve goals. Therefore, discourse is often assumed as a form of ideological practice or a reflection of a certain ideology (Darma, 2009). Then, the utterances produced by speakers in language activities are representations of their assumption of ideological concepts.

The tools used for analysis in CDA studies also involve elements that are out of the language. It is because, in CDA, language studies are more directed at studying language use. Fairclough (2001) states that the real point in CDA is more focused on the study of language and its use as a medium of communication. Fairclough (2001) also says that a critical view emphasizes more on the constellation of forces that occur in the process of production and reproduction of meaning (see also Brown & Yule, 1996). It means that the involvement of social aspects and its structure as a context significantly contributes to building the overall meaning of language. Concerning the three dimensions of Fairclough, Eriyanto (2001) explains that the individual and the writer's profession are discourse-forming factors, including educational factors, professional development of writers, political and economic orientations, and competence of writing skills possessed in conveying their writings.

Fairclough (2003) and Titscher et al. (2000) argue that language use is always simultaneously composed of social identity, social relations, and knowledge and belief systems, all of which are present since language is socially structured. Therefore, discourse text is seen as a social phenomenon, and language is a social practice. Texts have an ideational function that composes a knowledge system and an interpersonal function that creates a social identity. Fairclough (2003) signifies three dimensions in each discursive event: text (discursive practice, which includes the production and interpretation of texts) and social practice (where the analysis process is carried out according to the three-dimensional principle).

The level of discursive practice is the relationship between text and social practice. It means that text is a form of social activity, and the text indicates social change. Therefore, the search for text in a discourse cannot be separated from the context. Discursive practice is concerned with the socio-cognitive aspects of text production and interpretation. The analysis of discursive practice looks at the relationship between discursive events and the level of discourse, which is a problem of van Dijk's interdiscursivity (1993) and Fairclough (1993). Therefore, text analysis is the most crucial part of CDA and is based on three components, namely, description, interpretation, and explanation. The historical aspect of the formation of discourse is the primary key to seeing how the historical discourse is formed. The history of discourse formation has a vital role since discourse consumption aims to interpret discourse from the reader's point of view, as van Dijk (1993). It is also because socio-cultural practices in understanding discourse will be associated with elements outside the discourse, especially regarding the conditions of society at that time.

Ideology is a comprehensive idea that crystallizes and is universal and absolute (Kaplan & Robert, 2002). Meanwhile, Sobur (2004) defines ideology as a system of ideas expressed through language in communication. It proves that the spread and understanding of ideology to the public are never separated from language. A language is a place where various ideas and thoughts meet to convey thoughts, feelings, and desires to influence each other, dominate each other, and create a rival hegemony, even competing over ideas or problems (Marx, 2000). Thompson (2007) further states that in communication activities delivered through written language or text, ideology works through language, and language is a medium of social action. Therefore, ideology and language are two things that cannot be separated. Ideology has asymmetric power relations and dominant justifications; it is a representation system that dominates the human mind and achieves equality and common prosperity in society (Althusser, 2008). The ideological aspect of discourse has a close relationship with the three dimensions of Fairclough mentioned above in dismantling the ideology in a discourse text associated with social practice.

As a part of ideology, the concept of resistance ideology is characterized by the image of domination, doctrine, pressure, threats, and control over marginal groups (Mizuari, 2014). It means that the implementation of the resistance ideology as a form of compulsion due to the existence of doctrine, threats, pressure, and the like results in the emergence of a sense of injustice for the controlled group. Kress (1990) and Eriyanto (2001) stated that the resistance ideology is an ideology for marginal groups who fight against the dominant ideology.

2 METHODS

This research is qualitative descriptive research with data in the form of 81 pieces of written discourse from 79 documents in the form of letters to family and/or lovers, diaries, wills that have been recorded under the title *Kike Wadatsumi no Koe* as data sources. The data were identified based on three Fairclough dimensions to determine their type and function. Then, the data were analyzed and described based on the concept of the dimensions of representation, relation, and identity as well as their functions in the structure of the discourse text as a whole.

3 RESULTS AND DISCUSSION

Discourse text analysis was done by placing the three dimensions of Fairclough as the central point of the underlying study as follows.

3.1 *Representation*

As mentioned above, the aspect of representation is regarding how content is displayed in discourse (Fairclough, 2003). The content includes events, people, groups, situations, circumstances, or anything displayed and described in the discourse. The psychic picture presented by the author showed a decrease in morale in combat conditions due to pressure from opponents who have more robust equipment and war fleets. The psychological condition of the soldiers gave the impression of despair due to the increasingly depressing situation. This picture can be accepted if it is related to the context in which Japan decided to apply the *Kamikaze* attack strategy. As a form of panic, Japan was overwhelmed by the American attack, supported by much more sophisticated equipment and war fleets. The social conditions at that time indicated that Japan was approaching the point of defeat, and it was soon proven that Japan had surrendered. The description of these conditions can be seen in the following data excerpt.

> *遠い残雪のような稀みよ*、光ってあれ。例えそれが何の光であろうとも虚無の人をみちびく力とはなるであろう。全てを否定し、虚無の生活思想になり得たらどんなに愉快だろう。一刻一刻が*奈落への頂落の剎那*にある。
>
> "*Shine, desire likes the leftover snow far away. Likes whatever light it is, I think it will become a force that guides empty man. How can I enjoy it if I rebel and reject everything and become an empty fantasy of life? Minute by minute is just a moment of falling into hell*" (K.CH43.P.2)

悲しい護国の鬼達よ！すさまじい夜の春雷の中に、君達は銃剣を取り、遠ざかる俺達を呼んでいるのだろうか。大義のための戦。大義なんて何だ。医者の寝言に過ぎない。八月に入ってからは七七〇船団は台湾高雄を出て間もなく大島山丸（約一万五千トンの油）が魚雷により大爆発して二キロメートル四方を火の海として沈んでしまった。

"*O sad protectors of the country! In a summer storm, you take your bayonet eye, and will you call us who are now drifting apart? War for justice! What is justice! Nothing more than a fool's delirium! Since entering the eighth month, as many as 770 ships have left Gaoxiong Taiwan and soon will arrive at Ooshima Yamamaru (a tanker carrying approximately 15000 tons of oil) was suddenly blown up by missiles and all ships with a radius of 2 kilos in all directions drowned in the lake of fire*"(K. CH.43. P.5)

The condition of the declining fighting spirit possessed by the soldiers is implied from the diction 'leftover snow'. In Japanese society, snow is placed as something to be awaited. It makes someone happy and excited since it is considered to have aesthetic value, romantic value, and harmony because snow is considered a moment that wants togetherness. However, the diction 'leftover' on 'leftover snow' assumes that some of these values have passed or been lost. Furthermore, in the diction 'empty man' and 'empty mind'. the word 'empty' has the same meaning as 'not contained.' Emptiness describes the author's mental condition representing soldiers who no longer have a clear hope, purpose, and attitude. As a soldier who has the main task of guarding and maintaining his country's sovereignty, he is very bound by the command line of superiors and can only carry out all superior's orders and has no power to refuse. The social reality (the reality of war) is seen and felt by all soldiers. It creates a void in his soul towards hope and purpose. The author's statement reinforces the description of these conditions, which likens the conditions faced as destruction and torment. It is expressly conveyed through the expression, which states that the times he passed were felt to have 'fallen into hell'. This expression explicitly hints at the meaning of more destruction and torment. The diction 'fall' can be interpreted as describing a slumped condition, failure, and other unpleasant meanings. The combination of the word 'hell' has an impact on strengthening, emphasizing, and transcending the meanings mentioned above.

The discrepancy between the wishes and the Japanese government's doctrine of wanting to continue fighting until victory is achieved with the mental condition and spirit of the warriors, especially the author, can be seen in the form of the expression in paragraph 5. The author explicitly chooses the word 'devil' to call his fellow soldiers. The use of the word 'devil' is interpreted as a substitute for 'soldier' because of the phrase 'protector of the state'. Furthermore, it can be ascertained that those who have the main task of maintaining state sovereignty equipped with safety devices other than firearms as well as '*sangkur*' are soldiers. The use of this term indicates that psychologically, it can mean that the author has experienced pressure and depression. As a soldier who has been educated and nurtured his soul related to high discipline always to maintain his professional corps' honour, authority, and pride, the writer should not demean and insult his own company as 'devil'. It can be ascertained only in situations and conditions that are not appropriate for such degrading actions to be carried out by a soldier. The interpretation of the description of the author's psychologically depressed condition becomes stronger with the presence of the clause '...we are getting farther away'. Physically, the author is still in the camp since, at the time of the diary used as the subject of this study, the author was waiting for time because he had been prepared to carry out an attack using a single pilot fighter aircraft. That is, the meaning of the clause can be assumed that it is not the writer who is physically away, but the soul and heart of the writer can be made possible. This interpretation is based on the rules and procedures within the military sphere; that is, soldiers whom the state has assigned through the leadership are not allowed to refuse, let alone run away. This action can be considered as an act of discretion which will result in severe sanctions.

Through this understanding, the clauses used above describe a mental condition that can be possible to experience doubts and boredom due to various pressures that decrease sympathy and enthusiasm for military service. The condition of the writer's mental doubt becomes even more explicit if it is connected with the next sentence, namely, 'War for justice! What is justice! Nothing more than a fool's delirium!'. This sentence is more directed to the impression of 'apathetic' and 'pessimistic'. It is possible that the sentence was addressed to fellow soldiers. However, its role is more precisely directed to the Japanese government as the dominant party that makes policies and decisions on war actions. In this case, the clause 'Nothing more than the delirium of a fool' can be assumed as an expression of satire against the government for its policies. It is because, from that clause, it can be seen that the author felt pessimistic and doubtful about being able to win the war. This interpretation can be seen in the author's following sentence. The author provides an overview of the explosion of a fuel supply ship for war fleets named *Ooshima Yamamaru* belonging to the Japanese government to be bombed by fighter planes of American troops.

From the events presented by the author, it is not only a horrifying and terrible picture due to the ship's explosion; all ships within a radius of two kilometres became victims and even sank. Moreover, even other surviving war fleets would almost certainly not be able to operate and function due to the unavailability of fuel. In fact, fuel is an essential ingredient for a fleet to be used and worked. Therefore, if the war fleet as one of the primary weapons to fight the enemy and protect its safety cannot be operated or functioned, it can be interpreted as 'justice obtained through war'. As a result, the 'delusion of a fool' is close to reality.

The ideology of resistance conveyed by the author based on these data is more implicit. The message or resistance form can be implicit at the beginning of the paragraph (especially in paragraph 2) since the chosen diction uses more metaphorical figures of speech. The meaning conveyed by the author is not direct and not confrontational so that it can be interpreted as a form of hidden resistance. A more open form of resistance appears in the fifth paragraph. Although some of the selected dictions also use metaphorical figures of speech, semantically, the meaning of the chosen figure of speech is more assertive and extreme in character because the resistance conveyed is packaged in the form of swearing at the 'devil' diction and swearing in the phrase 'delusional fools'.

3.2 *Relation*

The relational analysis describes how the relationship between writers, audiences, and participants of discourse texts is displayed and described in the discourse of Fairclough (2001). Furthermore, Fairclough (2003) and Eriyanto (2001) stated that the concern point in the relational analysis is how the pattern of the relationship is shown in discourse?. These relationships are between the writer and the reader, the characters and the reader, and the writer and the characters. In every essay written in the form of discourse text, it is almost certain that every writer conveys an idea or idea to influence the reader in the hope that the reader can accept the author's ideas. To realize this goal, the writer's strategy is to use the greeting 'we' to display a picture of the relationship between the writer and the reader. The greeting word 'we' and 'our' found in paragraph 3-line 3,「*私達は生きぬばならない*」 'we have to stay alive', then in paragraph 6-line 3, ……*僕達の喜びは影のあるものになる相違ない*」 (…no more mistake, our happiness is only a shadow of an item). In paragraph 6 on the 13th line, even though the external structure of the discourse text does not appear, but semantically refers to the first person plural, namely 'us',「" *私達ば" 足元に自分の身に火をつけられている*」(If we light a fire at the base of our feet then actually, we light a fire in our own body).

The author's efforts to invite readers into the flow of the discourse text appear from the data above. By using 'we' or 'our' diction, the speech partner or reader is invited to enter into the text plot to have the same feelings and conditions as the author, namely a struggle of death, life under pressure, and the opponent attacking with the condition of strength under the opponent. Lexically, the word meaning 'we' or 'our' means that the writer and the reader are both included in the component area of its meaning.

Based on the data above, there is no explicit mention of the central figure who holds the dominant power since the diction used prefers to use symbolic languages such as metaphors, indirect sentences, and the like so that there is no relationship between the reader and the characters. However, by analyzing language functions involving components of the language context, it is possible to identify the characters involved in the structure of the discourse text being studied. The first is the Japanese government, namely the text element that has the dominant power. In this context, the government acts as the ruler and the reader as the people. The anxiety and fear of war events that readers feel reflect the actual conditions felt by the Japanese people, especially the parents and families of the soldiers who were assigned as members of the special offensive forces by the Japanese government. Implicitly, the relationship between the two can be seen in the following quote.

> 栄光ある祖国日本の代表的攻撃隊ともいうべき陸軍特別攻撃隊に選ばれ、身の栄光これに過ぐるものなきを痛感致しております......
>
> (*Selected and appointed as a member of the Special Assault Force, which is an assault force that represents the glorious country of Japan, actually only makes me feel glory for myself*...) (K.CH.H. 17.P.2)
>
> 理性をもって考えたなら実に考えられぬ事で、強いて考うれば、彼らが言うごとく自殺者とでも言いましょうか。精神の国、日本においてのみ見られる事だったと思います。
>
> (*Actually, I can't even think of what reason I should have. If I have to think about it, of course, I can call myself a suicide, as they say. The country of souls is something that can only be seen for the country of Japan*) (K .CH.H.19.P.2)

From those two quotations, the author does not explicitly present a central figure (person who has dominant power) by using only symbols, namely the use of 'state' diction and the subject component omission. The unique characteristic of the Japanese is the disappearance of the actor subject (decision-makers) who decided on the author's appointment to become a special attack force member. As a result, the relationship between the reader and the government is that the power structure in the political field is reflected in government policies, which is related to the decision-making of world war involvement.

Based on the data excerpt above, the relationship between the author and the characters is that the author fights with several figures. Both are against the government and or the leadership of the military corps. It is reflected in the text quote K.CH.H.17.P.2 and K.CH.H.19.P.2 through the analysis of the dimensions of relations and resistance to fellow soldiers, where the author uses the greeting 'O demon protector of the country who is sad' to refer to fellow soldiers as reflected in the text quote K.CH .43.P.5 through interpretation dimension analysis. On the other hand, the author also accepts one of the characters, namely fellow soldiers found in the quote K.CH.H.19.P.2. The author's acceptance is more of a comparative nature and strengthens the author's assumptions and ideology. Implicitly, through the structure of the discourse text, the author wants to convey a resistance form to acts of war by using the strategy of hitting fighter planes against enemy carriers.

3.3 *Identity*

Furthermore, identity dimension analysis is intended to find out how the author's identity, the object being discussed, and the participants or readers in the structure of the text. First, it was investigated by the author. Based on the identification results, it was found that there was a reference to the first pronoun 'I' and the enclitic '...I' as a reference to the author's self and ownership as in the data excerpt below. However, the Japanese sentence structure's personal pronoun 'I' does not always appear and is often lost, as in the following data.

例えそれが何の光であろうとも虚無の人をみちびく力とはなるであろう。全てを否定し、虚無の生活思想になり得たらどんなに愉快だろう。一刻一刻が奈落への頂落の刹那にある。

(*Like any light, I think that light will be able to become a force that guides empty humans. How can I enjoy it if I rebel and reject everything and become an empty dream of life? Minute by minute is just a moment of falling into hell*). (K.CH.43.P.2)

*自分*の理想の一端を受けて性質は*自分*の理想に向かって思いように妻教育に同化されてくれた大切な我が心の太陽です。

(*One of the essential characteristics of my mind is, I am committed to my own thoughts and as I think that, if I am concerned with my wife's education then, that is the eye of my heart*). (K.CH.53.P.3)

苦しさも空虚な悲しさも今は銀の小箱にそっと秘めて男らしく出発したいと思います。

(*The pain and sadness I currently store in a small iron box, and I want to depart mightily*). (K.CH.51.P.4)

未だ花弁を見せず、蕾のままで死んで行くものの一つのあり方であったかも知れない。

(*I've never shown my flower petals, I also don't know, maybe I'll just die while still in bud*). (K.CH.33.P1.B.1)

The pronoun 'I' in the structure of this discourse text refers to the author, meaning that this discourse text is indeed written by the author and is used to convey the author's ideas, thoughts, and feelings. Each essay is personal writing and describes the direction and state of the author's own mind. The author is given the freedom to cultivate and develop his thoughts which are then poured into an essay. This discourse text is also an essay that tells the reality of the lives of soldiers on the battlefield and is processed by developing their own ideas to compose a narrative that is considered to represent the thoughts and feelings of the author.

The use of '…I' enclitics which states the author's self-ownership, is raised by the author to emphasize that the thoughts, ideas, and desires are the author's property so that the flow of the text structure originating from the author seems to be stronger. The strategy of using '… I' enclitics is highly appropriate since it also shows the strength of the ideas presented in the text structure. Through the structure of the text, a strong impression of the author's personal identity can be built.

The structure of this discourse text also presents the plural first person greeting words 'we'. The diction 'we' refers to the meaning of the writer and other people (speech partners or readers) included as personas where both parties have the same position and status in the unity of the composition of the discourse text structure. It proves that the author makes an effort to attract readers who were initially outside the text flow structure into the text flow, where the writer and reader form a unity in the idea of text structure. Thus, both have the same role and concept of understanding in the structure of the discourse text they occupy, as in the following data.

「足元に*自分の身*に火をつけられている」

(*When we light a fire at the base of our feet, in fact, we light a fire in our own bodies*). (K.CH.P.6.B.13)

Seen from the text above, the author attempts to enter into the reader's ideology and assume that there are similarities between the writer and the reader to accept and agree with the ideas offered by the author. In fact, through the structure of the discourse text, the author has attempted to introduce an ideological concept to the reader without being realized or felt by the reader.

4 CONCLUSION

The overall theme built in the text of the discourse is the rejection of war actions that use special attack strategies, better known as *Kamikaze* attacks. Based on the text interpretation results, the writer is experiencing anxiety and doubt about the social conditions faced at that time. The doubts experienced were caused by the pressure of the situation, especially the strength condition of the opposing military forces, which were assumed to have more power, one of which was marked by the possession of more formidable and sophisticated war equipment and fleets. The pressure was also felt due to the policy of the Japanese government, which decided to continue the war until it achieved victory even though it was thought that it was not easy to materialize. The author's form of resistance is not in the form of physical activity but prefers to be actualized through written language that does not refer to direct meaning, namely the use of a figure of speech. The author's resistance ideology concept is taken from the experience and actual conditions experienced by the author. Then, the writer takes the arguments that are poured through language in written discourse text. It seems that there is an effort from the author to involve the speech partner or reader engaged in the structure of the text of the discourse and to incorporate the concept of the ideology of resistance that was initiated to the speech partner/reader by using the persona greeting word 'we;'. All of the personal and institutional characters, who have a high intensity of appearance presented in the text of discourse, are characters opposite to the author, and only characters who have a low intensity of appearance are acceptable to the author. This technique gives the impression that the author is more dominant in conveying ideas and concepts that build a complete and comprehensive text structure.

REFERENCES

Althusser, L. (2008). Tentang ideologi: Strukturalisme marxis, psikoanalisis. *Cultural Studies*.

Brown, G., & Yule, G. (1996). *Analisis wacana*. Jakarta: PT Gramedia Pustaka Utama.

Daeng, S. (2016). Pancasila alat perlawanan terhadap kolonialisme dan imperialisme. *Maritim News*. 30 May.

Darma, Y. A. (2009). *Analisis wacana kritis*. Bandung: Yrama Widya.

Eriyanto. (2001). *Analisis wacana: Pengantar analisis teks media*. LKiS Yogyakarta.

Fairclough, N. (1993). Critical discourse analysis and the marketization of public discourse: The universities. *Discourse & society, 4(2),* 133–168.

Fairclough, N. (2001). *Language and power*. Pearson Education.

Fairclough, N. (2003). *Analyzing discourse: Textual analysis for social research*. Psychology Press.

Gendeng, J. W. (2010). *Membongkar gurita geng Cikeas: Guyonan dan cerita*. Yogyakarta: Moncer Publisher.

Halim, R. (2014). *Linguistik kristis dan analisis teks: Suatu cadangan penelitian*.

Kaplan, D., & Robert A. M. (2002). *Teori budaya*. Translated by Simatupang, L. Yogyakarta: Pustaka Pelajar.

Kharismasari, A. (2006). Penerbang *Kamikaze* dan lambang bunga sakura dalam buku "*Kike Wadatsumi no Koe*", Universitas Negeri Surabaya (Unpublished thesis).

Kress, G. (1990). Critical discourse analysis. *Annual Review of Applied Linguistics*, 11, 84–99.

Marx, K. (2000). Peta pemikiran Karl Marx: Materialisme dialeksis dan materialisme historis. Translated by Ramly, A. M. LKiS Yogyakarta.

Muzairi, M. (2014). Pergeseran sistem kekuasaan dari Marxisme ke hegemoni dan politik media: Suatu kritik ideologi. *ESENSIA: Jurnal Ilmu-Ilmu Ushuluddin*, 15(2), 213–228.

Ojong, P.K. (2001). *Perang pasifik*. Jakarta: Kompas.

Richad, O. (2002). *Tokubetsu kougekika*. Heiwa Shinnen Kyoukai Kaijou Teikyou.

Shukry, A. S. M. (2013). A critical discourse analysis of Mahathir Mohamad's speeches on the "war on terror". *Intellectual Discourse, 21(2)*.

Sobur, A. (2004). *Analisis teks media: Suatu pengantar untuk analisis wacana, analisis semiotik, dan analisis framing*. Bandung: Remaja Rosdakarya.

Subandi, S., &. Mustofa, A. (2019). Ideological and hegemonic implicatures of Japanese male registers used by Japanese young women speakers: Gender based analysis and its implication in pedagogical domain. *Advances in Social Science, Education and Humanities Research*, 227, 93–97. https://doi.org/10.2991/steach-18.2019.21.

Thompson, J. B. (2007). *Analisis ideologi: Kritik wacana ideologi-ideologi dunia*. Translated by Yakin, H. Yogyakarta: IRCiSoD.
Titscher, S., Meyer, M., Wodak, R., & Vetter, E. (2000). *Methods of text and discourse analysis*. SAGE Publications Ltd, https://www.doi.org/10.4135/9780857024480.
van Dijk, T. A. (1993). Editor's foreword to critical discourse analysis. *Discourse & Society, 4(2),* 131-142.
Wolfowitz, C. (1991). *Language style and social space: Stylistic choice in Suriname Javanese*. Urbana and Chicago: University of Illinois Press.

Innovation on Education and Social Sciences – Maureen et al (Eds)

Altruism in *Serat Subasita*

O.D. Andriyanto, D. Darni, H. Subandiyah & S. Sukarman
Universitas Negeri Surabaya, Surabaya, East Java, Indonesia

M. Hardika
Surabaya European School, Surabaya, East Java, Indonesia

ABSTRACT: Javanese literary works have the content of local wisdom and intelligence of the community. As a representation of past knowledge and experience, *Serat Subasita* contains a lot of wisdom that has values and benefits in this era. This study focuses on altruistic behavior, such as tolerance, ethics, and empathy in Padmasusastra's *Serat Subasita*. The method used is qualitative with a pragmatic approach. Data was collected by literature study, there is related research, literature, local wisdom, and Javanese culture. The main data in this study are words and sentences in *Serat Subasita* which contain altruistic attitudes. The results of this study found that the altruism content of *Serat Subasita* includes tolerance, empathy, and ethics in social interaction of Javanese culture.

1 INTRODUCTION

Literary works are part of the wisdom of human thought. Moreover, literature is related to human life; both have relevance. Life as a source of ideas and ideas in literature along with the author's imagination with various building elements will reconstruct the literary work by the message the author wants to convey. The message in the story provides feedback to humans (readers) about how humans live and behave. It means that the reader can take the treasures of goodness in it, in the form of life harmony, wisdom, the harmony of nature, environmental ethics, equality, and various other perspectives in life that are considered to contain the goodness and benefits of life.

One of the Javanese literary works that contain many aspects of kindness, caring, empathy, and ethics is *Subasita Serat* written by Padmasusastra. *Serat Subasita* is a Javanese literary work published in 1914 in Surakarta. This Javanese text includes a lot about how Javanese people behave and behave ethically in front of other people. In detail, *Serat Subasita* contains an altruistic attitude that has respect, ethics in interacting with others, and the treatment of children and women. In general, altruistic behavior can be seen from the Javanese people's attitudes, which are full of tolerance, ethics, and empathy for others. This action is nothing but a way for harmonious interaction with others and can provide benefits.

Human values and aspects of empathy are often found in literary works. These values are a representation of human behavior, namely altruism. It is not easy to find this in life, but not in literature (Purabasari, 2019). The readers of literary works imitate stories in literary works that have benefits in life and interactions with the surrounding nature. Literary works contain a lot of spiritual values in the form of altruism that humans must care about living for others (Udayana & Indiatmoko, 2017).

Literary works cannot be separated from the linguistic elements used. The results of the power of imagination, critical power, reason with beautiful and artistic language media support the content of values in life. Moral teachings, love, inspire, and motivate to do good, high moral awareness, even spirituality (Udayana & Indiatmoko, 2017). The Javanese people have a religious nature. The outer and inner levels cannot be separated from the spiritual life of society and even emotional life

DOI 10.1201/9781003265061-30

in general. Community life refers to the existing norms and culture based on feelings, actions that function and are of value to the local community (Geertz, 2008).

Experts in ethical and moral studies state that there are good behaviors and attitudes to life in human interactions. The seeds of goodness are called altruism. Altruism has its roots in the Latin *alter*, which means 'other.' Thus, altruism is a view or attitude of life that focuses on kindness, caring, and other essential things that lead to people's welfare and happiness. Through this attitude, activities are born, actions that prioritize goodness, interests, and benefits for others (Mangunhardjana, 1997).

Altruistic behavior or activities are related to positive attitudes and good behavior, empathy in social life. In other words, individuals cannot be separated from one another. Altruism is a part of the culture and even encouraged in religion, regardless of place, time, and anyone. As part of a positive attitude, altruism needs to be habituated with full societal appreciation (Mesa, Aspin, & Rudin, 2020).

Ethics and norms cannot be separated from society and its life. The ethics and behavior of the community have distinctive characteristics so that they become the identity or marker of the social community. The content of ethics places humans at the highest point, namely the true philosophy of life. Behaving with self-skill and noble character leads to a harmonious life during community life. Exemplifying goodness in reality and classical Javanese literature also contains a lot of positive attitudes and benefits for life. This research focuses on the altruism aspect in Padmasusastra's *Serat Subasita,* which includes a lot of tolerance, ethics, and empathy in society. This research can provide benefits and implications for the wider community and also assist in growing character and strengthening national identity.

2 METHODS

This study uses a qualitative method with a pragmatic approach. The pragmatic approach in the literature is closely related to readers' reception through literary texts (Abrams in Sutopo, 2002). Data collection was carried out by a literature study with the primary reference being *Serat Subasita* or related research, literature, local wisdom, and Javanese culture. The primary data are words and sentences that have altruistic content. The steps in the research begin with identifying the data, grouping the relevant data with aspects of altruism, followed by data analysis.

3 RESULTS AND DISCUSSION

The values of goodness in *Serat Subasita* are closely related to human activities, which are clearly illustrated by depicting the author's past experiences. The following is a description of the charge of altruism in *Serat Subasita*.

3.1 *Tolerance*

> *"Mênawi wicantên sampun sêru-sêru...kalêbêt patrap digsura."* (Padmasusastra, 1914)
>
> "If you speak, do not be too loud ...including actions that are not following manners." (Padmasusastra, 1914)

Cultures around the world recognize that altruism plays an essential role in humanity. However, altruistic behavior is now rare due to economic, political, and social shifts (Yeung, 2006). The social aspect of Javanese society has distinctive characteristics, especially in language. The level of Javanese speech is still mostly well applied in society. In terms of speaking, Javanese people often say that you do not need to be too loud if you speak. It means that the Javanese still adhere to manners that are full of tolerance to maintain harmony.

> *"Mara tamu prayogi cêcala rumiyin, sagêd tampi utawi botên."*
>
> *"Mara tamu ingkang tanpa cêcala, bilih botên kapanggih ing jawi: asabawaa, dhèhèm-dhèhèm, mênawi wontên rencang, acanthela wêling, mawi amratelakakên namanipun. Salêbêtipun sawêg canthèl wêling, sampun lênggah rumiyin, ngadêk wontên ing latar kemawon ..."* (Padmasusastra, 1914).
>
> "If you are going to visit, you should make an appointment in advance; you can accept it or not."
>
> "A visit without an appointment, if you do not meet outside: say hello, if there is a servant, leave a message, by mentioning your name. When delivering a message, not to sit down first, just stand in the yard..." (Padmasusastra, 1914).

Society in general firmly adheres to the aspect of tolerance in life. Literature as a reflection of past lives deserves to be imitated by its good values. In *Serat Subasita*, it is explained that tolerance needs to be upheld when visiting. It is said that it is better to make an appointment before visiting. It is in line with the Javanese habit of prioritizing the '*empan papan*' attitude (situational). Myers (2009) states that altruistic behavior is categorized into empathy, belief in justice, responsibility attitude, self-control, and low ego. Altruism is influenced by mood, justice, empathy, situational, and sociobiological factors (Kamilah & Erlyani, 2017).

> *"Tatakraminipun tiyang katamuan, sadèrèngipun tamu mêdharakên pêrlunipun..."* (Padmasusastra, 1914).
>
> "The manners of people who have guests, before the guests tell their needs" ...(Padmasusastra, 1914).

Altruism is an attitude of help, sincerity in doing something that contains goodness in it, and even a responsibility to take care of others. Humans should be aware of their responsibilities, and the value of their humanism is based on a commitment to take action as a manifestation of their altruistic feeling towards others (Machmud, 2015). In *Serat Subasita*, etiquette is described when welcoming guests; guests should first convey their intentions and goals. None other than this is done to maintain communication with others and even tolerance (time, situation) before conveying the primary goal.

Humans are now challenged to have an attitude beyond their identity. It is because of the influence of diversity and the era of globalization. Ethical attitudes and obligations to others are no longer limited. It means that we cannot restrict moral actions; altruism has a significant role in advancing civilization in today's era. A person naturally has the nature of loving, giving to each other, and being willing to sacrifice, even considered a necessity and very important in life. However, altruism, both in the form of ideas and practices, is increasingly being ignored since it is considered only normative ethical. Altruism has never been used as a basis, even considered a basis for policy and intellectual thought (Robet, 2015).

3.2 *Ethics*

> *"Mênawi wicantênan sampun ngantos dipun wori gujêng, punika digsura, gumujêng kêdah angon kosok."*
>
> *"Mêgat wikalpa tiyang sawêg rêrêmbagan: digsura."* (Padmasusastra, 1914)
>
> "If you are talking, do not mix it with jokes. It is an act that is not following manners; joking must know the right time."
>
> "Cutting off the conversation of people who are discussing: not according to manners (insolent)." (Padmasusastra, 1914)

Moral and ethical actions are carried out to make it easier for someone when experiencing difficulties, difficulties, and suffering; in this case, humans are driven by their altruistic nature.

Altruism encourages humans to make moral/ethical actions as a natural human activity because they have a sense of empathy for others. Ethical principles drive moral actions that direct humans to obey (Jena, 2018). For example, one of the taboos in verbal interaction is cutting off the conversation partner or joking when the situation is not appropriate. It is far from the ethical values of the Javanese. In general, Javanese people must be able to control themselves and put ethics by moral teachings and norms that apply in society.

The human ability to think critically, rationally, and ethically is a human advantage. Humans eventually grow feelings of helping others and have behavior based on ethical attitudes that are following norms. Acting ethically and rationally, according to Peter Singer, is based on universally applicable moral principles. Altruism is the basis of moral behavior (ethical behavior) (Jena, 2018). Furthermore, Peter Singer argues that altruist behavior is based on ethical actions that are natural and not forced. Human altruism has biological characteristics, especially clan altruism (kin altruism) (Jena, 2018).

3.3 *Empathy*

> *"Mênawi wontên tiyang nêpsu dhatêng anakipun, sampun nimbrung tumut nyrêngêni utawi nutuh..."* (Padmasusastra, 1914)
>
> "If someone is angry with their child, do not join in scolding or bringing up ..." (Padmasusastra, 1914)

Basically, altruism contains noble moral values and can be used as an example for children. The value of altruism has recently become necessary due to the rampant moral crisis that makes people lose their sense of care and empathy for others. This phenomenon can be observed from the behavior of today's teenagers who like foreign cultures that are far from a reflection of the nation's culture (Asmaranty, Hasanah, & Suwignyo, 2019). In line with the existing data, self-control is also part of altruism, namely, not expressing anger and ego.

> *"Sampun udud wontên sangajênging tiyang èstri...nama botên ngajèni"* (Padmasusastra, 1914)
>
> "Do not smoke in front of women...very disrespectful." (Padmasusastra, 1914)

Prioritizing an attitude of responsibility by not harming others needs always to be done in life. The data stated that smoking in front of women is an act that is not polite, and is even considered disrespectful. It is far from being an altruistic act.

Adherents of altruism have the characteristics of a responsible attitude. In life, considering the consequences for oneself and others, society, country, and the world needs to be considered before acting. People who instill an altruistic attitude do not think about the consequences of their behavior, such as praise, raises/promotions, prizes, but focus on good deeds for themselves, others, and society. This attitude can be used as a reference and self-development and can focus on building a social life, a prosperous state (Mangunhardjana, 1997).

> *"Manawi sarêngan numpak sêpur, wontên tiyang èstri ngadêk botên uman papan, tiyang jalêr ingkang sampun lênggah rumiyin sacêlaking èstri wau, wajib ngawon lajêng ngadêk, panggenanipun dipun tawèkakên dipun kèn anglênggahi."* (Padmasusastra, 1914)
>
> "If at the same time riding the train, a woman is standing because she does not have a place, the man who has sat near the woman first must relent and then stand up; the place is offered to be occupied." (Padmasusastra, 1914)

Like the behavior of society in general, respecting others, especially women, can be done in various ways. It is illustrated that when a woman is standing on public transportation, someone (a man) should give his seat to a woman who needs it more. It is done as a form of respect and respect for others who are more in need.

A person can be categorized as having an altruistic attitude if he fulfills the following criteria: (1) has a sense of empathy, the ability to see the situation by feeling, caring, understanding others; (2) has a voluntary attitude, not wanting to get anything in return from others, being willing to sacrifice, behaving with values of honesty and fairness; and (3) having a desire to help others in the form of time, energy, and material (Myers, 2009). Altruistic behavior places a sense of empathy for others through the actions taken, like helping people selflessly, trying to give goodness by giving all useful things without expecting anything in return (Mesa, 2020).

4 CONCLUSION

Literary works are not just a medium or a means of entertaining, but the content in literary works contributes to strengthening character and authentic self-identity. *Serat Subasita* contains the author's experiences in the past, which of course are still very relevant to life in this era. Moreover, *Serat Subasita* is loaded with altruistic attitudes, such as human nature, which has positive characteristics, tolerance, ethics, and empathy to maintain interaction and harmonization in society.

REFERENCES

Asmaranty, P. Z., Hasanah, M., & Suwignyo, H. (2019). Pengembangan buku cerita berseri dengan tema altruisme untuk pembelajaran cerita rakyat. *Jurnal Pendidikan: Teori, Penelitian, Dan Pengembangan, 4(10)*, 1417–1426.

Geertz, C. (2008). *Local knowledge: Further essays in interpretive anthropology*. Basic books.

Jena, Y. (2018). Altruisme sebagai dasar tindakan etis menurut Peter Singer. *Respons: Jurnal Etika Sosial, 23(01)*, 59–82.

Kamilah, C., & Erlyani, N. (2017). Gambaran altruisme anggota komunitas 1000 guru Kalimantan Selatan. *Jurnal Ecopsy, 4(1)*, 33–40.

Machmud, S. (2015). Analisis nilai spiritual dalam novel Haji Backpacker karya Aguk Irawan MN. *Jurnal Humanika, 3(15)*.

Mangunhardjana, A. (1997). *Isme-isme dalam etika dari A-Z*, 186–187. Yogyakarta: Kanisius.

Mesa, N. M. R., Aspin, A., & Rudin, A. (2020). Pengaruh layanan bimbingan kelompok terhadap perilaku altruisme siswa. *Jurnal Ilmiah Bening: Belajar Bimbingan dan Konseling, 4(1)*, 35–44.

Myers, D.G. (2009). *Social Psychology*. New York: Mc Graw Hill.

Padmasusastra, K. (1914). *Serat Subasita*. Surakarta: Yayasan Sastra Lestari.

Purabasari, R. (2019). Building altruism through novels. In *Journal International Seminar on Languages, Literature, Arts, and Education (ISLLAE), 1(2)*, 334–338.

Robet, R. (2015). Altruisme, solidaritas dan kebijakan sosial. *MASYARAKAT: Jurnal Sosiologi*, 1–18.

Sutopo, H. B. (2002). *Metode penelitian sastra: Epistimologi model teori dan aplikasi*. Yogyakarta: Pustaka Widyatama.

Udayana, H., & Indiatmoko, B. (2017). Ekspresi cinta dan citra religius dalam novel Atheis karya Achdiat Kartamihardja. *Seloka: Jurnal Pendidikan Bahasa dan Sastra Indonesia, 6(2)*, 218–225.

Yeung, A. B. (2006). In search of a good society: Introduction to altruism theories and their links with civil society. *Civil Society Working Paper series (25)*.

Innovation on Education and Social Sciences – Maureen et al (Eds)
© 2022 copyright the Author(s), ISBN 978-1-032-20752-0
Open Access: www.taylorfrancis.com, CC BY-NC-ND 4.0 license

Author index